SAFETY AND QUALITY IN MEAT INDUSTRY
Innovative Approaches

The Editors

Dr. Pradeep Kumar Singh, graduated (BVSc and AH) from GBPUA&T, Pantnagar in 2003 and did masters (MVSc in Livestock Products Technology) from the same institute in 2005. He completed the in-service PhD (LPT) programme from LUVAS, Hisar in the year 2016 and has about 15 years of teaching/research/extension experience in different capacity at Apollo College of Veterinary Medicine, Jaipur (2005-08) and College of Veterinary Science and Animal Husbandry, Rewa, NDVSU (2008-till date). He has to his credit many publications and awards. He is actively involved in undergraduate and postgraduate teaching programme and his research area is livestock products like meat, milk, egg *etc.*

Dr. Samir Das did his BVSc and AH degree from GBPUA&T, Pantnagar, and then both Masters and PhD degree in Veterinary Public Health from IVRI, Izatnagar. He has 12 years of teaching and research experience in different capacity as Assistant Professor in Faculty of Veterinary and Animal Sciences, SKUAST-Kashmir; College of Veterinary Science and Animal Husbandry, Rewa, NDVSU and presently working as Scientist in ICAR since 2011. He has to his credit many international and national publications and is actively involved in the research of food borne and zoonotic diseases.

SAFETY AND QUALITY IN MEAT INDUSTRY

Innovative Approaches

— Editors —

Dr. Pradeep Kumar Singh

Dr. Samir Das

2022

Daya Publishing House®

A Division of

Astral International Pvt. Ltd.

New Delhi – 110 002

Cataloging in Publication Data--DK

Courtesy: D.K. Agencies (P) Ltd. <docinfo@dkagencies.com>

Safety and quality in meat industry : innovative approaches / editors, Dr. Pradeep Kumar Singh, Dr. Samir Das.

pages cm

Includes bibliographical references and index.

ISBN 9789354616556 (HB)

1. Meat--India--Safety measures. 2. Meat--Safety measures. 3. Meat--India--Quality. 4. Meat--Quality. 5. Meat industry and trade--India--Quality control. 6. Meat industry and trade--Quality control. I. Singh, Pradeep Kumar (College teacher of veterinary science and animal husbandry), editor. II. Das, Samir, editor.

LCC HD9428.I5S24 2021 | DDC 338.47664900954 23

Published by : **Daya Publishing House®**
A Division of
Astral International Pvt. Ltd.
– ISO 9001:2015 Certified Company –
4736/23, Ansari Road, Darya Ganj,
New Delhi-110 002
Ph. 011-43549197, 23278134
E-mail: info@astralint.com
Website: www.astralint.com

Preface

The book has been prepared for undergraduate and postgraduate students interested in exploring the latest techniques of safety and quality in meat sector. It may be of prime importance to students from livestock product technology and veterinary public health department. Additionally it may serve as a reference material for the food science and technology students having their interest area in muscle foods, or the industry people involved in safety and quality issues related to production, processing, packaging, storage and distribution of meat and meat products. The content of the book covers the latest trend/practices followed in production and further handling of quality products with major focus on safety principles. The book has covered all the major and related areas concerning production, meat speciation, regulations and standards, and other relevant aspects which can help in the enhancement of product features. During the last decade many new technologies have emerged and have shown their advantages to ensure meat safety and quality. Looking to their promising results, they have been discussed in this book to indicate their probable application on meat and meat products. This book has been written concisely but at same time authors have tried to address all the important issues/challenges by harnessing latest tools and techniques. The book is grossly woven around the latest development in the field of meat industry in relation to safety and quality. Hope the reader and industry will be benefited with the content of book.

Dr. Pradeep Kumar Singh

Dr. Samir Das

Contents

Introduction

The meat industry is growing with modernization witnessed in technology, but simultaneously facing a lot of challenges in its production, processing, storage and marketing phases; posing a huge threat to safety and quality. Consumers of nowadays are aware and demanding, hence industry needs to be equipped with latest tools and techniques. The approach initiates from the farm itself and continues till the product is consumed giving the famous notion of "farm to fork" management. The industry is going for organic meat, designer meat, genetically modified meat *etc.* as per the market demand and to keep a pace with quality they have to set new norms. The growing meat industry needs certain guidelines in the form of regulations and standards. Unless there is uniformity in standards available, it is difficult for the industry to follow it and also difficult for the monitoring agency to control the quality as per the standards. The industry has been moulding itself to achieve the HACCP concepts for ensuring safe and quality meat to consumers. It has to follow newer concepts of "traceability" to assure quality in the product value chain. It has to use traditional and modern day technologies *viz.*, molecular, proteomics to safeguard the interest of consumers. Once an animal is slaughtered and carcass has been obtained the approach of safety and quality shifts its balance from animal management to product (meat) management. Modern techniques like color and computer image analysis, near-infrared spectroscopy, *etc.* are useful in speedy analysis of large number of samples for nutritional quality. The industry is bound to follow the standards and methods for the detection of hazardous chemical residues, pathogens in the product, which otherwise can take toll on consumer's health. Hence, production of meat, free from residues and detection of pathogens is essential and many novel methods are under continuous development with improvement of sensitivity and the limit of detection. Demand for healthy, convenient and safe meat products with natural flavor and fresh appearance has led to the application of new preservation technologies like active packaging and intelligent packaging. There are several tools and instruments being used conventionally to check safety and quality, where biosensor based real time monitoring is going to be the future of industry. New dimensions of nano-materials have been added to the advanced processing techniques. The discussion in this book also focuses on different strategies for improvement of sensory quality of meat products with sensory testing facilities. The authors have tried to incorporate all the relevant aspect, essential for enhancing and supporting meat sector for ensuring safety and quality of product. Hope the book meets out the expectation of readers from the varied field involving academics and industry.

Dr. Pradeep Kumar Singh

Dr. Samir Das

Chapter 1

Designer Meat Production: A Nutritional Approach

Ankur Khare*, Ramesh Pratap Singh Baghel, Sunil Nayak and Vaishali Khare

College of Veterinary Science and A.H., N.D.V.S.U., Jabalpur, M.P.
**e-mail: ankur_khare29@yahoo.com; ankurkhare29@gmail.com*

ABSTRACT

Demand for functional foods is growing now days with the claims and a clear mandate of health and prevention of diseases as one of its main goals. Functional foods have developed a tremendous increase in interest of consumers due to their inherent ability of providing health benefits. They have been classified on the basis of potential health benefits and their ingredients' properties into different categories. For the development of functional meat there were some feeding strategies like improving fat content, enrichment with minerals, incorporation of vitamins and antioxidants, other added compounds/ingredients and reduction of some exogenous components that are unhealthy. In our country, still only few functional products especially of milk is available in the market but, meat based functional products are still under research only. Efforts are being taken to incorporate certain beneficial bioactive compounds in meat and market it as functional meat.

Keywords: *Functional meat, Nutritional manipulations, Human health, Bioactive components.*

1. Introduction

About two thousand years ago, Hippocrates mentioned that "Let food be your medicine and medicine be your food". Since from long time a clear relationship existed between the food people ate and their nutritional health status. There is also a proverb in Ayurveda "when diet is wrong, medicine is of no use, when diet is correct medicine is of no need". Present consumers are becoming more aware of relationships between diet and health and this has increased consumer interest in the nutritional values of foods. This is impacting on the demand for foods which contain functional components that play important roles in health maintenance and disease prevention. The meat industry is one of the most important in the world and, whether as a result of consumer demand or because of the ferocious competition in the industry; there is continuous research for new products.

Demand for functional foods is growing now days with the claims and a clear mandate of health and prevention of diseases as one of its main goals. Functional foods have developed a tremendous increase in interest of consumers due to their inherent ability of providing health benefits. Regulations were also implemented for the production and sale of functional foods because it has shown sustainable growth in the Asian markets. To maintain and improve health of the consumers, functional foods should be enriched with nutrients or other bioactive compounds. A working definition of functional food; "A food that beneficially affects one or more target functions in the body beyond adequate nutritional effects in a way that is relevant to either an improved state of health and well-being and/or reduction of risk of disease" (Yadav, 2013). In functional food, components are added or removed or modified by technological or by other means but they are in their natural form. Some of the qualities have also been advised for a functional food and they are as follows (Rinco n-Leo´n, 2003):

1. It is a food in its natural form.
2. It should be consumed as part of the daily diet.
3. It is having specific functions within body to regulate a particular process, such as enhancement of biological defence mechanisms, prevention and recovery from specific diseases and slowing of the aging processes.

As per WHO/FAO, the dietary patterns and lifestyle habits constitute major risk factors in the development of various diseases such as coronary heart disease, cancer, type 2 diabetes, obesity, osteoporosis and periodontal disease (WHO, 2003). Functional foods play vital role in eliminating such factors. But still the commercial success of such products eventually depends on taste, appearance, price, and health claim appeal.

2. Classification of Functional Foods

They are classified on the basis of potential health benefits and also on their ingredients properties. Health promoting components in such foods may be classified into different groups and one such classification has been given by Rinco´n-Leo´n (2003):

1. Dietary fiber
2. Oligosaccharides
3. Sugar/alcohols
4. Amino acids, peptides and proteins
5. Glycosides
6. Vitamins

7. Choline
8. Lactic acid bacteria
9. Minerals
10. Polyunsaturated fatty acids
11. Others (like phytochemicals and antioxidants)

3. Development of Functional Meat

Meat and meat products are potential source of high quality proteins containing all essential amino acids, fatty acids, vitamins, mainly B complex vitamins especially B_{12}; minerals, mainly iron and zinc of high bioavailability and manganese and other bioactive compounds (WHO, 2003). The only disadvantage is the lack of dietary fiber. Therefore, strategies were developed to reduce the health related risk factors of meat and to promote the health beneficial effects of meat. Strategies have been basically formulated to reduce the negative health implications of meat and meat product and/or increase the presence of beneficial compounds. Development of functional meat was based on two ways; Animal Production Practices and Reformulation Processes.

4. Animal Production Practices

It leads to a very challenging opportunity to modify the presence of natural bioactive components. Because the composition of animal tissues, or that of carcasses vary with many factors *i.e.* species, breed, age, sex, feeding type *etc.* The role of animal production practices depends on genetic strategies and feed management.

4.1 Genetic Strategies

Through this we can produce changes related to the production of healthy food for human diet which is also having economic and productive perspective.

- Selection and interbreeding
- Use of genetic markers for identification of carcass characteristics, fat content and fatty acid profile
- *In vitro* – meat production systems in labs.

4.2 Feed Management

It includes alteration in the diet of animals, to alter a particular component in its meat. More attention is given to lipids among all the bioactive compounds, for the development of healthier meat products due to its harmful health implications. Any compound can be enriched in the meat by supplementing it, in animal diet for example PUFA (polyunsaturated fatty acids). Feeding of many plant and marine sources (fish or algae) have been used to significantly increase PUFA levels in mutton, beef and pork (Raes *et al.*, 2004; Scollan *et al.*, 2006). Dietary supplementation has been used to enrich chicken, pork, beef and lamb with CLA (conjugated linoleic acid) (Raes *et al.*, 2004).

Muscle antioxidant level should also be increased through diet as there is a risk of increased per oxidation in muscle foods due to increasing unsaturated fatty acids. Farm animal dietary supplementation with vitamin E has increased the concentration of this compound in animal tissue to the extent that meat became a moderate vitamin E source. Supplementation with selenium, magnesium or iron has been employed to increase the concentration of these healthy elements in meat (Olmedilla-Alonso *et al.*, 2013). Lambs fed with concentrates having a rich source of carotenoids and fat soluble vitamins like retinol and tocopherol have been observed to contain significantly higher retinol and tocopherol levels in their body fat and meat products prepared from these lambs could contribute to health beneficial effects (Alvarez *et al.*, 2014).

4.3 Enrichment with Minerals

Enrichment of meat products with some of the essential minerals like selenium, calcium, zinc *etc.*, is becoming new trend in food industry, even though meat contain minerals like iron and has meat factor characteristics. In several broiler performance trials, organic Se supplementation showed a positive effect on weight gain and FCR (Feed Conversion Ratio) compared to controls (Spring, 2008). Birds supplemented with increasing levels of Se were better able to cope with the challenges of the reoviruses wasting disease. Several studies have also shown that supplementation with organic Se reduces drip loss in meat. Water holding capacity is also an important characteristic of meat, as it determines the level of exudative loss in packaging and during cooking, and the juiciness of meat. Organic Se exerts an antioxidant effect on the birds' cellular membranes and tissue structures resulting in less exudative losses from meat (Pan *et al.*, 2011). Organic Se (Selplex) at 0.1-0.3ppm can be added as anti-oxidants to the poultry diet (Surai, 2010).

4.4 Incorporation of Vitamins and Antioxidants

Meat is an excellent source of vitamins however, it is deficient in certain vitamins like vitamin C (ascorbic acid) and also has lower levels of vitamin E (tocopherol). Vitamin E was reported to improve phagocytic ability of the immune system (Boa-Amponsem *et al.*, 2000) in broilers. Konjufca *et al.* (2004) reported that supplemental α-tocopherol acetate enhances Fc-receptor-mediated macrophage phagocytic activity at early stages (up to 3 weeks) of broiler growth. Niu *et al.* (2009) observed that heat stress (23.9–38°C) severely reduced growth performance, feed intake, feed conversion and immune response of broilers, while dietary vitamin E supplementation improved the immune response of broilers under heat stress. Vitamin E was also reported to reduce the lipid oxidation (malondialdehyde concentration) in breast and thigh meats during refrigerated storage (Goñi *et al.*, 2007). For designer egg/meat production, Vitamin E at levels of 200-400mg/kg is supplemented in the diet.

Poultry meat is rich source of natural antioxidants like vitamin-E, Se, carotenoid pigments, flavonoid compounds, lecithin and phosvitin but at the same time, is highly susceptible to oxidative rancidity during storage. These antioxidants will protect the fat-soluble vitamins from oxidative rancidity. The designer meat, not only contain high levels of the above anti-oxidants but also contain synthetic anti-oxidant like ethoxyquin and anti-oxidants of herbal origin such as curcumin, cycopene, cuercetin and sulforaphene, depending upon the herbs used in the poultry diet (Narahari, 2005). Along with antioxidants like Vitamin E and Se, the enzymes like glutathione peroxidase, superoxide dismutase, catalaze constitute an integral part of antioxidant cellular enzyme system in omega-3 enriched products to reduce lipid peroxidation. The dietary supplementation of vitamin E is commonly used in commercial n-3 enriched products to mitigate the oxidation of n-3 FA, thereby preventing the formation of undesirable fishy flavour and warmed over flavour in refrigerated cooked and raw meat (Fraeye *et al.*, 2012). Besides these, other anti-oxidants as chemicals and herbs may be added, to prevent oxidative rancidity

5. Reformulation Processes of Functional Food

A large number of approaches are used to add or replace different bioactive components. Modification of meat formulation process includes use of such ingredients specifically designed with certain attributes that confer health-promoting properties. Olmedilla-Alonso *et al.* (2013) has classified reformulation processes of meat as follows.

5.1 Improving Fat Content

There are three ways for reformulation process of altering fat content in meat; reduce total fat and energy, decline in the level of cholesterol and alteration of fatty acid profiles (Jiménez Colmenero *et al.*, 2012). Reduction of fat is based on two main criteria, use of leaner meat and secondly, by adding water and other ingredients like carbohydrate based fat replacers or gums or protein based fat replacers with little or no calorific content (Kerry and Kerry, 2006). Conjugated linoleic acid (CLA) is having varied properties such as anti-oxidant, anti-carcinogenic, immune-modulative and anti-obesity properties and helps in regulating bone metabolism and reduces the risk of diabetes (Bernardini *et al.*, 2011). It is naturally present in the meat and milk of grass fed ruminants. Commercially produced CLA isomers have been injected into whole muscles to achieve enough levels of CLA to produce health benefits when consumed in smaller portions (Baublits *et al.*, 2007; Juarez *et al.*, 2009).

6. Other Added Compounds/Ingredients

Some other ingredients are also incorporated in meat for health benefits like sterols, taurine *etc.* These are incorporated into products either as purified compound or as non-meat ingredient like walnut, soybean, seaweed, carrot, and herbs which contain these compounds (Olmedilla-Alonso *et al.*, 2013). Some of the bioactive peptides have been identified from milk and meat and these peptides are having properties like antihypertensive, opioid, anti-oxidant and anti-thrombotic (Lafarga and Hayes, 2014). Castellano *et al.* (2013) observed that anti-hypertensive bioactive peptide (ACE inhibitory peptide) generated from porcine muscle protein by the action of lactic acid bacteria could be incorporated in preparing healthier functional meat.

7. Reduction of some Unhealthy Components

Cholesterol content of chicken meat contains about 60 mg per 100 g (USDA,

1991). Feeding dehydrated alfalfa free of choice also produce lean chicken breast meat as alfalfa is a good source of saponins which is hypocholesterolaemic in nature (Ponte *et al.*, 2004). A reduction of serum cholesterol has been reported in broilers fed with Lactobacillus culture (Jin *et al.*, 1998). Dietary supplementation of amino acids like glycine, lysine, methionine and tryptophan can decrease body fat deposition (Takahashi *et al.*, 1994). The carcass cholesterol levels can be significantly reduced by supplementing herbal plants and products like basil (tulsi), bay leaves, citrus pulp (nirangenin), garlic, grape seed pulp guar gum, roselle seeds, spirulina, tomato pomace (lycopene),and many more herbs in chicken diets will reduce the chicken cholesterol levels by 10-25 per cent. Canola oil, linseed oil, soybean oil and sunflower oil, reduced fat and cholesterol content in cockerel thigh and breast meat. Moreover, these substances act synergistically in reducing the cholesterol levels. Hence a combination of these supplements will be more beneficial, rather than a single substance.

7.1 Lean Meat Production

The present health conscious consumers prefer chicken meat with high protein, low fat and low cholesterol. In fact, chicken and yolk fats are not fats, but oil; because their saturated fatty acid content is only one third of the total fatty acid content, nearly 45 per cent is cardiac friendly monounsaturated fatty acids (MUFA) and the remaining are polyunsaturated fatty acids (PUFA). By dietary manipulations, lean meat with low carcass fat will lower the cholesterol levels and abdominal fat pad thickness; but this feed manipulation is not recommended as it will impart an undesirable 'fishy taint' to the meat. An increase in the lysine level in the pre-starter diets and methionine level in the finisher diets will yield lean breast meat in broilers.

8. Conclusion

In India, only few functional products especially of milk and egg are available in the market but, meat based functional products are still under research. Efforts have to be taken to incorporate certain beneficial bioactive compounds in meat and meat products and market those functional meat products. There are several laws and regulations regarding food market in the country but still there is dearth of significant law for regulation of functional foods. Now, a new set of regulations namely Food Safety and Standards Regulations, is under process which may help to integrate and stream line the several regulations covering the nutraceuticals, foods and dietary supplements.

REFERENCES

Alvarez, R.A., Mele´ndez-Martý´nez, A.J., Vicario, I.M., and Alcalde, M.J. (2014). Effect of pasture and concentrate diets on concentrations of carotenoids, vitamin A and vitamin E in plasma and adipose tissue of lambs. J. Food Comp. Anal., 36, 59–65.

Baublits,R.T., Pohlman,F.W., Brown Jr,A.H., Johnson,Z.B., Proctor, A., Sawyer,J. Dias-Morse, P., and Galloway, D.L. (2007). Injection of conjugated linoleic acid into beef strip loins. Meat Sci.,75, 84–93.

Bernardini, R.D., Harnedy, P., Bolton, D.,Kerry, J., ONeill, E., Mullen. A.M., and Hayes, M. (2011). Antioxidant and antimicrobial peptidic hydrolysates from muscle protein sources and by-products. Food Chem., 124,1296–1307.

Boa-Amponsem, K,. Price, S.E., Picard, M., Geraert, P.A. and Siegel, P.B. (2000). Vitamin E and immune responses of broiler pureline chickens. Poultry Science, 79(4): 466-470.

Castellano, P., María-Concepción, A.,Sentandreu, M.A.,Vignolo, G., and Toldrá, F. (2013).Peptides with angiotensin I converting enzyme (ACE) inhibitory activity generated from porcine skeletal muscle proteins by the action of meat-borne Lactobacillus. J. Proteom., 89,183–190.

Fraeye, I., Bruneel, C., Lemahieu, C., Buyse, J., Muylaert, K., and Foubert, I. (2012). Dietary enrichment of eggs with omega-3 fatty acids: A review. Food Research International, 48(2): 961-969.

Goñi, I., Brenes, A., Centeno, C., Viveros, A., Saura-Calixto, F., Rebolé, A., Arija, I., and Estevez, R. (2007). Effect of dietary grape pomace and vitamin E on growth performance, nutrient digestibility, and susceptibility to meat lipid oxidation in chickens. Poultry Science, 86: 508-516.

Jiménez-Colmenero, F., Herrero, A., Cofrades, S., and Ruiz-capillas, C. (2012). Meat and Functional foods. In: Y.H. Hui (ed), Hand book of meat and meat processing, (2nd ed), Boca Raton: CRC Press, Taylor and Francis group, pp.225-248.

Jin, L.Z., Ho, Y.W., Abdullah, N. and Jalaludin, S. (1998). Growth performance, intestinal microbial populations, and serum cholesterol of broilers fed diets containing Lactobacillus cultures. Poultry Science, 77, 1259-1265.

Juarez, M., Marco, A., Brunton, N.,Lynch, B. Troy, D.J., and Mullen, A.M. (2009). Cooking effect on fatty acid profile of pork breakfast sausages enriched in conjugated linoleic acid by dietary supplementation or direct addition. Food Chem., 117, 393–397.

Kerry, J.F., and Kerry, J.P. (2006). Producing low–fat meat products. In: C.Williams and J., Buttriss (ed.) Improving the fat content of foods: Woodhead Publishing Limited and CRC Press, LLC University College Cork, Cambridge, pp. 336-379.

Konjufca, V.K., Bottje, W.G., Bersi, T.K., and Erf, G.F. (2004). Influence of dietary vitamin E on phagocytic functions of macrophages in broilers. Poultry Science, 83(9), 1530-1534.

Lafarga, T., and Hayes, M. (2014). Bioactive peptides from meat muscle and by-products: generation, functionality and application as functional ingredients. Meat Sci., 98, 227–239.

Narahari, D. (2005). Nutrient manipulations for value added eggs and meat production. Conference of Indian Poultry Science Association and National Symposium-2005.

Niu, Z.Y., Liu, F.Z., Yan, Q.L., and Li, W.C. (2009). Effects of different levels of vitamin E on growth performance and immune responses of broilers under heat stress. Poultry Science, 88, 2101-2107.

Olmedilla-Alonso, B., Jiménez-Colmenero, F., and Sánchez-Muniz, J. (2013). Development and assessment of healthy properties of meat and meat products designed as functional foods. Meat Sci., 95, 919–930.

Pan, C., Zhao, Y., Liao, S.F., Chen, F., Qin, S., Wu, X., Zhou, H., and Huang, K. (2011). Effect of selenium-enriched probiotics on laying performance, egg quality, egg selenium content, and egg glutathione peroxidise activity. Journal of Agricultural and Food Chemistry, 59(21): 11424-11431.

Ponte, P.I.P., Mendes, I., Quaresma, M., Aguiar, M.N.M., Lemos, J.P.C., Ferreira, L.M.A., Soares, M.A.C., Alfaia, C.M., Prates, J.A.M., and Fontes, C.M.G.A. (2004). Cholesterol levels and sensory characteristics of meat from broilers consuming moderate to high levels of alfalfa. Poultry Science, 83, 810-814.

Raes, K., De Smet, S., and Demeyer, D. (2004). Effect of dietary fatty acids on incorporation of long chain polyunsaturated fatty acids and conjugated linoleic acid in lamb, beef and pork meat: a review. Animal Feed Science and Technology, 113, 199–221.

Rinco´n-Leo´n, F. (2003). Encyclopedia of Food Sciences and Nutrition. (2nd Ed). Elsevier Science Ltd, pp. 2827-2832.

Scollan, N., Hocquette, J., Nuernberg, K., Dannenberger, D., Richardson, I., and Moloney, A. (2006). Innovations in beef production systems that enhance the nutritional and health value of beef lipids and their relationship with meat quality. Meat Sci.,74,17–33.

Spring, P. (2008). Maximising meat quality with selenium. International Poultry Production, 15(5), 7-11.

Surai, P.F. (2010). Natural antioxidants in poultry nutrition: new developments. In Proceedings of the 16th European symposium on poultry nutrition (pp. 669- 675).

Takahashi, K., Konashi, S., Akiba, Y., and Horiguchi, M. (1994). The effect of dietary methionine and dispensible amino acid supplementation on abdominal fat deposition in male broiler. Journal of Animal Science and Technology, 65, 244-250.

USDA (U.S. Department of Agriculture) (1991). Composition of Foods, USDA Handbook 8.Washington, DC: U.S. Government Printing Office.

WHO. (2003). Diet, nutrition and the prevention of chronic diseases. Report of a Joint WHO/FAO Expert Consultation. WHO Technical Report Series 916, Geneva.

Yadav, M.R. (2013). Nutritional management through functional foods. Souvenir of 7th International Food Convention (IFCON), NSURE-Healthy foods, CFTRI, Mysore, pp. 359-362.

Chapter 2

Organic Meat Production and its Future

M. Raquib[1]*, S.K. Laskar[1] and S. Das[2]

[1]Department of Livestock Products Technology,
College of Veterinary Science, Assam Agricultural University,
Khanapara, Guwahati – 781 022, Assam
[2]Division of Animal Health, ICAR Research Complex for NEH,
Barapani –793 103, Umiam, Meghalaya
**e-mail: masuk1@rediffmail.com*

ABSTRACT

There has been a recent upsurge in the demand for organic food, among the consumers in the developed countries as well as among the elite and wealthy section of developing nations. With the ensured availability of food, the consumer demand increases for food with good taste and variety along with guarantee of food safety. The growing consumer preference towards organic food is due to increasing awareness about the ill effects of consuming foods having chemical residues in form of carcinogens, hormones, pesticides, drugs, antibiotics, etc. The organic products are acceptable in-spite of their premium prices, among the consumers with high purchasing power. The rising demand of organic products has been attracting the meat industry to initiate the venture for organic meat. It is in this context, an attempt is made to discuss the advantages of organic meat production, various mean and procedure to meet the industry standards.

Keywords: *Organic, Meat, Safe meat.*

1. Introduction

Since pre-historic times, humankind has been utilizing meat as a source of food and nutrition for their survival. Still most people in modern society enjoy meat in different forms, as a daily item in the diet or an item of delicacy, to not only satisfy their taste buds but also fulfill their nutritional requirements. However, the incidence of BSE (Bovine Spongiform Encephalopathy) epidemic in cattle, use of dioxin in animal feed, use of growth hormone and use of antibiotics as feed additives *etc.*, has raised a question on food safety. Food safety and quality is now an important issue discussed more strongly now than ever before on international platform. In lieu with the effects of environmental contamination of food, as well as with the increasing consciousness on animal welfare, the developed countries are searching for alternative livestock production systems, which could ensure safe food to them without disturbing environment and maintaining high standards of animal welfare. As a result, many consumers are seeking alternatives to conventionally produced meat which would be safe to eat and simultaneously will be safe to the environment. Organically produced meat is such an alternative to conventionally produced meat and the demand for this 'organic meat' is sharply increasing day by day in the so-called developed countries. Organic farming is a biodynamic farming system founded by philosopher Rudolf Steiner in Germany in 1924. The concept of organic farming had an interest in natural food, nutritional health and freshness. The Codex Alimentarius Commission Guideline (CAC/GL 32-1999) defines organic farming as an 'holistic production management system which promotes and enhances agro-ecosystem health including biodiversity, biological cycles and soil biological activity" and have established the guidelines and principles of organic production from the stages of production, processing, storage and transport to labeling and marketing. Lampkin (1990)-defined organic agriculture as, "a production system which avoids or largely excludes the use of synthetically compounded fertilizers, pesticides, growth regulators and livestock feed additives. To the maximum extent feasible, organic farming systems rely on crop rotations, crop residues, animal manures, legumes, green manures, off-farm organic wastes and aspects of biological pest control to maintain soil productivity to supply plant nutrients, and to control insects, weeds, and other pests".

FAO/WHO Codex Alimentarius Commission management system is the apex organization, which promotes and enhances health of agricultural ecosystem, which includes use of on-farm agronomic, biological and mechanical methods in exclusion of all synthetic off-farm inputs in the total farming activity involving biodiversity, biological cycles and soil biological activity. The main aim of organic farming is to establish and maintain soil-plant, plant-animal and animal-soil interdependence and to produce a sustainable agro-ecological system based on the local resources. Organic farming does not require and depend on the external resource inputs in any form of feed additives, fertilizer, pesticides, growth hormones, antibiotics *etc.* but the system rely on the available natural ingredients as input for feed, fodder, water by managing ecosystem in the best possible manner in a sustainable way to use the natural inputs only. The agriculture cultivations are incomplete without livestock component as they play a major role in agriculture cycle mainly as manure under the organic livestock production. Animal husbandry itself is an important component in developing an actual organic agriculture system but there are a number of challenges in developing organic agriculture and to carry forward for organic meat production, the challenges are even more. Thus, the establishment of organic animal/poultry husbandry requires a careful and gradual approach (FAO, 2015).

In short, 'organic meat' may be considered as the one which is obtained from domestic animals or poultry raised in an organic system, where the basic physiological needs like feed, water, medication of animals or birds are fulfilled through the naturally grown resources. Here, the behavioral needs of the animals are intended to satisfy through a congenial environment so that they can get adequate space for free activities and live more like a free animal or bird, growing in natural environment. The inputs required for farming needs to be grown in organic condition to certify the final product/output as organic one. These criteria may be economically tough to raise organic animals intended for organic meat production under such conditions. In case of organic animals, it may not be feasible to raise them under 100 percent organic condition, but largely efforts are made to pull off the maximum criteria. Animals are not caged, tethered or confined in buildings without adequate natural ventilation and lighting. The animals are given enough room for free movement and kept in appropriate herd size and flocks. Suitable and natural bedding materials, access to pasture and fresh water are essential. The health and vitality of the animals are maintained by sound nutrition and good management practices, prophylactic antibiotics should be avoided. All growth promoters and hormones are prohibited. Veterinary drugs are allowed only where there are no effective complementary treatments. Withdrawal periods after giving a veterinary drug are strict in order to prevent residues in meat (Srednicka-Tober *et al.*, 2016).For organic meat production to expand in a sustainable way, consumers must perceive it as good as conventional production regarding environmental quality and price. Therefore, possible future organic and conventional meat production need to be compared regarding production costs, land requirements, soil conservation, nature conservation, energy needs, and chemical requirements as well as the discharge of nitrogen and greenhouse gases.

2. Status of Organic Farming in World *vis-a-vis* India

Due to the increased health awareness and consumer preferences, organic land-area as well as organic livestock/poultry farming is increasing day by day throughout the world. Production and marketing of organic food is mainly confined to the developed countries like USA, EU, UK, and some developing countries including Argentina, Brazil, *etc.* Production of organic meat and meat products are one of the fastest growing sector in the developed countries (Kamihiro *et al.*, 2015). The reason behind its development in these countries is their food safety, consciousness and surplus production of these commodities. On the contrary, the scenario is reverse in most of the developing countries including India where population growth is a great problem and food security is a priority than food safety. Agricultural and Processed Food Products Export Development Authority (APEDA) declared that exports of organic food produce stood at Rs 5151 crore (2018-19) as against Rs 3,453 crore in 2017-18. The export volume in 2018-19 was 6.14 million tones (https: //www.grainmart.in/news/.). Organic products are exported to USA, European Union, Canada, Switzerland, Australia, Israel, South Korea, Vietnam, New Zealand, Japan *etc.* Almost all products exported from India are of plant origin, whereas, India has large number of livestock and poultry population in the world. Even a small shift from current conventional production to organic animal production can create a huge market to domestic consumption as well as export. Currently, 130 countries are producing certified organic products. Some developing countries like Argentina, Brazil and Mexico are now exporting organic animal products to the developed countries. Our country has a vast scope for promotion of organic farming in the export market, without compromising with the national food security as farming by tribals and under rainfed conditions is generally organic by default, since very little chemical inputs are used (NAAS, 2005).

3. Legal Requirements for Certification Process

The logo of organic itself adds value to a food commodity in today's market for privileged section of consumers. A food product has to fulfill certain legal requirements, prerequisite in order to label the product organic in nature. Certification is essentially aimed at facilitating and regulating sale of organic food products to the consumers, which are supposed to be superior in quality. In many countries, organic standards are formulated and supervised by the government. Countries like United States, the European Union and Japan have their own comprehensive organic legislation, and the term "organic" may be used only by certified producers while in countries without organic laws certification process is handled by non-profit organizations and private parties.

Countries	*Certifying Agency*
UK	UK Register of Organic Food Standards (UKROFS)
US	National Organic Program (NOP)
Sweden	Private corporation KRAV
Japan	Japanese Agricultural Standard (JAS)
Australia	Australian Quarantine and Inspection Service (AQIS)
China	China Green Food Development Center
India	Agricultural Processed Food Products Export Development Authority (APEDA) under Ministry of Commerce
International certifying bodies	
International Federation of Organic Agriculture Movements (IFOAM),	
Organic Crop Improvement Association (OCIA)	
Ecocert is an ***organic certification organization***, founded in France in 1991 and is based in Europe but conducts inspections in over 80 countries	

In India, APEDA under Ministry of Commerce is the controlling body for certification of commodities for export as organic. Till date there are no domestic standards for organic produce within India, and they are in process of development with 11 certification agencies been active to undertake certification process under National Programme for Organic Production (NPOP) and in 2006 this has been granted equivalence with European Union. For export consignment of meat products, the registered processing establishment has to apply for health certificate through Meat.Net online system where compulsory microbiological and other tests are carried out in government laboratories.

4. Organic Meat Production Practices

Organic meat production is considered as an ecologically balanced approach for production of meat animals. It promotes raising meat animals in a stress free zone using organic and biodegradable inputs from the ecosystem deliberately avoiding the use of synthetic inputs such as drugs, feed additives and genetically engineered breeding inputs, while ensuring the welfare of animals (Chander *et al.*, 2011; Chander *et al.*, 2012; Chander and Subrahmanyeswari, 2013). For sustainable organic meat production, the consumers must be convinced as regards to the quality of the organic meat products compared to the conventional meat products and they need to get satisfied with the premium price paid for the same. Reasonable comparative assessment between organic and conventional meat production in terms of cost of production, land requirements, environmental safety and chemical requirements are necessary before initiating the organic meat production practices in a particular ecosystem. In order to get certification for organic produce, it has to comply with the different laid down standards.

4.1 Animal Husbandry Management

The management of the animal should take into account the behavioural needs of the animals by providing sufficient free movement, sufficient fresh air and natural daylight according to the needs of the animals, protection against excessive sunlight, temperatures, rain and wind, enough lying and/or resting natural bedding to be provided, ample access to fresh water and feed according to the needs of the animals, adequate facilities for expressing behaviour in accordance with the biological and ethological needs of the species and access to open air and/or grazing appropriate to the type of animal and season welfare is not compromised. Poultry and rabbits shall not be kept in cages (Chander *et al.*, 2011; Patel *et al.*, 2017).

4.2 Conversion Period

Animal products may be sold as "product of organic agriculture" only after the farm or relevant part of it has been converted for at least twelve months and organic animal production standards have been met for the appropriate time. Regarding dairy and egg production, this period shall not be less than 30 days. Animals present on the farm at the time of conversion may be sold for organic meat if the organic standards have been followed for 12 months.

4.3 Newly Purchased Animals

When organic livestock is not available, the following age limits holds good for brought-in conventional animals:

- 2 day old chicks for meat production
- 18 week old hens for egg production
- 2 week old in case of any other poultry
- Piglets up to six weeks and after weaning
- Calves up to 4 weeks old which have received colostrum and are fed a diet consisting mainly of full milk.

- ✰ Breeding stock may be brought in from conventional farms at an annual rate not exceeding 10 per cent of the adult animals of the same species in the organic farm.

4.4 Breeds and Breeding

The certification programme shall ensure that breeding systems are based on breeds that can both copulate and give birth naturally. Artificial insemination, embryo transfer techniques, hormonal heat treatment, use of genetically engineered species or breeds are not allowed. Use of native breeds that suits to the agro-climatic zones are encouraged in terms of diseases resistance, hardiness and productivity of the animals.

4.5 Animal Nutrition

Landless animal husbandry system is not ideally suitable for organic livestock farming. At least more than 50 per cent of the feed should come from the farm unit itself or shall have to be produced in co-operation with other organic farms in the region. Wherein, if it is impossible to obtain certain feeds from organic farming then a percentage of feed consumed by farm animals to be sourced from conventional farm. The maximum percentages of such feeds shall be ruminants (dry matter intake) 15 per cent, non-ruminants (dry matter intake) 20 per cent and these has to be reduced within 5 years. Synthetic growth promoters or stimulants, synthetic appetisers, urea, farm animal by-products (*e.g.* abattoir waste), all types of excreta, feed subjected to solvent (*e.g.* hexane) extraction (soya and rape seed meal), pure amino acids and genetically engineered organisms or products thereof are to be excluded, however, vitamins, trace elements and supplements shall be used from natural origin.

4.6 Veterinary Medicine

The use of conventional veterinary medicines are allowed, when no other justifiable alternative is available. Where conventional veterinary medicines are used, the withholding period should be strictly adhered for at least double the legal period. Vaccinations are to be used only when diseases are known or expected to be a problem in the farm and where these diseases cannot be controlled by other management practices. However, genetically engineered vaccines are prohibited in organic production system.

4.7 Transport and Slaughter

The animals intended for organic production has to be handled gently without any stress. The use of electric pod and such instruments are prohibited during loading and unloading of animals in ramps. The other considerations that has to be taken during transportation includes fitness of the animal, quality and suitability of mode of transport and handling equipment, temperatures and relative humidity on the day of journey, hunger and thirst.

5. Organic versus Inorganic Meat

Consumers have a perception that organic food is more nutritious than non-organic foods; however, no systematic review comparing specifically the nutrient content of organic versus inorganic meat is available. The main driving force for consumers preference lie in the fact that they have a perception that organic livestock product contains higher concentration of nutritionally desirable compounds, thereby making them healthier. Meat is an important source of proteins, essential fatty acids (FA), minerals (F, Zn, Se, Cu) and vitamins (vitamin A, vitamin B_1, B_6 and B_{12}, riboflavin, folate, niacin, pantothenic acid). However, a comparative study on proteins, minerals and vitamins is lacking except some works related to fatty acid profile between conventional and organic meat has been published (Smith-Spangler *et al.*, 2012). Srednicka-Tober *et al.* (2016) reported significant differences in FA profiles between organic and conventional meat. Concentrations of SFA and MUFA were similar or slightly lower, respectively in organic compared with conventional meat. Larger differences were detected for total PUFA and n-3 PUFA, which were 23 per cent and 47 per cent higher in organic meat, respectively.

6. Challenges

Unlike the overall food industry, the organic food industry is still in its nascent stage after almost thirty years of its creation. However, there is continuous rise in demand despite higher price. The major challenges lies for the conversion of conventional farming to organic farming is the time taken to convert a conventional farm to organic one wherein the farm must lie fallow for three years before being considered organic and any postharvested products (meat, milk and egg) produced during that time must be sold as conventional. Secondly, organic farm operations are subjected to added fee and regulations. Thirdly, consumers of organic products tend to fall in high income group, thus the middle and lower income groups are reluctant to purchase at such an exorbitant price. Fourthly, as the farmers demand high organic premium to overcome the insecurities inherent in organic farming like low production and yield, there is shortage of supply inspite of the fact that there is continuous rise in demand despite higher price and lastly, the yield is uncertain and hence there is every risk on the part of the producers (Cooper *et al.*, 2007; Escribano, 2016).

7. Opportunities

In spite of challenges, there is tremendous scope for initiating organic meat production along with other livestock products. Initially, it may be difficult to capture the domestic market due to higher production cost and resulting high price of such products and the poor purchasing power of domestic consumers. Therefore, export oriented production adhering to strict quality norms may be initiated with proper help and technical support from APEDA. In this context, it has been found that already some developing countries like Argentina, Brazil and Namibia are producing and exporting organic meat to developed countries. Indian organic livestock sector should try to follow these ideal examples. Consumers from developed countries are quite capable and ready to pay the premium price for organic livestock products, as they are more health conscious. Besides, the native Indian livestock breeds have high disease resistance capacity and they are rarely exposed to pesticides, veterinary drugs and antibiotics. The huge indigenous livestock population thus, can provide a boost to organic meat production. Presently, in India, rapid urbanization, growing literacy and influence of electronic and print media has resulted in rising awareness among the consumers about food safety issues, mostly there is a growing concern about the ill effects of drugs, chemicals and pesticide residues in foods of vegetable as well as in animal origin. Under such stipulation, domestic consumption of organic foods, including livestock products is likely to be augmented. Up scaling of Indian traditional methods of livestock rearing and production and processing of certain traditional meat and other livestock products in organic way may open up tremendous opportunity for exporting such products to the developed countries. Even in the domestic market also, it is grossly evident that local/*desi* chicken, duck and other poultry meat and eggs, which are produced and reared under free range or backyard system, are sold at much higher price than the products from farm raised poultry and eggs. Thus, the rising consumer demand for good quality food products in the domestic market warrant the need for establishment domestic market for organic livestock products to fulfill the aspiration of local consumers particularly in urban areas.

8. Conclusion

In view of the high cost of production, sustainable organic food production is a complicated venture in a developing country like India. Considering huge population of the country living below poverty line, in reality food security is more important than food quality. However, on the other hand, organic food products like crops, fruits, vegetables, milk, meat and egg production may boost up the economy by generating employment, increasing income of the farmers and producers besides inspiring inventiveness. There is potential scope to develop an organic meat market channel to rich countries. India already has a well-organized export market for carabeef and mutton. Establishment of an export market for organic meat is very much feasible with suitable approach in this regard. In order to establish the domestic as well as export market for organic livestock products, well-organized approach is needed through research and development activities, creating awareness among the consumers regarding health benefits and economics of organic food production. Proper training for capacity building and embellishment of model organic livestock farms and processing units will be the necessary steps to initiate organic livestock production. The animal husbandry department and other government agencies along with researchers and academicians should come forward with sustainable technologies and create suitable marketing channels to provide adequate support to the farmers and entrepreneurs to carry on organic livestock and meat production.

REFERENCES

Chander M., and Subrahmanyeswari B. (2013). Organic Livestock Farming. Directorate of Knowledge Management in Agriculture, ICAR, New Delhi, 2013, pp. 293.

Chander, M., Rathore, R.S., Mukherjee, R., Mondal, S.K., and Kumar, S. (2012). Road map for organic animal husbandry development in India. In: Gerold Rahmann and Denise Godinho (Eds), Tackling the future Challenges of Organic animal Husbandry, Proc. 2nd Organic Animal Husbandry Conference, Hamburg, Trenthorst, 12-14 September, pp.59-62.

Chander, M., Subrahmanyeswari, B., Mukherjee, R., and Kumar, S. (2011). Organic Livestock Production: An Emerging Opportunity With New Challenges For Producers In Tropical Countries, India. Rev. Sci. Tech. Off. Int. Epiz, 30(3), 969-983.

Cooper, J., Niggli, U., and Leifert, C. (2007) Handbook of Organic Food Safety and Quality. Cambridge: CRC Press.

Escribano, A.J. (2016). Organic livestock farming – challenges, perspectives, and strategies to increase its contribution to the agrifood system's sustainability – a review. In: Organic Farming - A Promising Way of Food Production. Konvalina, P (Ed.). IntechOpen books.

FAO (2015). Animal Husbandry in Organic Agriculture. http: //www.fao.org/3/ CA2560EN/ ca2560en.pdf.

https: //www.grainmart.in/news/organic-food-exports-from-india-rise-by-50-to-inr-5151-crore-apeda/

Kamihiro, S., Stergiadis, S., Leifert C., Eyre, M. D., and Butler, G. (2015) Meat quality and health implications of organic and conventional beef production. Meat Sci., 100, 306–318.

Lampkin N. (1990). Organic farming. Farming Press, Ipswich. 1990, 1-4.

NAAS (2005). Organic Farming Policy 2005, Ministry of Agriculture, Department of Agriculture and Cooperation. http: //ncof.dacnet.nic.in/Policy_and_EFC/ Organic_ Farming_Policy_2005.pdf.

Patel, S.J., Kumar, U., Chaudhari, R.P., and Darji, S.S. (2017). Organic Farming in Animal Husbandry Practices. Indian Journal of Hill Farming., special issue, 55-60. http: //epubs.icar.org.in, www.kiran.nic.in; ISSN: 0970-6429

Smith-Spangler, C., Brandeau, M.L., Hunter, G.E., Bavinger, J. C., Pearson M., Eschbach, P.J., Sundaram, V., Liu, H., Schirmer, P., Stave, C., Olkin, I., and Bravata, D.M. (2012) Are organic foods safer or healthier than conventional alternatives? A systematic review. Ann Intern Med 157, 348–366.

Srednicka-Tober, D., Barański, M., Seal, C., Sanderson, R., Benbrook, C., and Steinshamn, H. (2016). Composition differences between organic and conventional meat: a systematic literature review and meta-analysis. Brazil Journal of Nutrition. 115(6), 994-1011. doi: 10.1017/S0007114515005073.

Chapter 3

Traceability in Meat Quality and Safety

R.P. Kolhe[1*] and K.N. Bhilegaonkar[2]

[1]KNP College of Veterinary Science, Shirwal, Dist. Satara, Maharashtra
[2]ICAR-IVRI Training and Education Center, Pune, Maharashtra
**e-mail: rpkolhe@gmail.com*

ABSTRACT

Meat traceability means the origin of meat should be known, its geographical location of farm, animal ID, slaughtering place, processing and packaging industry etc., so that if any issue arises in terms of quality, contamination and/or on legal front, the origin of meat can be traced. Systematic documentation and maintenance of records from farm to the processing plant will ensure the meat traceability. Consumer is willing to pay more for ensuring the safe products involving meat that can be traced to a genuine producer, it gives the assurance that the meat in food on the platter is authentic and had originated from the right place. The increase in the awareness of consumers is making traceability a happening in the meat industry.

Keywords: *Meat, Traceability, Meat quality, Meat safety.*

1. Introduction

Food is basic and fundamental requirement of every individual living on this planet. Moreover, availability and access to safe, wholesome and nutritionally enrich food is essential. There is direct relationship between food and health and therefore an implementation of Food Safety Management System (FSMS) in the food chain is mandatory. Ensuring food safety in the globalized food market is a challenging task and it can be only achieved through collaborative actions by different stakeholders in the entire chain of food business including consumers. It is not easy to quantify the actual human health and economic risks associated with unsafe food. Food borne pathogens and associated toxins including chemical contaminants may cause large array of illnesses right from diarrheal foodborne and waterborne illnesses to cancers (Fukuda, 2015). We all know that incidence of foodborne diseases has been increased several fold in recent years and many pathogens are of emerging nature. Foodborne diseases accounts to the significant morbidity and mortality in children and elderly immune-compromised people (Scott, 2003). If we look into the recent fact sheet of World Health Organization (WHO), globally approximately 600 million people suffer from foodborne illness and ~420000 succumb to death every year. Contaminated food is one of the important vehicles contributing in the diarrhoeal illnesses and deaths associated with it. Foodborne diseases create huge impact on the national development in terms of health, food trade and tourism. Thus production of safe and wholesome food should be taken as priority at national level under the provision of existing national and international food safety regulations. This will ultimately promote a healthy lifestyle and sustainable health through healthy food and nutrition (WHO, 2015).

In the globalized food business, it has become mandatory to produce and sell safe food as in compliance to the international food safety standards. For any nation to sustain in the international food trading business, deciding food safety policies and implementation of highest standards of food hygiene and safety would be a key to success. Ban on the export of various food commodities due to risk of pathogens, pest, mycotoxins, and toxic chemical residues has been experienced by many countries. Temporary ban on the few agriculture products of Indian origin has been imposed by European Union (EU) and Japan. Food traceability during entire food chain is basic requirement of numerous importing countries. Unsafe products need to be withdrawn from the food supply chain so that human health can be protected and promoted. This could be only achieved through effective food traceability system. Traceability is an important tool in food safety risk management and considered as a foundation of national food safety policy. The development and implementation of effective traceability system becomes mandatory since food risks are abundant. If food safety is compromised in the food production chain, it may account to the entry of pathogens, spoilage and risks associated with toxic substances of chemical nature. Anything above acceptable levels in food may pose a health risk to the consumer. Traceability is an essential risk-management tool that helps in the tracking, tracing and withdrawl of contaminated unsafe foods from the market. It is a foundation pillar of European Union's food safety policy (European Communities, 2007).

2. Description of Traceability

One can describe traceability in simple way as a system that ensures the accurate tracking and tracing of food in the food supply chain. Traceability should be established from manufacturing to consumers in order to monitor the food commodities through all stages of the supply chain. Traceability system has potential to identify the origin of the food product. It can point out the entry portals of hazardous substances (GS1 System, 2015). As per the European Union "traceability" means the ability to track any food, feed, food-producing animal or substance that will be used for consumption, through all stages of production, processing and distribution. In case of unsafe products if at all found, its proper identification in terms of its origin, source and marketing channels becomes a matter of prime importance. Withdrawal of such food or ingredients thereof could be alleviated by traceability. Globally, food trading is a multi million dollars business coming across several disputes. Traceability of food products during incidence of food safety hazard or foodborne outbreaks is more challenging. Traceability is an important way to assure food safety in agriculture and animal origin food chain.

It has been defined by several authors. The Codex Alimentarius Commission (CAC) has defined product tracing as "the ability to follow the movement of a food through specified stage(s) of production, processing and distribution". Tracing and tracking are the two major objectives of the traceability system (Petersen and Green, 2005). In the business flow, role of each traceability partners is very crucial. While implementing food traceability, linking of product flow (physical flow) with the information flow should be done by all the stakeholders involved in supply chain. Traceability enables each traceability partner to determine direct source and recipient of a traceable unit. That traceable unit may be a product or a logistics unit. Clarity and persistence of information related to food quality and safety across food chain is strongly supported by means of traceability (GS1 System, 2015).

A traceability system has been characterized by breadth, depth, and precision (Golan *et al.*, 2003). The breadth is amount of information recorded by the system and depth is how far backward and forward traceability is maintained, while precision is the ability of the system to pinpoint the original source of a problem. Benefits of the traceability are extensive (ISO 22005, 2007) and few of them are; tracing the material flow, stage wise production tracking, establishment of communication and coordination, improved productivity *etc.*

3. Traceability Requirements

Locating and prompt removal of defective or unsafe foods, pharmaceuticals or other products from table is major purpose of traceability. If effective traceability system is in place for tracking and tracing the meat and poultry products or any of the food commodities sold in international market, it will not only prevent the food borne disease outbreaks including food poisoning and food adulteration but it will also build confidence in consumers about the commitments and food safety policies of the manufacturing company. Due to increased risk of foodborne outbreak associated with *Salmonella* species, *Campylobacter* species, *E. coli O157: H7, Listeria monocytogenes, Staphylococcus aureus* (MRSA) *etc.* Safe and wholesome food is in great demand and thus it is essential for food establishments to produce foods with high quality. Integrated food safety management systems are being implemented by various food manufacturing organizations to achieve quality assurance (Aung and Chang, 2014).

This mechanism of quality assurance becomes more stringent and full proof with traceability. Establishments can follow the national and international requirements with traceability in an effective manner. Consumers perception is enhanced by the traceability system (Kher *et al.*, 2010). Traceability is becoming popular in the food business related to the foods of animal origin. Due to incidence of Bovine Spongiform Encephalopathy (BSE), dioxin in poultry feed, Foot and Mouth Disease (FMD) and melamine in milk, traceability has become integral part of animal origin food safety policies. Traceability is essential part of food safety auditing and also provides consumer assurance. Meat and poultry constitutes the major share of food commodity globally. Thus traceability requirements, importance, implementation and time to time up-gradation of traceability in meat food sector is important. Consumers are more knowledgeable and demanding, moreover increasing awareness of consumer on food safety issues demands food standards and practices that will help in tracing and tracking of meat products throughout the meat food chain.

Table 3.1. Requirements of a Traceability System for Food Products (Pigini and Conti, 2017)

Food Safety Control	Fast back and forward tracing in food crisis management that can cause health problems.
Food Quality Control	Possibility to monitor, control and time/geo-reference the complete chain. Quality preservation.
Information Security	Food authentication, fraud prevention, anti-counterfeit.
Public Authorities Control	Possibility to automatic and fast control by the public authorities.
Warehouse Optimization	Real-time location and identification of items in production and warehouse. Increment of the efficiency of the supply chain and logistics. Fast and accurate inventory.
Consumer Involvement	Increasing interest in certified quality, sustainability of the food production. Use of web and smartphone apps to buy the products.

Few countries have developed guidance documents on meat and poultry traceability to ensure consumers and traders about the best practices being followed for rapid identification, tracking and withdrawal of food lots if product is suspected or confirmed as unsafe for human consumption. For example, United States has developed traceability for meat and poultry implementation guide for trading partners like supplier, consumer, and government. Each traceability partner must be competent to identify the product source, its forward and back ward flow. Traceability can be used for a various purposes like product recalls, market withdrawals, regulatory compliance, public health trace backs, safety and quality assurance, and order management.

4. Recall Classification

A recall is a firm's voluntary removal of distributed meat or poultry products from food chain on the grounds of adulteration, misbranding, microbial risk, toxic residues *etc.* There are different sources through which a manufacture or a food safety agency can learn about the potential recalls. Manufacturer, distributor, retailer, laboratory findings, food safety department, export inspection council, consumer complaints; epidemiological data *etc.* are the examples of information for product recall (CRS, 2011). In food safety context, food recalls are classified based on risk posed by the product under question. There are three designated levels or classes of recalls as described below;

1. Class I recalls: These are serious recalls and probability of occurrence of health hazard is more severe. Pathogens like *L. monocytogenes, Salmonella* species, *E. coli O157: H7* are examples of Class I recalls.
2. Class II recalls: There is less probability of occurrence of health hazards. Presence of food allergens or foreign materials in small quantities is an example.
3. Class III recalls: A situation where food product will not cause harmful health hazards. Excess water in a food product could be an example of Class III recall.

Recalls of meat and meat products is not at all new for the world. USDA has recorded largest recall of frozen beef in 2008 which was a Class II type recall. In 2013, a recall due to substitution of beef with horse meat was recorded in Europe. Several poultry meat related recalls were also recorded in Europe in 2017(https: // www.stericycleexpertsolutions.). Therefore, traceability is very crucial part of the food safety system of meat and poultry industry. In the context of a food recall, importance of traceability can be summarized as below;

1. Identification of physical flow of raw material and finished product.
2. To create record keeping and documentation (FAO and WHO, 2012).

4.1 Food Recall Plan of FSSAI

All food business operators as prescribed in the regulation 7 of Food Safety and Standards (Food Recall Procedure) Regulations, 2017 must have an up to date recall plan as provided in Annex (Model Recall Plan)- I. A brief step by step procedure and its description are as below (FSSAI, 2017);

Assemble the recall management team

↓

Notify the authority

Identify all products to be recalled

Detain and segregate all products to be recalled which are in your firm's control

Prepare and distribute the information of recall including press release

Prepare the distribution list

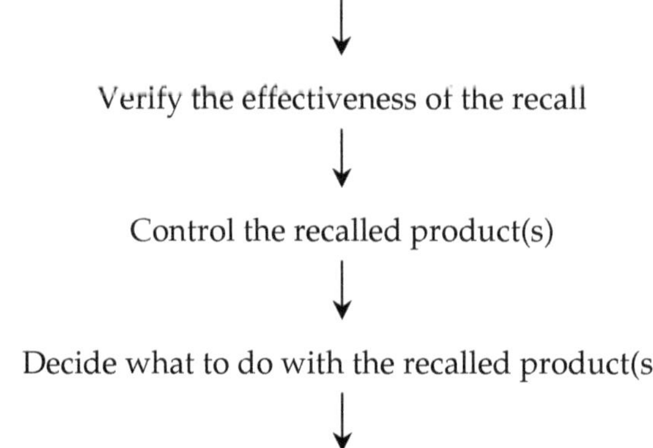

Fix the cause of the recall if the problem occurred at your facility

5. Role of Traceability

Traceability is of paramount significance in the supply chain of meat and poultry products, however achieving traceability in meat industry is challenging. Meat food traceability is crucial in the context of microbial food borne risks. The list of pathogens able to transmit through contaminated food including meat and poultry products is very extensive. Bovine Spongiform Encephalopathy, *Enterohemmorhagic E. coli,* Avian Flu, Foot and Mouth Disease, *Salmonella Enteritidis, Campylobacter jejuni, Listeria monocytogenes, Clostridium perfringens, Yersinia enterocolitica, Enteroviruses etc.* are examples of foodborne pathogens. To maintain transparency in international meat trade effective traceability system is needed which will reduce response time in case of food borne outbreaks. Traceability not only helps in tracking and tracing the food but also important in prevention of food safety hazards. There are several benefits of traceability system *viz.*, tracing food during entire food chain, knowing where food comes from, improving product recalls, meeting regulations and legal requirements and consumer trust (Aung and Chang, 2014; GS1 System, 2015).

6. Consumer Awareness

A comparative study on perception and knowledge of consumers of the EU, China and North America on traceability revealed that people are more concern about the food safety, however they are not much familiar to the traceability. Consumers inclination to pay higher for the quality food found varying in the different countries (Hansstein, 2014). Age, education, income, degree of awareness are important factors in the adoption of traceability. Traceability has been introduced in Croatia in 2007 for all food. Very low degree of awareness on meat traceability in Croatia was recorded (Cerjak, 2011). Lack of knowledge about traceability concept resulted mainly in difference regarding existence of this system. Recent study from Malaysia has shown positive attitude of respondents towards traceability systems for meat and meat products. It was also revealed that socio-demographic status like gender, race, income, education, marital status and household size has co-relation with preference for meat and meat products with traceability systems

(Nawi *et al.*, 2018). Findings of Zhaoa *et al.* (2010) from China concluded that many consumers never heard about traceability. Consumers did not completely believe in traceable food information provided by food suppliers. The low confidence in public policy and media prevented the food traceability system from quick promotion and development. In general, consumer's willingness to buy traceable food is low. If consumers are informed about the safety associated with traceability, their willingness to buy traceable foods will increase significantly.

7. General Methods Used for Meat Traceability

The beauty of traceability lies in the full proof system that can track every product from primary production till the consumers end (Aung and Chang, 2014). Tracing the origin of food is an important criteria for ensuring food quality. Thus, origin detection tools are regarded as important parts of the food traceability. Geographic, biological and analytical-based methods are used in determining the origin of the food (Dogu and Sireli, 2016). For livestock product assurance schemes to be effective, the animals must be reliably and uniquely identified. Animal identification is important part in the meat food traceability, however in many parts of the developing world, individual animal identification is not possible. Thus collective identification of group of units is possible. Animal identification and registration with unique identity scheme is essential to achieve meat food traceability (OIE, 2012). There are several methods used in food traceability.

7.1 Electronic Identification Devices

Electronic tags contain a radio frequency identifier (RFID). The RFID consists of a microchip and a coiled copper antenna. Full duplex (FDX-B) and half duplex (HDX) are the types of RFID being used at present. Current electronic tags are low frequencies (LF) which are operated in the frequency range 120-135 kHz. These are less sensitized to environment.

7.2 Identification Methods

Tattoo: These are frequently used on pigs, particularly white-skinned breeds. Pigs can also be tattooed on the ear. In all species the tattoo mark can be divided between the two ears *e.g.* the flock number on one ear and the individual number on the other. Tattoos have also been proposed as a means to identify poultry; the tattoo would be applied under the wing where there are few feathers.

Ear Notching: It may be used for on-farm management of pigs which is alternate to tattooing or ear tags: a special tool snips out a small piece of the ear flap to leave a permanent V-shaped notch. The position of the notch indicates a number and by using multiple notches it is possible to identify many pigs.

Ear Clipping: It is a similar procedure to ear notching but it is more frequently used term for livestock other than pigs.

Ear Tagging: It is the most common technique of marking livestock species like cattle, sheep and goats. It can be also used for identification of pigs. They may be electronic or non-electronic and mostly made of plastic. Ear tags are easy to apply, but each design requires a specific design of pliers.

Dewlap Tags: These are available for cattle herd management in the UK. The tag is inserted into the dewlap over the brisket.

Pastern (Leg) Bands: These are a non-invasive method in which a band is fitted around the animal's lower leg. It may be electronic or non-electronic. As an alternative to ear tags and ruminal boluses, pastern bands are allowed under EU regulations.

Ruminal Bolus: It is a bullet-shaped, high density ceramic container with an electronic identifier. It is designed to lodge in the rumen of the animal. Ruminal boluses are more expensive than ear tags.

Hot Branding: It is the application of a red hot branding which causes a permanent mark through scarring of the skin.

Freeze Branding: It is permitted for cattle as well as equines and is now more common than hot branding. The branding iron is cooled in liquid nitrogen, dry ice or similar coolant and applied to a shaved area of the animal's skin.

Micro-chipping: Electronic transponders are inserted under the skin and read by the electronic reader. Micro-chipping is widely used for companion animals and equines.

7.3 Future Technologies

Barcode: Use of printed barcodes on the ear tag is possible. They are permitted for cattle in the EU and are used on ear tags in the Republic of Ireland.

Biometric identification: It is more precise and most authentic method of animal identification. Scanning of rectum, muzzle, facial recognition, immunological labelling and DNA analysis have been suggested. The biometric pattern cannot be manipulated.

Bioactive immunological labeling: The system ('ImmunoTrack') was developed in Germany by Responsif GmbH. This uses highly antigenic peptide sequences with appropriate adjuvants to induce strong peptide-specific antibody responses in cattle or pigs. Specific antibodies are detected in blood or meat juices. This approach is helpful in encoding the origin, breed of animals (Gonzales-Barron and Ward, 2005).

Ultra High Frequency (UHF) Tags: These tags could store up to 96 bits of information and had a frequency range of 860-960MHz.

RFID: This technology is based on the radio frequency identification. Radio frequencies are used to identify product or object. RFID (radio frequency identifier) is combined with tags, readers, middle-ware, and application system. It is automatic technique having many advantages, like a large amount of data storage, can read and write, farther messaging distance, stronger air-penetration, read faster, long service life, better adaptability of environment *etc.* Compared with the popular bar code, RFID technology speed up the exchange of information and helps to save on production and distribution costs. The main application of RFID tags is for tracing livestock, like cattle.

GIS: Geo-based monitoring technique has come up as new methods in recent years. They are generally used for traceability of food-agriculture and agricultural products. Mineral isotope and GIS (Geographical Information System) are the main tool of the geographic based applications.

QRcode: The QRcode is an extension of barcode which is used in the final product. Using smart phone one can read the code. It is used as a fast link to the company web site. It is popularly used in various sectors due to storage capacity and good readability. Barcodes are the most cost efficient form of storing traceability coding.

NFC: Near Field Communication is a wireless close-range connectivity technology derived from RFID technology. It allows data trade between two devices. NFC has been widely integrated in mobile devices such as smartphones, tablets and notebooks. This allowed the development of applications directly addressed to the end user.

EAN.UCC: International Numbering Association Universal Code Council (EAN. UCC) system is an extensively used coding system for product traceability which provides a sequence of numbers designated to identify unit, locations and services. It also includes Global Trade Identification Number (GTIN) used to identify trade items.

7.4 Molecular Methods for Traceability

Molecular technology has revolutionized the techniques of identification of biological substances targeting genomic markers. In DNA-based methods specific DNA sequences are used as markers. Animal and plant identification becomes more reliable and easy by molecular techniques. They are of two types *viz.* hybridization-based markers, and Polymerase Chain Reaction (PCR)-based markers.

Species-specific profiles of DNA can be exposed by DNA hybridization with restriction enzymes. PCR using targeted primers and restriction fragment length polymorphism using restriction enzymes is very popular in meat species identification. Amplified product is separated by electrophoresis (Galimberti *et al.*, 2013). In PCR-RFLP, species-specific mitochondrial markers are targeted. In microarray cytochrome b-derived probe is done to identify forensic samples of meat. DNA barcoding using cytochrome b is a potential tool for meat traceability. Other analytical and biological methods include ELISA (Enzyme Linked Immunosorbent Assay), MS (Mass Spectrometry), IRMS (Isotope Ratio Mass Spectrometry), ICPMS (Inductively Coupled Plasma Mass Spectrometry), GS-MS (Gas Chromatography Mass Spectrometry), NMR (Nuclear Magnetic Resonance Spectroscopy) IR (Infrared Spectroscopy), ASR (Atomic Spectroscopy), VS (Fluorescence Spectroscopy), HPLC (High Performance Liquid Chromatography), GC (Gas Chromatography) and CE (Capillary Electrophoresis) (Dogu and Sireli, 2016).

7.5 Labeling Compliance

Proper labeling of packed food items is very significant in the food safety management. Labeling of foods must be in compliance to the Food Safety and Standards (Packaging and Labelling) Regulations, 2011 for the foods manufactured in India. Product specific requirements related to the packaging and labelling are mentioned in the said regulations. Class of food, nutritional information, declaration of food additives, manufactures address, quantity packed, date, expiry, country of origin, lot/batch number *etc.* are the mandatory instructions which must be furnished on labels. EU regulations are stringent and as per the EC (Labelling of Beef and Beef Products) Regulations, 2002 it is mandatory to have a traceability system in place from slaughterhouse to the various points of sales.

8. Global Perspective

EU countries, *viz.* Austria, Belgium, Denmark, Finland, France, Germany, Ireland, Italy, Netherlands, Sweden, and the United Kingdom, fall under the mandatory regulation of EU Legislation 178/2002. New Zealand and Brazil implemented a mandatory traceability and identification system for livestock. Mandatory requirements for tagging and identification of cattle, sheep, and goats are in place in Australia. It is very important part of the meat industry in Canada and all animals are tagged as a part of animal identification and meat traceability. United States require total document of traceability and enforcement for registration. Russia also requires complete assemblage of pesticide usage and prohibited to genetically modified organisms (GMO). Two dimensional bar code ear tagging is used for livestock in China (Dandage *et al.*, 2017). Beef traceability system is mandatory system in Japan (Jin and Zhou, 2014).

A study was conducted by Souza-Monteiro *et al.* (2004) to compare the economic significance of mandatory and voluntary traceability systems in the EU, Japan, Australia, Brazil, Argentina, Canada, and the United States in terms of the systems' breadth, depth, and precision. There is difference in the approaches of establishing food traceability system by each nation. The EU and Japan have mandatory traceability. Backward and forward traceability is achieved in case of beef right from farm to the retail market.

In Australia and Brazil traceability is presently mandatory for exported beef only. Traceability is voluntary in USA. Animal identification for all the animals moving from farm is mandatory in Canada. It is mandatory in Argentina only for the beef which is intended for export. The mandatory systems in the EU and Japan are the deepest. Variations are found in the meat traceability systems adopted by each nation. In Australia, abattoirs are not linked to the retail meat shops or stores, but details of the containers which are exported is registered. Brazil, Japan, Australia, and the EU have the comprehensive systems for meat traceability, whereas, Canada and Argentina have easy traceability systems in terms of breadth. Meat traceability system in Japan, EU, Australia, and Brazil is very precise and individual animals and their farms of origin can be linked with beef products. DNA based approach is used in Japan. Electronic tagging is an important part of Australian beef traceability.

9. Indian Perspective

An extensive review of Indian perspective on food traceability has been given by Dandage *et al.* (2017). There is no any obligatory system of traceability however, Indian government is working with FSSAI, APEDA, GS1 India, NABARD, FPO, ITC's eChaupal *etc.* for developing the traceability system. Existing product identification technologies; use of Alphanumerical-codes, Hologram, Barcode, RFID-tags, GI-tag *etc.* are the examples of existing systems of product identification in India.

Pioneering work in the area of food traceability is done by APEDA in India. APEDA has set up a food traceability mechanism for agricultural products. APEDA helps to establish linkages between different stake-holders through 'farm-to-fork' monitoring so as to guarantee quality and food safety. 'Grapenet' is developed by APEDA for Indian grapes sector. 'Anarnet' is also developed for quality assurance of pomegranate export. 'Tracenet' is another web based system of traceability which is highly user friendly. This is implemented since 2010 for the organic products. 'Peanutnet' is also developed for control of risk associated with aflatoxins in peanuts (APEDA, 2013).

In case of export of meat and meat products, APEDA issues Health Certificate to the meat processing establishments registered with them. This can be done through Meat.Net Online System. Health certificate is issued for each of the export consignment of meat products. It has become mandatory in India that each consignment subjected for export must be examined for chemical and microbial tests in the approved government laboratories.

10. Conclusion

The countries that have already adopted traceability systems have benefited in the domestic as well as international market. Understanding the benefits of traceability, this system will become an integral feature of international market of food products. For detail information and guidelines for establishment of traceability system in meat and poultry supply chain, GS1 Global Meat and Poultry Traceability Guideline must be referred. This guide is useful for the establishments wish to implement traceability at their place. The GS1 Guideline consists of seven documents - starting with Part 1 the GS1 System. Parts 2 to 5 are separate sections covering the Beef, Lamb, Pork and Poultry Sectors. Consumers' awareness on traceability must be promoted. Collaborative approach and co-operation from different stakeholders *i.e.* producer-processor-distributor-consumer and political will is essential in traceability implementation in food sector.

REFERENCES

APEDA. (2013). Traceability. Retrieved from http: //apeda.gov.in/apedawebsite/ index.html.

Aung, M.M. and Chang, Y.S. (2014). Traceability in a food supply chain: Safety and quality perspectives. Food Control, 39, 172-184.

Cerjak, M., Kljusuriæ, J.G. and Mesiæ, Z. (2011).Traceability in Croatian meat sector: are consumers aware of it?. Croatian Journal of Food Technology, Biotechnology and Nutrition 6 (3-4), 123-128.

CRS;Congressional Research Service (2011). The USDA's Authority to Recall Meat and Poultry Products 7-5700. www.crs.gov

Dandage, K., Badia-Melis, R., and Ruiz-García, L. (2017). Indian perspective in food traceability: A review. Food Control, 71, 217-227.

Dogu, S. O., and Sireli, U. T. (2016). Determination tool of origin in the food traceability. J. Food and Health Sci., 2(3), 140-146.

EC;European Communities (Labelling of Beef and Beef Products) (Amendments) Regulations, 2002 (S.I.No.485/2002).

European Communities (2007). Food Traceability. (www.ec.europa.eu/food/sites/).

FAO and WHO (2012). FAO/WHO guide for developing and improving national food recall systems, Food and Agriculture Organization of the United Nations and World Health Organization Rome.

FSSAI (2017). Food Safety and Standards (Food Recall Procedure) Regulations, 2017, Guidelines for food recall.

Fukuda, K. (2015). Food safety in a globalized world. Bull. World Health Organization. 93, 212.

Galimberti, A., Mattia, F.D., Losa, A., Bruni, L., Federici, S., Casiraghi, M., Martellos, S., and Labra, M. (2013). DNA barcoding as a new tool for food traceability. Food Res. International, 50, 55–63.

Golan, E. B., Krissoff, F., Kuchler, K., Nelson, G. P., and Calvin, L. (2003). Traceability in the US Food Supply: Dead End or Superhighway. Choices, 2,17-20.

Gonzales-Barron, U., and Ward, S. (2005). Review of biometric and electronic systems of livestock identification. The BioTrack Project: Development of a protocol for biometric-based animal tracking and tracing. Report to the Department of Agriculture and Food for Ireland, Dublin.

GS1 System (2015). Global Meat and Poultry Traceability Guideline, Part 1. The GS1 System.

Hansstein, V. (2014).Consumer Knowledge and Attitudes towards Food Traceability: A Comparison between the European Union, China and North America Francesca. International Conference on Food Security and Nutrition IPCBEE Vol. 67, IACSIT Press, Singapore DOI: 10.7763/IPCBEE.

https: //www.stericycleexpertsolutions.co.uk/recall-index/.Stericycles second quarter 2017 Recall and Notification Index

ISO-22005. (2007). Traceability in the feed and food chain- General principles and basic requirements for system design and implementation. (http: //www.iso. org/iso/catalogue_detail?csnumber=36297).

Jin, S., and Zhou, L. L. (2014). Consumer interest in information provided by food traceability systems in Japan. Food Qual. Pref., 36,144-152.

Kher, S. V., Frewer, L. J., Jonge, J. D., Wentholt, M. Davies, O.H. Luijckx, N.B.L., and Cnossen, H.J. (2010). Experts' perspectives on the implementation of traceability in Europe. British Food J., 112 (3), 261-274.

Nawi, M. N., Basri, H. N., Kamarulzaman, N. H. and Shamsudin, M. N. (2018). Factors influencing consumers' preferences towards meat and meat products with traceability systems in Malaysia. Int. Food Res. J., 25(2),S157-S164.

OIE (2012). Terrestrial Animal Health Code. World Organisation for Animal Health. Paris.

Petersen, A., and Green, D. (2005). Seafood traceability: A practical guide for the US industry.

Pigini, D., and Conti, M. (2017). NFC-Based Traceability in the Food Chain. Sustainability 2017, 9, 1910; doi: 10.3390/su9101910.

Scott, E. (2003). Food safety and foodborne disease in 21st century homes. Canadian J. Infect. Dis., 14 (5), 277-280.

Souza-Monteiro, M., Julie, A., and Caswell J.A. (2004). The Economics of Implementing Traceability in Beef Supply Chains: Trends in Major Producing and Trading Countries, Working Paper No. 2004-6. (http: //www.umass.edu/resec/workingpapers).

World Health Organization (2015). Estimates of the global burden of foodborne diseases (http: //www.who.int/foodsafety/areas_work/foodborne-diseases/ferg/en/).

Zhaoa, R., Qiaoa, J., and Chena, Y. (2010). Influencing factors of consumer willingness-to-buy traceable foods: An analysis of survey data from two Chinese cities. Agriculture and Agricultural Science Procedia, 1, 334–343.

Chapter 4

Recent Techniques to Assess Meat Quality in Carcass and Meat Cuts

Ruma Devi[1]* and Praneeta Singh[2]

[1]Assistant Professor, Department of Livestock Products Technology, College of Veterinary Science and Animal Husbandry, NDUA&T, Kumarganj, Ayodhya, U.P.

[2]Assistant Professor, Department of Livestock Products Technology, College of Veterinary and Animal Sciences, G.B. Pant University of Agriculture and Technology, Pantnagar, U.K.

**e-mail: rumadevi.2@gmail.com*

ABSTRACT

To ensure production of quality meat products, it is equally important to have meat having desirable features. The access of quality meat is the first and foremost approach and it starts from the farm itself where an animal is procured for meat production but the real assessment starts after the animal is slaughtered and meat in the form of carcass or meat cuts is obtained/produced. It is desired to have accurate, fast and non-invasive method for assessing technological and sensory characteristics. The crucial qualitative attributes that need to be determined are nutritional value, texture and appearance. Similarly the other aspects of meat involving physico-chemical features, can be evaluated by using the latest techniques and tools and preferably the non invasive one to have a holistic picture for consumers' aid and to have best quality in the products produced.

Keywords: *Meat, Quality, Texture, Sensory.*

1. Introduction

Meat animals have traditionally been assessed for their value by assessment of their live contours and appearance. From the last 30 years, pig, buffalo, sheep and goat have been assessed for economic value depending on the carcass attributes under a grading methodology. Carcass "quality" defines the preliminary features of lean which affect the palatability mainly in the form of juiciness, tenderness, and flavour. Meat quality is mainly affected by different physiological processes that occur not only during the growth of the animal but even after slaughter. The consumers' evaluation of meat quality is dependent on its colour, juiciness, tenderness and meat flavour, which influences their decision during meat purchase repeatedly. The purpose of assessing the meat quality is to provide to the consumer tasty, wholesome and safe meat at a rational cost. Assessment of meat quality is also very important for preparation of good quality meat products.

Meat industry poses major emphasis to gather consistent information on meat quality throughout the production process, so that guaranteed quality meat products may be made available to consumers. For the above need, it requires accurate, fast and non-invasive method for assessing technological and sensory attributes. In recent days, many methods have been developed, and they principally determine the meat quality characteristics. Most of these methods are invasive, meaning that they are difficult to implement on-line. In meat industry, the crucial qualitative attributes that need to be determined are nutritional value, texture and appearance. Variability in raw meat leads to highly variable products which are marketed without an oversee level of quality. This problem is enhanced when the industry is not able to satisfactorily characterize the high level of quality and cannot therefore market meat products with a certified quality standard, which is an essential condition for withstand and development of any modern meat industry. Different meat quality parameters like texture, nutritional value and appearance have varying importance regarding consumer satisfaction, tenderness is observed as one of the most important variable which is positively correlated with taste and juiciness (Naganathan *et al.*, 2008). Fat is another important quality attribute, though the consumers prefer leaner meat due to health concerns but a minimum quantity is required to conserve the juiciness and flavour. Ideally these attributes should be evaluated by consumer panels in order to ultimately satisfy the customers. This is however an ambitious task, which is very difficult to arrange and is usually not practicable due to time and financial limitations (Jackman *et al.*, 2009).

The basic traits relate to nutritional content like fat, protein, fibers, vitamins and minerals, mainly iron. Another key criterion is safety. The food must be clean in relation of agro-chemical residue, pesticide residue, heavy metals, pathogenic and spoilage micro-organisms, and any other substance which causes a potential health risk. The other side of quality deals with "functional" attribute which is related to the sensory quality of appearance and taste (Grunert, 1997). The difference in functional attributes is related to the biological diversity of the animals from which meat is obtained. This chapter will focus on the recent techniques currently being studied and used in laboratories proposed for meat quality evaluation.

2. Mechanical Methods

Since the thirties, mechanical methods to detect texture and sensory quality have been extensively used. They are of two types; one is invasive method like traction, compression and shearing, all these require sampling. Another one is non-invasive method like direct or resonance tests, performed on intact muscles.

Table 4.1. Summary of Instrumental Mechanical Methods

Instrumental Mechanical Methods	*Specification*	*Quality Parameter*	*Meat Quality*	*Merit*	*Demerit*
Warner–Bratzler shear force test (WBSF)	Tenderness	Sensory attributes	Functional attributes	—	Destructive, anisotropic
Slice shear force test	Tenderness	Sensory attributes	Functional attributes	More strongly associated with tenderness than WBSF	Destructive, anisotropic
20 per cent and 80 per cent of meat sample deformation	Tenderness	Sensory attributes	Functional attributes	Assess (20 per cent) myofibrillar proteins and (80 per cent) intramuscular connective tissues separately	Destructive, anisotropic
Armor tenderometer	Tenderness	Sensory attributes	Functional attributes	Movable, non-destructive	Invasive
Torque tenderometer	Tenderness	Sensory attributes	Functional attributes	Movable, non-destructive	Invasive
Tendertec penetrometer	Tenderness	Sensory attributes	Functional attributes	Movable, non-destructive	Invasive
Ultra sound	Texture, collagen content, fat content	Sensory and nutritional characteri-stics	Functional and basic attributes	On live animals, on whole carcasses,3D	
Ultrasonic spectral analyses					
Ultrasonic elastography (or transient elastography)	Local viscoelastic properties	Sensory attributes	Functional attributes	Non-invasive,3D	

2.1 Instrument Measurements

The traditional instrumental methods which are employed to measure the tenderness of meat is basically based on compressive and shear forces produced during chewing are:

2.1.1 Warner-Bratzler Shear Force (WBSF) Testing

WBSF test do not give clear view in interpretation of results but WBSF testing is always used as a standard technique for comparing the other techniques (Lorenzen *et al.*, 2010). It measures the maximum shear force. WBSF measurement differ between raw and cooked meat (Tornberg, 1996). Although the precaution is taken at the time of orienting the measurement probe due to diversity in muscle fiber direction (Stephens *et al.*, 2004) and it is destructive and time consuming method but WBSF remains most widely used instrument technique for the measurement of meat toughness. For the improvement of this method, use flat blades (Slice Shear Force) instead of the V-shaped blade used with WBSF.

2.1.2 Compression Tests

In compression tests as deformation increases, on raw meat, the three structural components that are intramuscular connective tissue, myofibrillar proteins and perimysium play an important role in mechanical resistance. This initiated the formation of complementary mechanical test to evaluate intramuscular connective tissues and myofibrillar proteins respectively, measuring strain at 80 per cent and 20 per cent of meat sample deformation (Lepetit and Culioli, 1994).

2.1.3 Tenderometers

Armor tenderometers are portable instrument having six sharp needles, that gives good result on raw-meat toughness by trained sensory panel (Stephens *et al.*, 2004). The "Torque Tenderometer" from MIRINZ, New Zealand, and the Tendertec Mechanical Penetrometer from the Australian Meat Research Corporation give a good correlation with meat toughness (Belk *et al.*, 2001).

The major disadvantage of these shear strength and compression measurements is that they are not just time consuming but also bound to the lab.

2.2 Ultrasound Methods

Ultrasound refers to sound which cannot be detected by human ear. The term 'Ultrasound' applies to all acoustic energy with the frequency above human hearing (20000 Hz or 20 KHz), below this frequency, sound can be heard by the human ear and is known as 'infrasound'. Tissue imaging or live animal evaluation frequencies range from 1 to 10 MHz (Amin, 1995). The composition of carcass can be assessed on all species of livestock by the use of ultrasound technology. Wild in 1950 (Houghton and Turlington, 1992) first described the use of ultrasonics as a nondestructive and humane means of measuring fat and muscle in live animals. The food industry offers manifold possibilities for the use of ultrasonics. Acoustic waves are propagated through materials. Hence, the acoustic properties of materials can be compared to its macroscopic composition and structure. It uses to measure the small deformations in rheological properties at low frequencies known as mechanical spectroscopy. However, ultrasonic waves with higher frequency can give valuable information in meat. Ultrasonics can have frequency from 20 kHz to 10 MHz for instrumentation in foods. These waves are mechanical waves which causes a series of mechanical disturbances that propagate as stresses and strains in the physical bonds of the material. Thus the nature of bonds and masses of molecules present in material affect the absorption or speed of transmission of these waves. The low powered ultrasonic waves should be distinguished from high powered ultrasound waves which are used for homogenization and cell disruption.

The benefits of ultrasonics are (1) it gives easy and quick measurement (2) applicable in optically opaque food materials (3) easy adaptability for automated process control. By using the two methods of ultrasound: ultrasonic spectral analysis and ultrasonic elastography or "transient elastography" we can measure the functional quality of muscle food.Ultrasonic wave propagation in meat not only depends on the composition (*e.g.* lipid and water content) but also depends on the structure (*e.g.*, organization of connective tissue, inclination of muscle fibers). As reported by Monin (1998), ultrasonic measurements give a good prediction of meat texture on whole carcass and on live animals, while at the same time being inexpensive and non-invasive. Animal tissues behave like a viscoelastic material means they present as fluid viscosity properties and solid elasticity properties. Since acoustic wave propagation is directly connected to these mechanical properties, following the tissue propagation of acoustic waves could be a solution for assessment of local viscoelastic properties. It can be done by way of an echographic system, through technique called "transient elastography". Supportive with ultrasound analysis, transient elastography is an exceptional and non-invasive technique for assessing the local mechanical properties of muscle tissues. It consists in an ultrasonic transducer, which is applied at the surface of muscle tissue, combine with an ultrasonic pulse echo method. The basic idea is that the low frequency vibrations generated by the transducer itself induce a low frequency motion of the scatterers inside the media. This motion can be detected and measured with the conventional echographic system, and via an inverse problem resolution, this motion can yield the local viscoelastic properties.

3. Optical Methods

3.1 Spectroscopic Methods

Spectroscopy is a technique that uses the synergy of energy with a sample to perform a wave dissemination and the study of radiation, absorption and more generally of any interactions between electromagnetic radiation and matter is termed as spectroscopy. The data which is obtained from spectroscopy is termed as spectrum. A spectrum is a frame of the intensity of energy detected, alternatively the wavelength (or mass or momentum or frequency *etc.*) of the energy. A spectrum can be used to gain information about atomic and molecular energy levels, chemical bonds, molecular geometries, synergy of molecules and related process. Usually spectra are used to determine the elements of a sample (qualitative analysis). Spectra may also be used to assess the amount of material in a sample (quantitative analysis).

Optical spectroscopy offers a panel of useful methods for on-line characterization because of its non-contacting possibilities and because of the fibre-optical components which make it easy to design portable devices. It has been commonly investigated in the area of meat science as a method of gaining structural information. Polarized light gives further organizational data and are therefore often used for these applications. Below is a summary of the main techniques developed and used in meat science and some times in the biomedical field, where tissue organization is also of importance. There are many different types of energy sources as from low frequencies to high frequencies, optical spectroscopy covers near-infrared (NIR), infrared (IR), visible, and ultra-violet (UV) (including fluorescence). Spectroscopic techniques are broadly used for muscle food quality.

Table 4.2. Summary of Optical Techniques

Optical Spectroscopy	*Specification*	*Quality Parameter*	*Meat Quality*	*Merit*	*Demerit*
Infrared spectroscopy	Structure of molecules	Chemical composition	Functional Attributes	Non-contacting, rapid	Need complex data analysis
Near infrared spectroscopy	Structure of molecules, instrumental texture, sensory tenderness, pastiness, crusting, juiciness, discrimination between fresh and frozen-thawed products	Physico-chemical characteristics	Functional Attributes	Non-contacting, rapid	Need complex data analysis
Raman spectroscopy	Structure of molecules, interaction of molecules, structure of proteins, water activity, fat organization,water holding capacity, instrumental texture, tenderness	Nutritional, sensory and chemical characteristics	Functional as well as basic attributes	Non-contacting, rapid, can be performed in vivo using optical fiber, small sample portion	Need complex data analysis
Visible spectroscopy-colorimetry	Colour, sarcomere length, myofilaments organization, PSE, tenderness, water holding capacity, drip loss, collagen content, fish freshness	Sensory characteristics	Functional attributes	Often performed with polarized light	Need complex data analysis
Fluorescence spectroscopy	Tryptophan microenvironment, connective tissue content, palatability, chewiness, fish freshness, tenderness, myofiber organization, collagen destructuration with heating, aging, sarcomere length	Sensory characteristics, Chemical composition, Health protecting characteristics, Nutritional composition	Functional and basic attributes	Non-contacting, rapid, can be perormed in vivo using optical fiber, often performed with polarized light	Often need extrinsic fluorophore probe, high temperature sensitivity

3.1.1 Infrared Spectroscopy

The infrared absorption spectrum of a matter is sometimes called its molecular finger print. Infrared (IR) spectroscopy is a spectroscopic process that utilize source of energy as infrared zone of the electromagnetic spectrum (from about 800 to 2500 nm). Infrared spectroscopy is based on the basic that the chemical linkage in organic molecules absorb or release infrared light when their vibrational state changes. Indeed, although infrared spectroscopy gives direct molecular-level information, research shows that it can be successfully used to determine macroscopic structural changes associated with meat or muscle structure. IR spectroscopy is one of the most powerful analytical techniques which is available to final meat product evaluation because of its ease to use, speed and versatility.

3.1.2 Near Infrared Spectroscopy

Frequently expanded to the visible region is under research in several laboratories accompanied with evaluating its prospective use for meat structure control. Most of these laboratories are working on eating meat quality, focusing on instrumental texture, sensory tenderness, pastiness, crusting and juiciness.

3.1.3 Raman Spectroscopy

Raman spectroscopy is also a vibrational spectroscopic technique used in condensed matter physics, biomedical applications and chemistry to study rotational, vibrational and other low-frequency modes in a technique. It depends on inelastic scattering of monochromatic light, generally from a laser in the visible, IR, or near-UV spectra. It gives similar but complementary information to IR spectroscopy. Raman spectroscopy can be performed in vivo by utilizing optical fiber technology. Furthermore, this spectroscopy technique has several advantages compared to traditional methods, since it is a direct and non-invasive technique that requires only small sample portions. The evolution of anon-line probe designed and optimized for conducting Raman measurements in both industrial and laboratory conditions. Developing this type of Raman spectrometer in the biomedical area has covered the procedure for its development in food science. Moreover, the components contributing to Raman scattering in muscle include collagen and elastin, both well known for their role in meat structure. At the protein level, Beattie *et al.* (2006) demonstrated the ability of Raman spectroscopy to measure changes in secondary form and then to be useful for determining textural properties such as meat tenderness. N-IR FT-Raman spectroscopy was also utilized to explore protein structure changes in food at the time of heating. Ellis *et al.* (2005) proposed an approach that would aid food regulatory bodies to rapidly identify meat and poultry products, focusing the potential use of Raman spectroscopy for rapid evaluation of food adulteration and discrimination between both species and distinct muscle groups within these species.

3.1.4 Visible Spectroscopy and Colorimetry

This field of biophysical methods overlay the visible spectra (generally expanded to near-UV or N-IR regions) and the CIE L*a*b*colour space as objective and non-destructive methods for the tissue characterization. There has been a great deal of biomedical research into tissue characterization using these techniques. As this research is often focused on gaining structural information, it merits coverage here due to the potential for use in muscle and therefore meat structure evaluation. The use of visible light to access muscle structural information is not a new concept. In the area of meat science, early detection of pale, soft and exudative (PSE) meat is a main potential application of visible spectroscopy and colorimetry for both poultry meat and pork. It can be used to predict toughness, water holding capacity and drip loss in salted comminuted chicken breast meat by visible spectroscopy with polarized light (Swatland and Barbut, 1999).

3.1.5 Fluorescence Spectroscopy

Fluorescence spectroscopy is a type of electromagnetic spectroscopy which analyzes fluorescence from a sample. It entails using a beam of light, commonly UV light, which excites the electrons in molecules of specific compounds and element, resulting to eject lower-energy light. In fluorescence spectroscopy, the electrons are first excited, by absorbing a photon of light, from its ground electronic state to one of the various vibrational states in the excited electronic state. Concussion with other molecules induce the excited molecule to drop vibrational energy until it reaches the lowest vibrational state of the excited electronic state. The molecule then drops back down to one of the different vibrational levels of its ground electronic state, emitting a photon in the process. As molecules can fall down into any of various vibrational levels in the ground state, the emitted photons will have various energies and frequencies. Therefore, analyzing the different frequencies of light which ooze out in fluorescent spectroscopy, beside with their relative intensities, makes it possible to determine the structure of the different vibrational levels. Tryptophan is one of the important intrinsic fluorescent probe which can be used to estimate the nature of the tryptophan microenvironment. Proteins which have no tryptophan can be connected to an extrinsic fluorophore probe. Meat and meat products are known as opaque samples, for that front face fluorescence used, and because these products having tryptophan, this method has been used in muscle and meat science to examine sample structure without using extrinsic fluorophore probes. As we know that meat tenderness is directly linked to the amount of connective tissue in meat. Since connective tissue is the best intrinsic fluorophore in meat or muscle, its intensity of fluorescence should be a good marker of meat or muscle tenderness.

3.2 Imaging

3.2.1 Microscopic Imaging

Meat and meat product structure has been widely controlled by microscopy. It can be divided into two fields: "optical microscopy" and "electron microscopy".

3.2.1.1 Optical Microscopy

Optical microscopy offers the simplest techniques which obtain magnified images of biological tissues. This field includes a large range of methods which have been used to know property of meat and meat product structures. Methods can be differentiating in a simple way depending on whether samples have been

Table 4.3. Summary of Microscopic and Other Techniques

Microscopy	*Specification*	*Quality Parameter*	*Meat Quality*	*Merit*	*Demerit*
Optical microscopy	Collagen typing, collagen myofilaments organization, fat configuration, myofiber typing, difference between fresh and frozen-thawed products, myofiber spacing, Z line degradation, sarcomere length, endomysium structure, myofiber diameter, density, organization, PSE specific proteins recognition	Physico-chemical characteristics and sensory quality	Functional quality	Selective analysis, 3D reconstruction with confocal microscopy	Sample preparation: thin cuts
Electron microscopy	Structure of proteins, myofilaments organization changes, connective tissue configuration, perimysium and endomysium structure, Z lines degeneration, I band breaks, attachment of myofibrils with sarcolemma	Chemical constitution	Functional quality	Superior resolution and magnification, observe samples without freezing or dehydration with environment scanning electron microscopy	Sample preparation need cryofixation, dehydration, embedding, or staining with heavy metal
Macroscopic imaging	Collagen: arrangement and content, lipids organization, tenderness and juiciness, fat content	Sensory, chemical, nutritional attributes	Functional and basic properties	Easy to use	–
Impedance measurement	Membranes soundness, pH, fat content, tenderness aging, differentiation between fresh and frozen-thawed products	Sensory and physico-chemical attributes	Functional quality	Non-destructive, inexpensive	Incursive

prepared in thin cuts or not. The non-thin-cut samples were used for very early phase contrast measurement allowing the detection of I and A bands in muscles.

3.2.1.2 Histology

Histology is a widely used method for observing biological tissues at the microscopic level, particularly as a tool for controlling meat texture in food science. The technique may or may not require tissue staining with specific dyes. However, histology always needs very thin sample cuts. In biological microscopy, it is almost always necessary to enhance contrast by using specific dyes to make certain biological components more visible during histological observation. The histo-chemical quality of a muscle sample, such as composition of fibre type, fibre area, glycolytic and oxidative ability and lipid and glycogen capacity are the factors which have been found to influence meat properties. Similarly, histochemistry, with various other staining protocols such as Sudan black-B, myosin ATPase has been used to categorize fibres.

3.2.1.3 Confocal Laser Scanning Microscopy

It is a fluorescence method as it helps in obtaining high-longitudinal resolution optical images. Confocal laser scanning microscopy is an evolution of the more traditional fluorescence microscopy, its fundamental property being the ability to produce point-by-point in-focus images of thick specimens, allowing 3D reconstructions of complex tissues. This method depends on fluorescence, so samples commonly need to be treated with fluorescent dyes to create targets visible, but there is no need for thin cuts. In meat science for examination of changes in fresh and cooked pork muscle during ageing confocal laser scanning microscopy is applied.

3.2.1.4 Electron Microscopy

By the application of electron beam for illumination of a specimen and form an enlarged image which leads to image observation. It has a greater resolution capacity than with optical microscopes. The coupling of an X-ray probe enhances the effort of the electronic microscope for performing the X-microanalysis.

3.2.1.5 Scanning Electron Microscopy (SEM)

SEM gives images with more depth-of-field, yield a characteristic 3D display which gives greater penetration into the surface structure of a biological sample. Dehydration, cryofixation, embedding (in resin), or staining (with heavy metal) of samples is required in scanning electron microscopy. It is a high-performance device for analyzing process-related changes in ultra structure of meat.

3.2.1.6 Transmission Electron Microscopy (TEM)

Electrons are passed through the sample in transmission electron microscopy (TEM), its resolution is more than SEM. In this technique the sample can be stained with heavy metals for the improvement of the image quality. Samples are prepared in very thin slices after that this sample is placed on a grid for the examination, making the method difficult to implement. This method is used for the examination of the Z line changes during the process of tenderization by proteolytic enzymes and by calcium chloride following osmotic dehydration of culled animal meat.

3.2.2 Macroscopic Imaging

Visual inspection is used extensively for quality assessments on meat products applied to processes running from initial grading to consumer purchases. Numerous laboratories have investigated the possibility of using image-based meat quality evaluation. In meat science, image analysis consists in analyzing the texture of images produced from muscle and meat sections at one or more wavelengths. This technique makes it possible to distinctly highlight the lipid and collagen structures of muscular tissue.

4. Dielectric Methods

4.1 Impedance Measurement

Electric impedance is the attribute of a material to confront the flow of electric current. When this property is not dependent on the frequency of the current, it qualifies as resistance; otherwise, as it is the case with biological tissues, the impedance has a resistive component and a capacitive component. In systematic terms, biological tissues are composed of cells that are surrounded by an extracellular liquid. The cell membrane acts as an insulator at low frequencies, behaving like a capacitor. Biological tissue, particularly meat, has anisotropic impedance. The impedance of the meat reduces rapidly with rigor and continues to decrease, albeit much more slowly, during storage. Electric impedance is used for a broad range of purposes in meat technology:

4.1.1 Detection of Frozen Meat

In the 1970s, it was shown that frozen meat samples have very weak impedance (Sale, 1974). However, we now know that this measurement cannot certify that a meat has been frozen, because meat aged for very long can present similarly low impedance.

4.1.2 pH

Most of the research on impedance is used for controlling the abnormal drop in pH during post mortem changes. The evaluation of water holding capacity of pork is one of the major problem, being affected in pale soft exudative (PSE) meat having low pH and highly exudative characteristics. This change makes meat unsuitable for processing. Similarly, dark firm dry (DFD) meat with high pH in case of beef is the another major issue. These two defects are associated with modifications of membranes and extracellular fluids which finally affect the meat electrical properties. For compensation of the inaccuracy of pH measurements electrical measurements have been used. Most of the studies in this field have focused on detecting the defects early, *i.e.*, within 45 min to 1 hour post-slaughter. The problem in detection of PSE meat during rigor set is that meat involves rapid pH and temperatures fall during this period, whereas the associated metabolic changes will only affect its structure and thus its electric properties later on. Once the ultimate pH is reached, impedance (conductivity) is more capable of detecting PSE meat.

4.1.3 Fat Content

Fat is a non-conductor which affects the impedance of tissues. Electric impedance methods can obtain remarkable results. A simple electric conductivity measurement on a carcass immediately after slaughter can be associated with anatomical data to give fat content with remarkable accuracy (R2 =0.95). This could be explained by the fact that there is no membrane or extracellular compartment modifications occurring immediately after slaughter and the measurements are made at a stable temperature.

4.1.4 Tenderness

A study (Byrne *et al.*, 2000) related to electrical properties of muscle after cooking to tenderness as assessed by WBSF and attempted to establish a link between the electrical properties and the mechanical resistance of meat. There was no direct connection between meat tenderness and straight forward electrical measurements. This is due to the fact that connective tissue, which plays a crucial role in tenderness, has similar impedance to muscle fiber and thus cannot be detected by electrical measurements.

4.1.5 Ageing

Ageing involves meat-tenderizing biochemical and physicochemical processes. These processes include the action of endogenous proteases on muscle fibre structure, progressive increase in membrane water permeability, and the weakening of connective tissues. Faure *et al.* (1972) set out to evaluate state of maturation by quantifying these effects. They proposed an approach based on the ratio of low-frequency impedance to high-frequency impedance, which decreases during refrigerated storage.

4.2 Microwave Characterization

The microwave has been employed for heating food in many applications; thawing, cooking and disinfection purposes. Recently, however, there has been an emergence of sensor systems based on the interaction of low power electromagnetic microwaves with biological matter. The microwave frequency ranging from 0.3–300 GHz, observes the dielectric properties of biological tissues, closely correlated with water content and state. In particular, dielectric relaxation spectroscopy determines the molecular motion response of polar molecules (mainly water) to a weak external alternative electric field. As electric field frequency increases, the polar molecule can no longer rotate with the electric field, this frequency is known as 'relaxation frequency'. Dielectric properties change markedly around this relaxation frequency. The technique has been explored for measuring water activity in protein gels. Water activity is closely related to water binding, which itself is linked to water holding capacity of meat and also with pale soft exudative and dark firm dry phenomena. Moreover, fraudulently added water in meat products can be detected because of its higher value of water activity.

Table 4.4. Summary of Microwave Techniques

Microwave	*Specification*	*Quality Parameter*	*Meat Quality*	*Merit*	*Demerit*
Microwave measurement	Water activity, ageing, fish freshness	Physico-chemical characteristics	Functional attributes	Potentially non-contacting	–

5. X-ray Measurements

X-rays have long been used in medicine and others areas. Its principle is based on the measurement of the depletion of the penetrating energy. For bone, lean meat and fat have different attenuation properties and so depending on the level of penetrating energy; it should be possible to obtain quantitative measurements. For discrimination of fat, bone and lean meat multiple technology tools using X-ray beams at different energy levels have been developed according to the energy attenuation measured.

Table 4.5. Summary of X-ray Measurement

X-ray Measurement	*Specification*	*Quality Parameter*	*Meat Quality*	*Merit*	*Demerit*
X-ray measurement	Amount of fat, changes in myofilament structure	Nutritional characteristics and chemical composition	Basic characteristics	-	Ionizing radiation is used

6. Magnetic Resonance Methods

6.1 Nuclear Magnetic Resonance (NMR)

NMR is used for the description of many products which include muscle food. The high costs involved do make it currently difficult to consider installing NMR systems on production lines. The tool nevertheless has a broad range of research applications, particularly for product assessment, and can be seen as a reference method given the richness of measurements obtained: the diffusion coefficient and the relaxation time being the most useful. Nuclear magnetic resonance is basically based on the absorption and ejection of energy in the range of radio frequency of the electromagnetic spectrum.

The major interest of this technique is that it can solve many product control problems during production. NMR sequences run on bovine samples can produce images where the water and lipid signals are selected, it makes feasible to identify the various components of the conjunctive network. This technique is further enhanced by "susceptibility" imaging that makes it possible to identify the conjunctive network fibers whose thickness is much lower than the dimensions of the NMR image. Fat content investigation by NMR needs selecting an array of impulses to be applied to the product. The outcome focus the versatility and practicability of the technique, since the equipment involved is compact and the method can be equally well deployed for controlling food composition as for checking food quality. NMR imaging (MRI) produces a morphological image of a sample that differentiating the bone, fat and lean meat. Sample elements can be differentiated by differences in water content and in water mobility in various biological elements. Water content and water mobility are variables that can be studied by measuring particular NMR parameters (proton density, relaxation time, T1, T2,T_2, diffusion coefficient, *etc.*). For the test range of 5–15 per cent fat content, actual fat content was determined with very good accuracy. It is thus feasible to use this technique to describe the connective tissue structure of the perimysium (Bonny *et al.*, 2001; Laurent *et al.*, 2000).

6.2 Magnetic Resonance Elastography

Like transient elastography, magnetic resonance elastography (MRE) measures local viscoelastic properties by following an acoustic wave in a biological tissue. The technique takes advantage of MRI image quality, and microscopic MRE, which is made possible with high-resolution MRI. The technique provides the complementary microscopic and macroscopic datasets on intramuscular conjunctive tissue structures, both of which are necessary for predicting sensory tenderness. Although a high number of studies have already been conducted, the field of research is vast, and meat scientists still have years of exciting work ahead before offering to meat industry a cheap, robust, reliable, portable, rapid, universal magnificent meat quality sensor.

Table 4.6. Summary of Magnetic Resonance Methods

Magnetic Resonance	*Specification*	*Quality Parameter*	*Meat Quality*	*Merit*	*Demerit*
NMR spectroscopy and imaging	pH, water activity, water content, water holding capacity, denaturation of connective tissue, cooking losses, fat content, PSE, DFD, collagen content, collagen organization, myofiber typing	–	Basic attributes	Precise, 3D restoration	Costly
Magnetic resonance elastography	Local viscoelastic character	–	–	3D restoration of mechanical character	Costly

7. Conclusion

The uses of all above technologies continue to increase and serve a valuable role in our society. The entire meat chain is working together as never before to improve the quality of the products they place before their consumers. The biggest roadblock is the availability of accurate information upon which to base decisions on breeding and selection, as well as production and management. Technology is sorely needed to identify critical control points for the origin of meat quality problems in the industry. Once these are distinguished, the possibility for rapid enhancement in the quality and uniformity of meat is improved. Among the numerous techniques which have been proposed for meat quality evaluation on the fresh intact product, very few meet even potentially, the requirements of industry.

When instrumentation has advanced enormously at the laboratory level, permitting a better understanding of the mechanisms underlying meat quality, application of this knowledge to quality evaluation in the meat industry remained very limited. Among the techniques which have been recently described, I think that meat quality evaluation at large-scale, the most important methods are: ultrasonic measurements, for the inspection of potential texture on live animals and whole carcasses, mainly because of their non-invasive character and low cost; for meat joints and cuts, image processing and NIR spectroscopy, image processing possess proven capability to estimate basic acceptability attribute, that is colour and marbling; it is totally non-invasive and obviously, use of this technology could greatly improve quality control in meat industry. NIR spectroscopy possesses a large potential range of utilization, from toughness prediction to frozen-thawed meat detection. However, more information on its real capabilities in industrial conditions is still needed. As Swatland *et al.* (1994) noted it: "introducing new technology into meat industry, at the level of slaughtering, meat cutting, and distribution, is not easy".

REFERENCES

Amin, V. (1995). "An introduction to principles of ultrasound." Iowa State University. *Study Guide*, 1-34.

Beattie, R. J., Bell, S. J., Borggaard, C., and Fearon, A. (2006). Prediction of adipose tissue composition using Raman spectroscopy: Average properties and individual fatty acids. Lipids, 41, 287–294.

Belk, K. E., George, M. H., Tatum, J. D., Hilton, G. G., Miller, R. K., Koohmaraie, M., Reagan, J.O., and Smith, G. C. (2001). Evaluation of the Tendertec beef grading instrument to predict the tenderness of steaks from beef carcasses. Journal of Animal Science, 79(3), 688–697.

Bonny, J. M., Laurent, W., Labas, R., Taylor, R., Berge, P., and Renou, J. P. (2001). Magnetic resonance imaging of connective tissue: A non-destructive method for characterising muscle structure. Journal of the Science of Food and Agriculture, 81(3), 337–341.

Byrne, C. E., Troy, D. J., and Buckely, D. J. (2000). Postmortem changes in muscle electrical properties of bovine M-longissimus dorsi and their relationship to meat quality attributes and pH fall. Meat Science, 54(1), 23–34.

Ellis, D. I., Broadhurst, D., Clarke, S. J., and Goodacre, R. (2005). Rapid identification of closely related muscle foods by vibrational spectroscopy and machine learning. Analyst, 130(12), 1648 1654.

Faure, N., Flachat, C., Jenin, P., Lenoir, J., Roullet, C., and Thomasset, A. (1972). Contribution à l'étude de la tendreté et de la maturation des viandes par la méthode de la conductibilite électrique en basse et haute fréquence. Revue De Médecine Vétérinaire, 123, 1517–1527.

Grunert, K. G. (1997). What's in a steak? A cross-cultural study on the quality perception of beef. Food Quality and Preference, 8(3), 157–174.

Houghton, P.L., and Turlington, L.M. (1992). Application of ultrasound for feeding and finishing animals: A review. Journal of Animal Science, 70, 930-941.

Jackman, P., Sun, D. W., and Allen, P. (2009). Automatic segmentation of beef longissimus dorsi muscle and marbling by an adaptable algorithm. Meat Science, 83, 187–194.

Laurent, W., Bonny, J. M., and Renou, J. P. (2000). Muscle characterisation by NMR imaging and spectroscopic techniques. Food Chemistry, 69(4), 419–426.

Lepetit, J., and Culioli, J. (1994). Mechanical-properties of meat. Meat Science, 36(1-2), 203–237.

Lorenzen, C. L., Calkins, C. R., Green, M. D., Miller, R. K., Morgan, J. B., and Wasser, B. E. (2010). Efficacy of performing Warner-Bratzler and slice shear force on the same beef steak following rapid cooking. Meat Science, 85(4), 792–794.

Monin, G. (1998). Recent methods for predicting quality of whole meat. Meat Science, 49(1), S231–S243.

Naganathan, G. K., Grimes, L. M., Subbiah, J., Calkins, C. R., Samal, A., and Meyer, G. E. (2008). Visible/near infrared hyperspectral imaging for beef tenderness prediction. Computers and Electronics in Agriculture, 64(2), 225–233.

Sale, P. (1974). Instrument for detecting frozen meat by electrical conductance. Revue Generale du Froid, 65(4), 343–348.

Stephens, J. W., Unruh, J. A., Dikeman, M. E., Hunt, M. C., Lawrence, T. E., and Loughin, T. M. (2004). Mechanical probes can predict tenderness of cooked beef longissimus using uncooked measurements. Journal of Animal Science, 82(7), 2077–2086.

Swatland, H. J., and Barbut, S. (1999). Sodium chloride levels in comminuted chicken muscle in relation to processing characteristics and Fresnel reflectance detected with a polarimetric probe. Meat Science, 51(4), 377–381.

Swatland, H. J., Ananthanarayanan, S. P., and Goldenberg, A. A. (1994). A review of probes and robots: implementing new technologies in meat evaluation. Journal of Animal Science, 72, 1475–1486.

Tornberg, E. (1996). Biophysical aspects of meat tenderness. Meat Science, 43, S175–S191.

Chapter 5

DNA Based Meat Speciation Methods to Ensure the Quality of Meat and Meat Products

Ajit Pratap Singh[1]*, Samir Das[2], P.K. Singh[3] and Akshay Garg[4]

[1]Associate Professor, Animal Biotechnology Center, NDVSU, Jabalpur, Madhya Pradesh
[2]Scientist, Division of Animal Health, ICAR Research Complex For NEH Region, Umroi Road, Umiam – 793 103, Meghalaya
[3]Assistant Professor, Department of Livestock Products Technology, College of Veterinary Science and A.H., Rewa, Madhya Pradesh
[4]Assistant Professor, Department of Vet. Microbiology, College of Veterinary and Animal Science, SVPUAT, Merrut, Uttar Pradesh
**e-mail: ajitvet123@gmail.com*

ABSTRACT

DNA based speciation of the meat ensures rapid affirmation for the quality of meat to the consumer. DNA acts as a biomolecule for ascertaining adulteration of meat. Many a time there is adulteration or mislabeling of meat product with meat from unwanted species. Foodstuff should be authentic and trust between producer and consumers needs to be established on scientific evidences. The producer or retailer should not subject a fraudulent product hiding the identity, its origin, its composition etc. to the consumer. Identification of animal species in commercial meat products is important with respect to export, international norms, economic and sanitary issues. Consumers have a preference of meat from specific species because of cultural, taste, religious issues (beef is a taboo for Hindus and similarly pork is taboo for Muslims), and many sect of people prefer goat meat rather than sheep etc. Consumer has all the right to select the species of meat as per their preference. Wild animals of endangered species sometime enter in the meat market which needs to be authenticated for forensic report. With the advent of modern techniques these issues can be resolved by DNA based speciation techniques. The meat products adulterated with low cost products can also be detected by the isolation and identification of DNA from that product.

Keywords: *DNA, Meat, Quality, Speciation.*

1. Introduction

Meat and meat product industry is growing with a very fast pace in India and globally. With increased demand and consumption of meat products, nowadays food safety and quality are a major concern. It has a great impact on opinion of public, if food contamination, poisoning and adulterations are reported by the media. There is an increasing demand for the improvement of quality controls of food; hence researchers are focusing towards the development of reliable molecular tools for food analysis and meat speciation. The molecular techniques play a pivotal role in implementation of different social and conservation acts like prevention of cow slaughter acts of different states of India, wild life conservation act, PFA acts of India and some other similar acts of the world (Singh and Sachan, 2010). Fraudulent practices are prevalent in meat industry, where adulteration practices involving inferior quality meat mixed to superior quality meat to increase the profit margin leading to a risky approach affecting consumers. These practices are not new and existed since a long time; the first case of fraudulent substitution was recorded in thirteenth century A.D. at Florence in Italy (Thornton, 1968). A common fraud in the meat industry in which uninspected meat is substituted for meat that has undergone inspection and been branded as satisfactory. The other frauds are the substitution of meat of another species, *i.e.*, horse for beef especially in Britain and Ireland, kangaroo meat in beef in Australia, cat for chicken or rabbit, goat meat for mutton, mutton for venison, dog meat and cat meat for chevon (Kangehte *et al.*, 1986) in other countries including India. Meat adulterations as high as 25-30 per cent of the meat sold in India is documented (Bhat *et al.*, 2015). These practices are more common in comminuted meat products. For detection of meat species in adulterated meat; several techniques are there starting from simple physical tests to recent and reliable molecular techniques.

The advent of molecular technique has led to the development of a diverse array of assay for quality control of meat and meat products. Rapid analysis using DNA hybridization and amplification techniques offer more sensitivity and specificity in results than conventional methods; additionally the recent techniques are faster than the conventional one. Several techniques have achieved the high level of automation too, facilitating their application for a routine sample assays. The text here is planned to produce a vivid picture of different molecular techniques, helpful in quality control of meat and meat products.

2. DNA Based Molecular Techniques

There are different DNA based methods for the quality and safety assurance of meat and meat products. DNA is a molecule of choice for species specifications due to its stability during heating and processing. The techniques are applicable even in rendered meat products, genetically modified foods *etc.*

2.1 DNA Hybridization Technique

It is a qualitative or semi-quantitative technique of meat species speciation in which species-specific DNA sequence is detected (Ebbehoj and Thomsen, 1991). For this purpose probes prepared from DNA or cloned DNA are hybridized with target DNA and detected by colour development. During the early development of DNA sequence analysis, genomic DNA was used as a species specific probe and was hybridized to DNA extracted from meat samples. The subsequent development of probes derived from species-specific satellite repetitive DNA sequences has greatly improved the specificity of the assay, now making it possible to detect admixtures that contribute as little as 5 per cent or less to a product (Wintero *et al.*, 1990). However, as DNA hybridization is a laborious exercise and later with advent of species specific polymerase chain reaction, which is more specific, sensitive and can be routinely performed, this technique got replaced for better results (Kumar *et al.*, 2013).

2.2 Polymerase Chain Reaction (PCR)

PCR is a rapid method because in this technique we can obtain multiple copies of specific piece of DNA sequence *in vitro* and it has high degree of selectivity and sensitivity. The species specific primers are used for amplification of target gene. PCR amplifies a target DNA sequence in an exponential phase, being capable of detecting even a single copy sequence from a single cell sample. It is a qualitative test for meat species specification. The usual PCR reaction is carried out in thermal cyclers in which cyclic temperature-time are variables. The pre-requisite to start a PCR reaction is the need for the sample nucleic material (DNA/RNA) which can be isolated by different methods usually, the simplest of which is hot-cold lysis where the sample is boiled for 5 min followed by quick chilling which basically breaks the cell and releases the DNA. The second method is a conventional method where the samples are lysed with combination of chemical compounds like SDS, protenase-K, EDTA, lysozyme *etc.*, which lyses the cell and DNA is released but this solution contains protein and fat as impurities which is further removed through phenol, chloroform solutions. Phenol is used to remove protein and chloroform for fat impurities and chilled absolute alcohol is used for precipitation of DNA. The third method is commercial kit based which includes lysis solution and spin column for removing the impurities. All the methods have their own advantage and disadvantage and were used as per the downstream work. The simplest hot-cold lysis is easy to perform and economic but contains other impurities which hinders many downstream assays. The DNA obtained from this crude method usually works for species-specific PCR. The conventional method using various chemicals gives good yield for performing most of their downstream reaction but is lengthy and requires technical knowledge and skills. The kit based methods are easy but costly; the yield is also lower than conventional chemical methods. The DNA so obtained by kits of good quality can be used for most of the downstream PCR assays.

This alternative DNA detection system of PCR amplification used for meat speciation mainly targets the mitochondrial cytochrome b gene and even subsequent cleavage by a restriction enzyme gives rise to a species-specific pattern on an agarose gel. This method does not requires the development of species-specific probes and, because it is PCR-based, is most suitable for critical samples in which DNA is largely degraded (Chikuni *et al.*, 1990). This is a good technique for detection of adulteration as low as 0.1 per cent (Janssen *et al.*, 1998).There are two main techniques for amplification of genetic marker, *i.e.*, mono-locus-specific primers for amplification of a concrete DNA fragment (*e.g.*, simplex PCR) and multi-locus amplification of non-targeted DNA (*e.g.*, RAPD).The wild pig and domestic pig can be differentiated by the ARMS –PCR technique (Jadav *et al.*, 2016).

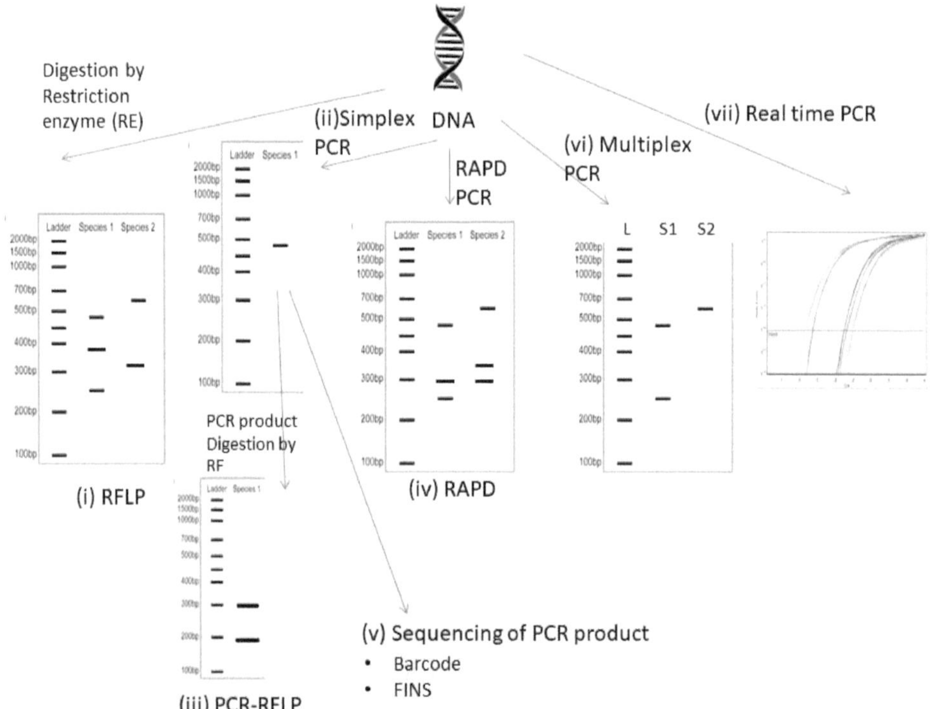

Figure 5.1. Representative Diagram of Various PCR Based Methods for Meat Speciation.

2.2.1 Species-Specific PCR

These are mainly based on amplification of target DNA by the use of species specific primers. The prerequisite of this test is the prior knowledge of nucleotide sequence to be used as a target. This is one the most common and simplest PCR method where the isolated DNA is put into a PCR tube usually 0.2 or 0.5 ml size with a reaction mixture containing DNA polymerase, primers, nucleotide mixture (A,T,G,C), buffers (containing magnesium sulphate, tris-hydrochloride, gelatin, potassium chloride *etc.*) and rest DNAase-RNAase free water. The usual volume is 20 to 50 microlitre for the whole reaction and the PCR assay is performed in a thermal-cycler with condition optimised like initial pre-denaturation at 95°C for 5 min, followed by 25 to 40 cycles of three steps denaturation at 95°C for 1 min, annealing at X°C (X here need to be calculated as per temperature of melting of the primer used which is usually 50-65°C) for 1 min and extension 72°C for 1 min and then a final one cycle of 72°C for 5 min extension. This is an example of PCR program which can be changed as per the standardization of laboratory protocol. This allows multiplication of the targeted DNA sequence to millions of copy which is separated in a gel electrophoresis apparatus and visualized on gel-documentation after staining with ethidium bromide (an example shown in Figure 5.2). With the advancement of science and technology, the PCR assays have become more efficient, robotics have made handling large number of samples with convenience and efficiency. PCR mastermix have been made, leading to more convenient way of sample processing, where it requires only the primer, sample DNA and nuclease free water to be added. Future may further enhance the technical interventions making the test more handy and reliable.

Table 5.1. PCR Amplification Programme

Step No	*Process*	*Temperature*	*Time*	
1.	Initial Denaturation	95°C	4minutes	
2.	Cycle Denaturation	95°C	30seconds	35 cycles
3.	Annealing	59°C	30seconds	
4.	Extension	72°C	35seconds	
5.	Final Extension	72°C	7minutes	
6.	Hold at 4°C			

In this technique of simplex PCR, the species-specific DNA in femto grams (fg) and pico grams (pg) can be detected in both processed and unprocessed meat samples by using targeted amplification of rRNA genes (12, 16, 18 S), actin-multigene which is highly conserved in eukaryotes, cytochrome-b, satellite DNA, mitochondrial D-loop, cytochrome oxidase-II, growth hormone gene, melanocortnin gene, myofibrillar components, satellite-I DNA *etc.* using polymerase chain reaction. By PCR technique very old samples even of more than 100 million years can also be identified (Girish and Nagappa, 2009). Mitochondrial DNA is one of the most preferred and used gene for differentiation of meat species. Mitochondrial D-loop PCR specific for beef can detect adulteration with even as low as 0.1 per cent of other species of buffalo, sheep, goat, pig and chicken and in term of DNA as low as 0.1 picograms for both cooked, microwaved and raw meat can be detected showing the efficacy of species –specific PCR (Karabasanavar *et al.*, 2017). In another study for Kebab, the PCR targeting mitochondrial cytochrome-b (cyt-b) gene could detect beef, mutton and chicken effectively without being affected by cooking and addition of non-meat ingredients (Hassan *et al.*, 2019).

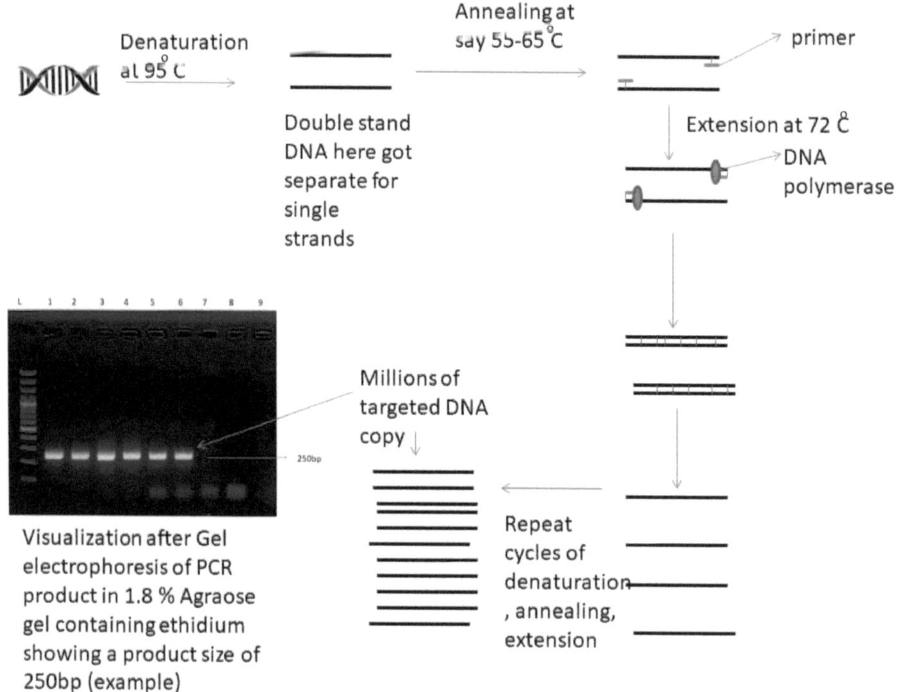

Figure 5.2. An Illustrative Flowchart Showing Polymerase Chain Reaction.

2.2.2 Multiplex PCR

In this technique many targets are simultaneously amplified which helps in detection of many species in a short period of time. Many such multiplex PCR have been developed for simultaneous detection of different meat species. Multiplex PCR usually starts with standardization of few sets of simplex PCR individually and then these sets are combined in one single tube with further standardization in PCR format making it multiplex PCR. Ali *et al.* (2015) used 5 set of species-specific primers targeting mitochondrial ND5, ATPase 6, and cytochrome-b genes which could differentiate the targeted cat, dog, pig, monkey and rat meat where they were assessed for specificity with 15 meat, fish species and 5 plant species. They were able to detect 0.01–0.02 ng DNA for raw and 1 per cent for processed meatball. Many other such multiplex PCR are available which could efficiently detect and differentiate various meat species (Kitpipit *et al.*, 2014; Hou *et al.*, 2015; Bhat *et al.*, 2016; Alikord *et al.*, 2017; Li *et al.*, 2019).

2.2.3 Random Amplified Polymorphic DNA Fingerprinting (RAPD)

It is a modified PCR technique in which DNA fingerprints are generated in a very short period of time which can be visualized on gel electrophoresis. In this method prior knowledge of DNA sequencing is not required but known standard has to run each time. In this technique DNA extraction from sample is the first step and then it is amplified with specific short length primers (generally 9-10 mer) and DNA fragments so obtained are visualized, recorded and compared with standards (Calvo *et al.*, 2001). Chai *et al.* (1997) developed RAPD for differentiation of bird meat. Martinez and Yman (1998) developed for elk, kangaroo, reindeer, buffalo and ostrich, as well as some domestic meat species. Arslan *et al.* (2005) developed RAPD for detection of meat from wild boar, bear, camel and domestic species. Many other such RAPD based meat speciation methods are available. RAPD is relatively easy and economical to use but there is a need for post- PCR analysis of band pattern which requires software and skill for the analysis.

2.2.4 PCR-restriction Fragment Length Polymorphism (PCR-RFLP)

It uses both the techniques, PCR and RFLP, here conserved targeted gene is amplified and resultant PCR product is digested with specific restriction endonuclease enzyme to get a restriction pattern or fingerprints. This is also suitable method for identification of even degraded DNA with apomorphic sites (Bauer *et al.*, 1987; Borgo *et al.*, 1996; Girish *et al.*, 2005). The first step in this technique is extraction of DNA then these DNA fragments are amplified with PCR technique using specific primers. It is followed by digestion with restriction endonucleases and the band patterns are observed by gel electrophoresis. It is suitable method for the analysis of very low amount of meat (1 mg). It is a faster and more sensitive method than DNA hybridization. By this technique two species of animal can be differentiated even after heating at 120°C. In this technique proper identification of male and females can be done by using PCR and genomic DNA extraction from raw muscles (Lahiri *et al.*, 1992).

2.2.5 Polymorphism Techniques

PCR-single strand conformation polymorphism (PCR-SSCP) is one of such promising DNA based polymorphism technique which is used for meat speciation. Here polymorphic DNA is used for amplification through PCR followed by denaturation by high temperature and formamide and subsequently these denaturated PCR amplicons are separated by non-denaturing polyacrylamide gelelectrophoresis (PAGE) for visualization of DNA bands through silver-staining method. Csikós *et al.* (2015) performed PCR-SSCP and could obtain meat speciation for undeclared species. Tisza *et al.* (2015) used PCR-SSCP targeting mitochondrial 12S rRNA for differentiation of several poultry species, *viz.*, chicken, guinea fowl, pheasant, turkey, goose, duck and muscovy duck. PCR-SSCP is a relatively economic technique than other sequencing based or real-time based methods.

2.2.6 PCR Product Sequencing

In this technique sequencing of a particular gene is carried out to know the nature of gene responsible for particular meat species specificity. The sequence so obtained can be analysed into different form to known their similarity, dissimilarity, phylogenetic distance between various meat species. Many downstream techniques of PCR sequencing using targeted gene are used for such purpose in meat speciation such as DNA barcoding, forensically informative nucleotide sequencing (FINS) *etc.* DNA barcoding basically targets PCR amplification of small standardized fragment say mitochondrial cytochrome oxidase I gene followed by sequencing and to observe the match so produced to reference sequences or DNA barcodes (Singh *et al.*, 2014). FINS is a technique that combines DNA sequencing and phylogenetic analysis. In this technique meat samples are identified on informative nucleotide sequences basis. Actually PCR amplification and sequencing of conserved gene is one of the first technique for meat species identification. Among them mitochondrial DNA is highly conserved and gene on it Cytochrome-b and 12S-r RNA used for meat species identification can be exploited for the meat species speciation. FINS had been used for differentiation of beef, buffalo meat, mutton and chevon (Murray *et al.*, 1995; Girish and Nagappa, 2009).

2.2.7 Real Time PCR

Is a quantitative PCR (qPCR) where PCR products are detected and monitored during amplification in the same reaction vessel with the help of fluorescent compounds. All the stages of amplification are visible on the screen and hence identification, detection of the product during its amplification of the targeted DNA is possible (Ghatak and Gill, 2009). The amount of DNA is measured after each amplification cycle via fluorescent dyes that yield increasing fluorescent signal in direct proportion to the number of PCR amplicons generated. Data collected in the exponential phase of the reaction yield quantitative information on the starting quantity of the amplification target. The fluorescent-based methods used in real-time PCR can be classified into two categories *viz.*, TaqMan Probe-based and DNA intercalating dye based like SYBR Green. TaqMan probe-based RT-PCR are more specific but is expensive, and much more difficult to design and optimize. Alternatively, dyes are sequence-independent and bind into the minor groove of dsDNA. SYBR Green I is the most common florescent dye used in this category. In this technique quantification of the meat species may also be possible. The traditional PCR detects meat adulteration even in smaller quantity but if there is a requirement to quantify the extent of adulteration then there is a role of Real Time PCR. Sawyer *et al.* (2003) was able to detect beef in lamb and also able to quantify the amount of its addition by using Real Time PCR. He used universal primer to amplify all the common meat species and additionally he used a species specific primer for beef labelled with probe. The amplifications are compared for the cycle number when they are first detected to the reference standards of known species content, ultimately determining the percentage of a particular species in a mixed sample.

2.2.8 Digital PCR (dPCR)

Also known as droplet digital PCR (ddPCR) is a novel quantitative PCR method similar to qPCR or Real time PCR that provides a sensitive and reproducible way for detection of nucleic acid (DNA or RNA) and quantification. Digital PCR is more precise than traditional PCR but needs expert handling like qPCR. The dPCR reaction contains the traditional PCR component and uses the Taqman based chemistry for recording the quantification. In dPCR, the PCR reaction is carried out in thousands of droplet, at nano-scale simultaneously, which increases the efficiency of this PCR in detection. The ddPCR reaction contains the mastermix, primer, probe, and after addition of the target DNA the reaction is divided into thousands of droplet by droplet generator and the PCR cycles similar to traditional PCR is done followed in ddPCR machine and later product is visualized in droplet reader.The ddPCR does not require internal control and standard curves and had better efficiency than RT-PCR (Taylor *et al.*, 2017). It was used to detect chicken fraction in sheep or goat keeping many other meat species as control *viz.*, pork, duck, goose, pigeon, beef, horse, dog, rabbit, turkey and donkey (Ren *et al.*, 2017).

2.3 DNA Sequence Bar Coding

The accurate species identification by DNA sequence barcoding relies on a suitable DNA barcode, which refers to a standardized sequence (usually less than 1,000bp) of the genome. The barcode should have universality so that it can be easily amplified from diverse species, and it should contain few insertions or deletions to facilitate sequence alignment. Also, its mutation rate must be sufficient to generate a barcoding gap, which means the maximum intraspecific variation is less than the minimum interspecific distance (Hebert *et al.*, 2003). For animal identification, the most broadly used barcode marker is mitochondrial cytochrome c oxidase subunit I (COI), which is highly conserved across species employing oxidative phosphorylation for metabolism (Hebert *et al.*, 2003). Numerous studies have shown that COI-based DNA barcoding can delimit diverse animal species, indicating the high rates of sequence change at species level and constraints on intraspecific divergence in COI sequence. For analyzing fresh and well-preserved animal tissues, a full-length barcode such as a 658-bp region of COI gene is recommended as its PCR amplification and sequencing can be feasible.

3. Conclusion

Several methods are available for meat species specifications but no single method can be applied in all cases. The physical, chemical and anatomical methods are more suitable for raw meat while minced or comminuted meat requires sophisticated techniques. An array of DNA based methods have been discussed in the chapter. These DNA based methods continuously evolve with molecular science advancements. The same has lead to several techniques for speciation starting from DNA hybridization, DNA fingerprinting, species specific PCR, Multiplex PCR, sequence analysis, real time PCR, digital PCR *etc.* The DNA based methods have now become fast and accurate and have been utilised in vetro-legal and adulteration investigations. The acceptance of these methods have led to solve the pending cases/ issues without any delay. These efforts are ultimately giving a sense of surety to consumer for a safe and quality meat products.

REFERENCE

Ali, M.E., Razzak, M.A., Hamid, S.B., Rahman, M.M., Amin, M.A., Rashid, N.R. and Asing. (2015). Multiplex PCR assay for the detection of five meat species forbidden in Islamic foods. Food Chemistry, 177, 214–224.

Alikord, M., Keramat, J., Kadivar, M., Momtaz,H., Eshtiaghi, M.N. and Rad, A.H. (2017). Multiplex-PCR as a rapid and sensitive method for identification of meat species in Halal-meat products. Recent Patents on Food Nutrition and Agriculture, 8(3),175-182.

Arslan, A., Ilhak, I., Calicioglu, M., and Karahan, M. (2005). Identification of meats using random amplified polymorphic DNA (RAPD) technique. Journal of Muscle Foods, 16 (1), 37-45.

Bauer, V.C., Teifel-Greding, J., and Liebhardt, E. (1987). Species identification of heat denaturized meat samples by DNA analysis. Archives Lebensmittel Hygiene, 38, 149-176.

Bhat, M.M., Salahuddin,.M., Mantoo, I.A., Adil,S., Jalal, H. and Pal, M. A. (2016). Species-specific identification of adulteration in cooked mutton Rista (a Kashmiri Wazwan cuisine product) with beef and buffalo meat through multiplex polymerase chain reaction. Veterinary World, 9(3), 226–230.

Bhat, M.M., Jalal, H., Para, P.A., Bukhari, S.A., Ganguly, S.,Bhat, A.A., Wakchaure, R and Qadri, K. (2015). Fraudulent Adulteration/Substitution of Meat: A Review. International Journal of Recent Research and Applied Studies, 2, 12 (5),22-33.

Borgo, R., Soulie-Crosset, C., Bouchon,D., and Gomot, L. (1996). PCR-RFLP analysis of mitochondrial DNA for identification of snail meat species. Journal of Food Science, 61, 1-4.

Calvo, J.H., Zaragoza, P. and Osta, R. (2001). Random amplified polymorphic DNA fingerprints for identification of species in poultry pate. Poultry Science, 80, 522-524.

Chai, K.M., Huat, L.C., Thai, C. S., and Phang, S.T.W. (1997). Random amplified polymorphic DNA (RAPD) fingerprint profiling of domestic and game birds. Asia-Pacific Journal of Molecular Biology and Biotechnology, 5(3), 173-182.

Chikuni, K., Ozutsumi, K., Koishikawa, T. and Kato, S. (1990). Species identification of cooked meats by DNA hybridization assay. Meat Science, 27, 119-128.

Csikós, A., Tisza, A., Ádám, S., Gabriella, G., András J., and Levente, C. (2015). Species identification in meat and cheese products by PCR-single strand conformation polymorphism (PCR-SSCP) and DNA sequencing. Aweth, 11(2), 78-83.

Ebbehoj, K.F. and Thomsen, P.D. (1991). Differentiation of closely related species by DNA hybridization. Meat Sci., 30, 359-366.

Ghatak, S., and Gill, J.P.S. (2009). Molecular techniques for detection and typing of food borne pathogens from livestock products. Proceedings of the Recent Developments in Postharvest Processing and Value Addition in Livestock Produce, Oct. 22-Nov. 11, CIPHET, Ludhiana, pp. 52-57.

Girish, P.S. and Nagappa, K. (2009). Molecular techniques for meat species identification. Proceedings of the Recent Developments in Postharvest Processing and Value Addition in Livestock Produce, Oct. 22-Nov. 11, CIPHET, Ludhiana, pp. 57-61.

Girish, P.S., Anjaneyulu, A.S.R., Viswas, K.N., Shivakumar, B.M., Anand, M. Patel,M. and Sharma, B. (2005). Meat species identification by Polymerase Chain Reaction-Restriction Fragment Length Polymorphism (PCR-RFLP) of mitochondrial 12S rRNA gene. Meat Science, 70, 107-112.

Hassan, K.I., Ali, B.A.M. and Mohammed, N.A. (2019). Detection of the origin of animal species in kebab meat using mitochondrial DNA based - Polymerase Chain Reaction (mtDNA-PCR). Iraqi Journal of Veterinary Sciences, 33 (1), (39-43).

Hebert, P. D. N., Ratnasingham, S., and DeWaard, J. R. (2003). Barcoding animal life: cytochrome c oxidase subunit 1 divergences among closely related species, Proceedings of the Royal Society B Biological Science, vol. 270, supplement 1, pp. S96 S99.

Hou, B., Meng, X., Zhang, L., Guo, J., Li, S., and Jin, H. (2015). Development of a sensitive and specific multiplex PCR method for the simultaneous detection of chicken, duck and goose DNA in meat products. Meat Science, 101, 90-94.

Jadav, K. K., Rajput, N., Singh, A. P., Sarkhel, B. C., and Srivastava, A. B. (2016). A species-specific ARMS PCR test for detection of Indian wild pig DNA. Indian Journal of Animal Sciences, 86(6), 67-69.

Janssen, F.W., Hagele, G.H., Buntjer, J.B and Lenstra, J.A. (1998). Species identification in meat using PCR-generated satellite probes. Journal of Industrial Microbiology and Biotechnology, 21, 115-120.

Kangehte, E.K., Gathuma, J.M. and Lindqvist,K.J. (1986). Identification of species of origin of fresh, cooked and canned meat and meat products using antisera to thermostable muscle antigens by Ouchterlony's double diffusion test. Journal of the Science of Food and Agriculture, 37, 157-164.

Karabasanavar, N., Girish, P. S., Kumar, D and Singh, S. P. (2017). Detection of beef adulteration by mitochondrial D-loop based species-specific Polymerase chain reaction. International Journal of Food Properties, 20(sup2), 2264-2271.

Kitpipit, T., Sittichan, K. and Thanakiatkrai, P. (2014). Direct-multiplex PCR assay for meat species identification in food products. Food Chemistry, 163,77-82.

Kumar, A., Kumar, R. R., Sharma, B. D., Gokulakrishnan, P., Mendiratta, S. K., and Sharma, D. (2013). Identification of Species Origin of Meat and Meat Products on the DNA Basis: A Review. Critical Reviews in Food Science and Nutrition, 55(10), 1340–1351.

Lahiri, D. K., Bye, S.,Nurnberger, J. I., Hodes, M. E., and Crisp, M. (1992). A non-organic and non-enzymatic extraction method gives higher yields of genomic DNA from whole blood samples than do nine other methods tested. Journal of Biochemical Methods, 25,193–202.

Li, J., Li,J., Xu, S., Xiong,S., Yang, J., Chen, X., Wang, S., Qiao, X. and Zhou, T. (2019). A Rapid and Reliable Multiplex PCR Assay for Simultaneous Detection of Fourteen Animal Species in Two Tubes. Journal of food chemistry, 295,395-402.

Martýnez, I., and Yman, I.M. (1998). Species identification in meat products by RAPD analysis. Food Research International, 31(6-7), 459-466.

Murray, B.W., McClymonts, R.A., and Strobeck, C. (1995). Forensic identification of ungulate species using restriction digests of PCR-amplified mitochondrial DNA. Journal of Forensic Science, 40, 943-951.

Ren, J., Deng, T., Huang, W., Chen, Y., and Ge, Y. (2017). A digital PCR method for identifying and quantifying adulteration of meat species in raw and processed food. PLoS ONE, 12(3), e0173567.

Sawyer, J., Wood, C., Shanahan, D., Gout, S. and McDowell, D. (2003). Real-time PCR for quantitative meat species testing. Food Control, 14, 579–583.

Singh, V.P. and Sachan, N. (2010). Collection and dispatch of meat samples for vetro-legal cases. Training Manual of Veterinary Officers of U.P. Govt. under ATMA Scheme, Department of Extension DUVASU, Mathura, pp: 63-67.

Singh, V.P., Pathak, V., Nayak, N.K., Verma A K., and Umaraw, P. (2014). Recent Developments in Meat Species Speciation-a review. Journal of Livestock Science, 5, 49-64.

Taylor, S.C., Laperriere, G. and Germain, H. (2017). Droplet Digital PCR versus qPCR for gene expression analysis with low abundant targets: from variable nonsense to publication quality data. Science Reproduction, **7**, 2409 doi.org/10.1038/s41598-017-02217-x.

Thornton, H. (1968). Textbook of Meat Inspection. 6th Edn., Tindall and Cassel, Bailliere, London.

Tisza, Á., Csikós, Á., Simon, Á., Gulyás, G., Jávor, A., and Czeglédi, L. (2015). Identification of poultry species using polymerase chain reaction-single strand conformation polymorphism (PCR-SSCP) and capillary electrophoresis-single strand conformation polymorphism (CE-SSCP) methods. Food Control, 59,430-438.

Wintero, A.K., Thomsen, P.D., and Davies, W. (1990). A comparison of DNA-hybridization, immunodiffusion, countercurrent immunoelectrophoresis and isoelectric focusing for detecting the admixture of pork to beef. Meat Science, 27, 75-85.

Chapter 6

Protein Based Techniques of Meat Authentication

R.R. Kumar* and Preeti Rana

ICAR-Indian Veterinary Research Institute,
Izatnagar, Bareilly – 243 122, Uttar Pradesh
**e-mail: dr_rajivranjan@yahoo.com*

ABSTRACT

With increase in the demand of meat and meat products, and the profitability in this business, the malicious addition of undesired meat by seller to mimic the meat preferred by consumer is growing. Here comes the role of meat authentication, where the consumer gets satisfaction that the meat which he is consuming is the meat he is intending from that particular species. Because of religious issue, a Hindu person doesn't eat beef, and similarly, Muslim sect does not consume pork. Some prefer chevon instead of mutton but sellers try to mix mutton, if it is available cheaper. Many non-farm animals' meat also seems to get entry like dog meat. Consumer use to have a specific preference for a particular species and it hurts the religious sentiment, ethos, taste etc. if they unintentionally ate such adulterated meat. Hence, to provide the consumer the right meat, it is important that the meat inspector or forensic people adopt the methods which can differentiate meat of various species. The meat authentication can be done by observing anatomical differences, appearance of meat, by molecular methods and through protein based methods. Here, we will be discussing the various meat authentication methods through protein based techniques which are easy to perform, cost effective and can give concrete results in short time.

Keywords: *Meat authentication, Protein, Meat adulteration.*

1. Introduction

Global issues related to meat are most importantly concerned with authentication of meat. With ever rising demand for the meat the problem of adulteration has taken to its toll due to large economic benefits and shortage of supply as reported in European horse meat scandal in 2013.Adulterated food can be defined as food incompatible with the declaration of the seller. Advances in science have led to greater sophistication in adulteration, and specific sensitive analytical methods are needed to detect the adulterating substances (Fennema and Tannenbaum, 1996). Considerable number of incidences has occurred on global scale such as substitution of domestic animal meat with wild animals in Kenya, kangaroo and horse meat added to beef in Australia, sale of mutton as goat meat in India and chicken and pork frequently in beef sausages (Barai *et al.*, 1992). Therefore, there is dire need of authentication by employing various techniques. Authentication plays an important role in determining the wholesomeness, labelling, avoiding unfair competition, assessing the value of meat and providing assurance to consumers against any adulteration. Increased awareness among consumers for their health has led to demand of more inclusive information on the origin, composition and safety of the foods (Grunert, 2002; Taljaard *et al.*, 2006; Verbeke and Ward, 2006). Health reasons and religious taboos are other equally important aspects for which meat authentication is needed as for the consumer even traces of the objectionable meat are unacceptable. Analysis of protein composition and nucleic acids, play an important role in meat identification.

2. Basics of Protein Based Techniques

Immunological methods involve antigen–antibody interactions for species. Species determination on heated/autoclaved meat samples can also be achieved by raising antibodies against denatured proteins. (Muldoon, *et al.*, 2004; Kim *et al.*, 2005). A general advantage of ELISA techniques are that commercial sticks for species determination can be applied on-site and results can be obtained in about 15 min (Muldoon *et al.*, 2004). Detection of proteins in meat products is not restricted to immunoassays. Liquid chromatography (LC) methods with focus on protein profiles have been published and include qualitative detection of meat from a variety of animal species (Ashoor and Osman, 1988; Chou *et al.*, 2007). Also quantitative results obtained from LC analysis of known unheated chicken and turkey mixtures with a LOD of 1 per cent (w/w) have been published (Ashoor and Osman, 1988). If animal material consists of pure pork, lamb, chicken, or beef, dipeptide profiles can be used to identify the exact species (Aristoy and Toldra, 2004). Protein profiles from ground meat mixtures have successfully been studied by isoelectric focusing and a subsequent multivariate data analysis distinguished between beef, pork, and turkey (Skarpeid *et al.*, 1998).

2.1 Electrophoretic Techniques

Electrophoresis is the motion of dispersed particles relative to a fluid under the influence of a spatially uniform electric field and used in laboratories in order to separate macromolecules based on size. The technique applies a negative charge so proteins move towards a positive charge. The interpretation of results is done on the basis of protein band mobility. Most common method of protein based analysis for fish species identification. Primary structure of meat is made up of protein making it ideal subject for electrophoresis for species identification on basis of protein separation. Protein composition differs between species and depends on other factors such as tissue type, storage time of the product and technological processes employed to manufacture. Hofmann (1985) had analysed the aqueous extract of mammalian muscles by IEF (isoelectric focusing) and SDS-PAGE (sodium dodecylsulphate polyacrylamide gel electrophoresis) for species identification. Protein profiles from ground meat mixtures have successfully been studied by isoelectric focusing and a subsequent multivariate data analysis distinguished between beef, pork, and turkey (Skarpeid *et al.*, 1998).

3. DNA versus Protein Based Techniques

Genetic methods are the most specific and sensitive methods for food components authentication but they require expensive laboratory equipment and a certain degree of expertise. Immunological assays provide an alternative by reducing the test time and cost. Enzyme-Linked Immuno Sorbent Assay (ELISA) has been the most widely used technique for regulatory purposes in detecting food authenticity because of its specificity, simplicity and sensitivity, among other advantage. Primary structure of some meat proteins is relatively resistant to processing and skeletal muscle proteins are both species and tissue-specific, presenting them as a good candidate for use of specific muscle proteins and peptide markers for meat authentication (Buckley *et al.*, 2009, 2013; Sentandreu and Sentandreu, 2011; Montowska and Pospiech, 2012). Peptidomic analysis is extremely helpful for species identification as well as assessment of the quality of the product.

4. Techniques in Trend: Chromatographic, Electrophoretic, Immunological

4.1 Chromatographic Methods

Chromatographic methods involve the separation of a mixture of substances into individual constituents on the basis of determination of properties of a given constituent/protein and then measurement of their quantities which takes place during the flow of the mobile phase (gas or liquid) along a specially designed surface of the stationary phase (liquid, gel or porous solid body) made up of substances exhibiting sorption capabilities or capable of exerting other effects on flowing substances. Chou *et al.*, 2007 elaborated a high-performance liquid chromatography method with electrochemical detection (HPLC-EC) that allows differentiation of 15 meat species commonly consumed in Taiwan *viz.*: cattle, pig, goat, deer, horse, chicken, duck, ostrich, salmon, cod, shrimps, crabs, scallops, bullfrog, and alligator revealing unique electrochemical profiles. Mass spectrometry (MS) based proteomic methods only quite recently entered the field of meat authentication. In general, specific proteins or peptides (generated after tryptic digest) are detected in MS methods. Notably, the primary structure of proteins is, in general, quite stable against processing (Buckley *et al.*, 2013). Furthermore, a limited degree of protein degradation is often less critical for the detection of species-specific peptides than DNA shearing for the PCR detection, as fragmentation events within the relatively short peptide sequences are less likely compared to degradation of the longer DNA templates. Ponce-Alquicira and Taylor (2000) employed electro spray mass spectrometry coupled with collision-induced dissociation (ESI-CID-

MS/MS) to identify cow, pig, horse, and sheep meats on the myoglobin basis. The ESI-MS-MS spectra myoglobin extracted from the above four meat species were compared with the amino acid sequence of the entire protein differentiating sheep and beef meat from each other and from pig and horse meat. Simó *et al.* (2004) elaborated a physically adsorbed polymer coating for capillary electrophoresis-mass spectrometry (CE-MS) to successfully determine lysozyme content, in order to detect minced beef adulteration by addition of chicken egg white.

4.2 Electrophoretic Methods

The electrophoretic methods are based on the separation of proteins in the electric field following their extraction from the muscle tissue and later placed on special media. The IEF method consists in the separation on a gel which is characterised by the pH gradient resulting from the addition of ampholyte. Individual proteins move towards the pH value which is in agreement with their isoelectric point (*pI*). In the case of meats of such animal species as: cattle, horses, pigs, sheep, roe deer, kangaroos, camels, brown bears, rabbits, hares, chickens, ducks and ostriches the set of myoglobin bands is characteristic and they can be identified on the basis of myoglobin patterns (Hofmann, 1997). The IEF technique can be employed for the identification of related animals with low myoglobin content, for example, a chicken and turkey, using pseudo-peroxidase staining with the aid of a mixture composed of odianizidine and hydrogen peroxide. Some to several minutes after the treatment, red-brown bands appear on the gel (Montowska and Pospiech, 2007). The IEF method based on the myoglobin analysis is also effective in the situation of mixtures composed of different species. When the IEF method is used for the species identification of meats subjected to thermal treatment, the obtained results are influenced by the height of the temperature during the technological process. Although high temperature does not affect the relocation of the isoelectric points, the bands are not sharp and this may make the assessment more difficult. Therefore, it is recommended to conduct the analysis on one gel onto which the examined samples as well as the reference sample should be placed. The SDS-PAGE (Sodium Dodecyl Sulphate Polyacrylamide Gel Electrophoresis) method makes it possible to analyse proteins which dissolve poorly in solvents other than SDS solutions. This method can be applied to analyse proteins which do not dissolve in urea-containing solutions. The method can be utilized for protein quality analyses, for example, proteins of various species of fish or for the quantitative analysis which takes into account different degrees of binding of the applied dye by individual proteins.

4.3 Immunological Procedures

These procedures are based on very specific antigen antibody interactions, and can be used to detect and determine particular analytes without prior separation in complex mixtures such as biological fluids or food extracts.

4.3.1 Agar Gel Immunodiffusion (AGID)

This technique is carried out by placing a few millilitres of aqueous extract of soluble proteins from a meat sample in wells cut in a shallow layer (2-3 mm) of agarose gel. Antisera raised against the test species are placed in an adjacent well.

The plate is then incubated to allow antibodies and antigens to diffuse together. If both are present, a visible opaque band will appear in the gel between the wells, and results are available within 24 h. It is not, however, possible to distinguish sheep from goat, or horse from donkey, by this method. Subjective detection limits in raw beef are: sheep, 5 per cent; pork, 10 per cent; horse, 10 per cent; turkey, 15 per cent; kangaroo, 20 per cent; beef itself could be detected at the 2 per cent level in pork.

4.3.2 Enzyme Linked Immuno Sorbent Assays (ELISA)

It used as a fast qualitative analysis systems on the production site (test-stripe based) or as a quantitative method (Depreux, 2000). ELISA can be performed for processed samples and, on pork mixed with heterologous meat can have a detection limit of 0.5 per cent (w/w) for pork. Liu *et al.*, 2006 developed monoclonal antibody-based sandwich ELISA for porcine skeletal muscle detection in raw and thermally treated meat articles with limit of detection 0.05 per cent pork in chicken as well as pork in soy-based feed and 0.1 per cent pork in beef mixtures. In ELISA the antibody - antigen interaction is usually carried out in a monomolecular layer immobilised on the surface of a plastic microtitre plate. At least three forms of ELISA have been used in meat speciation work. The first to be applied was a straight forward form of ELISA (sometimes called indirect ELISA), in which all the soluble protein components of an extracted meat sample are absorbed directly on to the ELISA plate., double-antibody sandwich (DAS) version of ELISA which have the advantage of improved specificity and sensitivity for detection of components at low concentrations, require two species-specific antibodies. A competitive ELISA is a third variant of the system, which has been developed for meat species identification. Janssen *et al.* (1987) showed that dot-ELISA could be used to identify samples that contain prohibited constituents (especially proteins). They developed a procedure that could detect wheat gluten, whey protein, casein, ovalbumin and soybean protein in heated meat products at 0.1 per cent of adulterant. This technique could reduce labor costs when used to screen samples. Demeulemester *et al.* (1991) used an improved dot- ELISA to detect whey proteins in meat products, particularly in liver pates.

5. Novel Approach

Meat and meat product authenticity can be investigated much more comprehensively using two-dimensional electrophoresis (2-DE) allows for separation of thousands of proteins in one gel simultaneously. During the first dimension, proteins are separated with the assistance of the native technique (most frequently IEF) depending on their isoelectric point (*pI*) while the second dimension uses SDS-PAGE to separate proteins depending on their molecular weight (*MW*). The combination of the above-mentioned techniques result in the development of 2-D maps on which proteins are visible as spots with positions determined by their *pI* and *MW*. This technique can be used to separate simultaneously up to 10 000 proteins as well as to identify and determine quantitatively spots that contain less than 1 ng protein. Meat and meat product authenticity can be investigated much more comprehensively using two-dimensional electrophoresis (2-DE), which has been developed in recent years (Lopez, 2007).

Proteomic based method has been developed for the detection of chicken meat within mixed meat preparations. The procedure is robust and simple, comprising the extraction of myofibrillar proteins, enrichment of target proteins using OFFGEL isoelectric focusing, in-solution trypsin digestion of myosin light chain 3, and analysis of the generated peptides by liquid chromatography-mass spectrometry/ mass spectrometry (LC-MS/MS). Using this approach, it was possible for example to detect 0.5 per cent chicken in pork meat with high confidence (Sentandreu *et al.*, 2010).

At the present time, mass spectrometry (MS) techniques play a key role in protein and peptide analyses in food products. The combination of electrophoretic, enzymic, and chromatographic techniques with mass spectrometry increased significantly the analysis efficiency of complex protein mixtures. A mass spectrometer separate molecules of analyte according to their mass-to-charge (*m/z*) ratio

6. Detection of Non-animal Protein

Several proteomic approaches based on identification of peptide biomarkers have been already reported to reveal food composition. Chassaigne *et al.* (2007) developed a proteomic approach for detecting the main allergens present in peanuts, identifying some peptide biomarkers specific for the allergenic proteins. In the meat industry, soy protein is the most widely used vegetable protein, due to its biological value, its properties as an emulsifier, stabilizer and its capacity to increase water-holding capacity and improve the texture of the final product. With its hydrating capacity, soy protein can considerably decrease the final cost of the product to the benefit of the manufacturer.

Typical cases of intentional meat adulteration involve the substitution or addition of animal proteins (normally cheaper varieties) or plant proteins (such as soyabean or grain derivatives) not declared as such in the ingredient list (Flores-Munguia *et al.*, 2000). Plant-derived ingredients such as soya and wheat have attracted attention as meat substitutes for decades due to their cheaper prices compared to animal-derived components and because their addition in small quantities can enhance the technological characteristics of the final products (water binding capacity, texture) (Vanha *et al.*, 2009). Meat samples were analysed for the presence of soya using a commercial quantitative enzyme-linked immunosorbent assay (ELISA). The antibodies utilised in this ELISA are capable of detecting both raw and processed soya. Undeclared soya was detected with the ELISA in five of the 41 (12 per cent) mince samples, with three of these testing for levels exceeding 1000 mg kg soya.

Proteomic tools have been successfully applied to solve the problem of fish authentication. Mazzeo *et al.* (2008) carried out a molecular profiling strategy based on the use of MALDI–TOF MS as a rapid screening technique to identify protein profiles characteristic of the different fish species. The study of parvalbumins allowed the identification of specific biomarker proteins for unambiguous discrimination between them. These authors highlighted the importance to select suitable protein biomarkers capable of verifying product authenticity in both fresh and processed fish products. In another work, Carrera *et al.* (2007) carried out a classical proteomic approach including protein extraction, 2-DE gel electrophoresis, in-gel digestion of protein spots and de novo analysis of the generated peptides by MALDI–TOF and LC–ESI–MS/MS capable to identify specific peptides of the different fish species of the Merlucciidae family. A similar approach was developed to characterize peptides capable to differentiate between shrimp species of commercial interest (Ortea *et al.*, 2009).

The proteomic approach developed by Leitner *et al.* (2006) comprising prefractionation of soybean glycinin proteins, trypsin digestion and identification of peptides by LC–ESI–MS/MS has proved to efficiently detect the addition of soybean proteins to processed meat products, which could also be used for quantitative purposes. Sentandreu *et al.* (2010) unambiguously differentiated between chicken and turkey meats using species-specific peptides obtained from trypsin digestion of myosin light chain 3 (MLC-3).

7. Future Perspectives in Meat Authentication

Traditional strategies used to assess meat authenticity are based on methods such as chemometric analysis of a large set of data analysis, immunoassays or DNA analysis. The identification of peptide biomarkers specific of a particular meat species, tissue or ingredient by proteomic technologies open a new area in field of identification of species due to its high discriminating power, robustness and sensitivity.

8. Conclusion

Analysis of protein/peptide biomarkers using proteomic technologies as a way to assess meat authenticity still has limited use in meat science as compared to other established methodologies reviewed here such as chemometrics, immunoassays or DNA analysis. Protein based techniques provide a promising alternative because of their high discriminating power, comparable to that of DNA-based analysis, robustness and sensitivity. As in DNA analysis, discrimination is made at sequence level, being suitable to differentiate between closely related species whereas proteomic approach displays more robustness facing at the major limitations of DNA analysis in both quantitative determinations and degradation of DNA in highly processed samples. From this perspective, the primary amino acid sequence of marker peptides would be considerably more resistant to food processing than DNA sequences. The possibility to use routine mass spectrometry equipment would make implementation of this proteomic approach attractive and affordable for control laboratories.

REFERENCES

Aristoy, M. C., and Toldra, F. (2004). Histidine dipeptides HPLC-based test for the detection of mammalian origin proteins in feeds for ruminants. Meat Science, 67(2), 211–217.

Ashoor, S. H., and Osman, M. A. (1988). Liquid chromatographic quantitation of chicken and turkey in unheated chicken-turkey mixtures. Journal – Association of Official Analytical Chemists, 71(2), 403–405.

Barai, B. K., Nayak, R. R., Singhal, R. S., and Kulkarni, P. R. (1992). Approaches to the detection of meat adulteration. Trends in Food Science and Technology, 3, 69-72.

Buckley, M., Collins, M., Thomas Oates, J., and Wilson, J. C. (2009). Species identification by analysis of bone collagen using matrixassisted laser desorption/ionisation timeofflight mass spectrometry. Rapid Communications in Mass Spectrometry, 23(23), 3843-3854.

Buckley, M., Melton, N. D., and Montgomery, J. (2013). Proteomics analysis of ancient food vessel stitching reveals> 4000yearold milk protein. Rapid Communications in Mass Spectrometry, 27(4), 531-538.

Carrera, M., Cañas, B., Piñeiro, C., Vázquez, J., and Gallardo, J. M. (2007). De novo mass spectrometry sequencing and characterization of species-specific peptides from nucleoside diphosphate kinase B for the classification of commercial fish species belonging to the family *Merlucciidae*. Journal of Proteome Research, 6(8), 3070-3080.

Chassaigne, H., Nørgaard, J. V., and van Hengel, A. J. (2007). Proteomics-based approach to detect and identify major allergens in processed peanuts by capillary LC-Q-TOF (MS/MS). Journal of Agricultural and Food Chemistry, 55(11), 4461-4473.

Chou, C. C., Lin, S. P., Lee, K. M., Hsu, C. T., Vickroy, T. W., and Zen, J. M. (2007). Fast differentiation of meats from fifteen animal species by liquid chromatography with electrochemical detection using copper nanoparticle plated electrodes. Journal of Chromatography B, 846(1–2), 230–239.

Demeulemester, C., Lajon, A., Abramowski, V., Martin, J. L., and Durand, P. (1991). Improved ELISA and dotblot methods for the detection of whey proteins in meat products. Journal of the Science of Food and Agriculture, 56(3), 325-333.

Depreux, F. F. S., Okamura, C. S., Swartz, D. R., Grant, A. L., Brandstetter, A. M., and Gerrard, D. E. (2000). Quantification of myosin heavy chain isoform in porcine muscle using an enzyme-linked immunosorbent assay. Meat Science, 56(3), 261-269.

Fennema, O. R., and Tannenbaum, S. R. (1996).Introduction to food chemistry. In O. R. Fennema, (Ed.), Food Chemistry, (pp. 1-15). Marcel Dekker, New York.

FloresMunguia, M. E., BermudezAlmada, M. C., and VázquezMoreno, L. (2000). A research note: Detection of adulteration in processed traditional meat products. Journal of Muscle Foods, 11(4), 319-325.

Grunert, K. G. (2002). Current issues in the understanding of consumer food choice. Trends in Food Science and Technology, 13, 275-285.

Hofmann, K. (1985). Principal problems in the identification of meat species of slaughter animals using electrophoretic methods. In: Biochemical identification of meat species; Patterson R.L.S., Elsevier Applied Science Publishers: London and New York, 9–31.

Hofmann, K. (1997). Nachweis der TierartbeiFleisch und Fleischerzeugnissen. 2. Mitteilung. Fleischwirtschaft, 77(2),151-154.

Janssen, F. W., Voortman, G., and De Baaij, J. A. (1987).Detection of wheat gluten, whey protein, casein, ovalbumin, and soy protein in heated meat products by electrophoresis, blotting, and immunoperoxidase staining. Journal of Agricultural and Food Chemistry, 35(4), 563-567.

Kim, S. H., Huang, T. S., Seymour, T. A., Wei, C. I., Kempf, S. C., Bridgman, C. R., Momcilovic, D., Clemens, R.A., and An, H. (2005). Development of immunoassay for detection of meat and bone meal in animal feed. Journal of Food Protection, 68(9), 1860–1865.

Leitner, A., Castro-Rubio, F., Marina, M. L., and Lindner, W. (2006).Identification of Marker Proteins for the Adulteration of Meat Products with Soybean Proteins by Multidimensional Liquid Chromatography– Tandem Mass Spectrometry. Journal of Proteome Research, 5(9), 2424-2430.

Liu, L., Chen, F. C., Dorsey, J. L., and Hsieh, Y. H. P. (2006).Sensitive Monoclonal Antibodybased Sandwich ELISA for the Detection of Porcine Skeletal Muscle in Meat and Feed Products. Journal of Food Science, 71(1).

Lopez, J. L. (2007). Two-dimensional electrophoresis in proteome expression analysis. Journal of Chromatography B, 849(1), 190-202.

Mazzeo, M. F., Giulio, B. D., Guerriero, G., Ciarcia, G., Malorni, A., Russo, G. L., and Siciliano, R. A. (2008). Fish authentication by MALDI-TOF mass spectrometry. Journal of agricultural and food chemistry, 56(23), 11071-11076.

Montowska, M., and Pospiech, E. (2007).Species identification of meat by electrophoretic methods. ACTA Scientiarum Polonorum Technologia Alimentaria, 6(1), 5-16.

Montowska, M., and Pospiech, E. (2012). Myosin light chain isoforms retain their speciesspecific electrophoretic mobility after processing, which enables differentiation between six species: 2DE analysis of minced meat and meat products made from beef, pork and poultry. Proteomics, 12(18), 2879-2889.

Muldoon, M. T., Onisk, D. V., Brown, M. C., and Stave, J. W. (2004). Targets and methods for the detection of processed animal proteins in animal feedstuffs. International Journal of Food Science and Technology, 39(8), 851–861.

Ortea, I., Canas, B., and Gallardo, J. M. (2009). Mass spectrometry characterization of species-specific peptides from arginine kinase for the identification of commercially relevant shrimp species. Journal of Proteome Research, 8(11), 5356-5362.

Ponce-Alquicira, E., and Taylor, A. J. (2000). Extraction and ESI–CID–MS/MS analysis of myoglobins from different meat species. Food Chemistry, 69(1), 81-86.

Sentandreu, M. A., and Sentandreu, E. (2011).Peptide biomarkers as a way to determine meat authenticity. Meat Science, 89(3), 280-285.

Sentandreu, M. A., Fraser, P. D., Halket, J., Patel, R., and Bramley, P. M. (2010). A proteomic-based approach for detection of chicken in meat mixes. Journal of Proteome Research, 9(7), 3374-3383.

Simó, C., Elvira, C., González, N., San Román, J., Barbas, C., and Cifuentes, A. (2004).Capillary electrophoresismass spectrometry of basic proteins using a new physically adsorbed polymer coating. Some applications in food analysis. Electrophoresis, 25(13), 2056-2064.

Skarpeid, H. J., Kvaal, K., and Hildrum, K. I. (1998). Identification of animal species in ground meat mixtures by multivariate analysis of isoelectric focusing protein profiles. Electrophoresis, 19(18), 3103–3109.

Taljaard, P. R., Jooste, A., and Asfaha, T. A. (2006).Towards a broader understanding of South African consumer spending on meat. Agrekon, 45, 214-224.

Vanha, J., Hinkova, A., Slukova, M., and Kvasnicka, F. (2009).Detection of plant raw materials in meat products by HPLC. Czech J. Food Sci. Vol, 27(4), 234-239.

Verbeke, W., and Ward, R. D. (2006). Consumer interest in information cues denoting quality, traceability and origin: an application of ordered probit models to beef labels. Food Quality and Preference, 17, 453-467.

Chapter 7

Proteomics a New Era in Meat Quality Control

Anjay[1]*, Purushottam Kaushik[1] and Ajeet Kumar[2]

[1]Assistant Professor, Department of Veterinary Public Health and Epidemiology,
[2]Assistant Professor, Department of Veterinary Biochemistry,
Bihar Veterinary College, Bihar Animal Sciences University,
Patna – 800 014, Bihar
**email: dranjayvet@gmail.com*

ABSTRACT

The application of proteomics in the area of meat science has gained popularity due to advancement in characterization of protein and its association with bioinformatics tools. Proteomics can be valuable tool to characterize meat quality traits such as tenderness, juiciness, flavour and odour. Further, its application in determination of proteins and development of protein biomarker related with deterioration of meat quality during natural process of rigor mortis, detection of adulteration or substitution of meat or meat protein provide a vast future prospect of proteomics in meat science.

Keywords: *Meat, Meat quality, Proteomics*

1. Introduction

Meat is a vital source of proteins in the diet of non-vegetarian human and the quality characteristics of meat such as juiciness, tenderness, flavour and odour are influenced by the biological and genetic traits, variant of live food animals. The quality of meat is also directly related with the structure and function of constituent proteins. The methodical characterization of whole proteins that are expressed in a cell or tissue is called as Proteomics. Nowadays, proteome based tools are gaining importance for monitoring and optimizing meat quality.

The term "proteome" or "proteomics" was formerly used by Marc Wilkins and colleagues in 1995 which refers to the complete protein pool of an organism encoded by the genome. Proteomics allow various set of proteins analysis at a time in contrast to traditional methods which allows single protein identification. There are vast collections of proteins in diverse cells and tissues that exhibit variations in response to various conditions. Hence, the entire set of protein categorization in a multi-cellular organism is a vast task due to comprehensive expression of protein.

The four major components namely separation, identification, characterization and quantification of proteins direct the proteome analysis. The experimental approaches of proteome has been extended from the conventional identification of proteins to functional proteomics. The functional proteomics deals with the quantification and identification of diversely expressed proteins among different circumstances, identification and characterization of proteins interactions, and post translational modifications.

Proteome characterization includes two approaches one is mapping proteomics and global proteomics that reflect the methodical efforts for classification of whole protein complements of cells, tissues and organisms. While, second one comparative proteomics and expression proteomics refer to relative studies that are based on corresponding quantitative protein analyses of expression patterns. The molecular markers or biomarkers can be identified by expression proteomics. Now a day, the identification of biomarkers has gained importance in the field of biological sciences, as use of biomarkers pave a way of its extended applications, including the methods involved in quality meat production and processing (Pan *et al.*, 2009).

Various proteomic approaches have been developed for isolation and identification of proteins using 2D electrophoresis, mass spectrometry and others, which may be applied in conjunction with bioinformatics. In the meat science the proteomics technologies are used for identification of meat and study of proteome changes as a result of genetic variations, pre-slaughter conditions, post-mortem changes, and changes associated with meat quality traits (Bauchart *et al.*, 2006; Hollung *et al.*, 2007).

2. Techniques in Proteomics

Protein is a highly complex substance existing in all living entities and involved in various chemical processes of life. The comprehensive biochemical diversity of protein makes proteomics as very challenging task. The complete genome of any animal species contains as many as 20,000 genes, which may generate 5 or 6 different messenger RNA. Every messenger RNA species are translated into proteins that are processed in many ways to generate 8 to 10 diverse forms of protein. Therefore, the genome may potentially produce approximately 1.8 million different proteins (Jensen, 2004). In muscle cells, 80 per cent myofibril proteins are constituted by myosin heavy chain, titin, actin and nebulin, while other proteins are present only in limited copies. Hence, a perfect proteomics method must have high-throughput potential which can be able to detect as many as protein products in a sensitive, reproducible and measurable methods as well as provides the detail information about the alterations of the individual detected proteins.

Among various proteomic approaches, a gel-based two-dimensional gel electrophoresis (2DE) is the most frequently used approach. In this proteins are separated in primary dimension as per isoelectric points, followed by separation as per molecular weights in the secondary dimension followed by visualization on gel after staining with either commasie brilliant blue, silver or fluorescent dye. Each spots on gel related to one protein type and each cell or cell condition is linked with the precise association of spots. In last few years, fluorescent staining of protein gets more popularity in proteomics research as a consequence of its wide dynamic range with high sensitivity. With the help of this method, we can perform proteome analyses with multiple labeling, using different fluorescent probes on the same gel. This method is also referred as 2D-difference gel electrophoresis (Tonge *et al.*, 2001).

In spite of various advantages, this method shows various limitations for analysis of complex protein. The first limitation is related with the physicochemical properties required for protein separation *i.e.*, isoelectric point and molecular weight. The basic proteins (extreme isoelectric points), and high (>150 kDa) and low (<10 kDa) molecular weight proteins are generally not detected by 2DE. Another limitation is related with hydrophobic proteins that are either not extracted in the sample loading buffers or precipitated during electrophoresis. These limitations of 2DE act as a barrier during investigation of numerous muscle structural proteins having molecular weight >150 kDa, such as myosin heavy chain, titin and nebulin. There are also some technical limitations *viz.*, methodological, laborious and prolonged, hinder to produce its utilization.

In other proteomic approaches the mass spectrometer (MS) has long been an indispensable tool in chemistry. In this initially a vacuum system used to ionize the molecules and the ionised molecules are then placed into an electric and/or magnetic field. The paths of the ionised molecules in the field are considered as function of their mass-to-charge ratio, (m/z) which is used to infer the mass (M) of the molecule with very high precision. Matrix-assisted laser desorption/ionization mass spectrometry (MALDI-MS) has also been effectively used to evaluate the mass of broad range of macromolecules. In this proteins are placed in a light-absorbing matrix and ionized with a short pulse of laser light followed by desorption into a vacuum system. Electrospray ionization mass spectrometry (ESI-MS) is also an equally important method, in which a charged needle kept at a high electrical potential is used to pass the analytes solution. This leads to dispersion of the analytes solution into a thin film of charged microdroplets. The solvent around the macromolecules evaporates quickly, resulting into entering of the charged macromolecular ions into the gas phase. The m/z of the molecule can be analyzed in the vacuum chamber.

Current advancement in MS technology has led to the development of various MS-based proteomic methods. In a classical MS-based proteomics, the proteins are initially cleaved into peptides with the use of enzymes and subjected to fractionation by means of single- or multidimensional separation. These fractionated peptides are further assessed by MS analysis annotated through a web based database search. The peptides quantification in proteomics based on MS primarily depends on the differential isotopic labeling of 2 or more samples. In proteomics based on

MS, iTRAQ (isobaric tag for relative and absolute quantitation) and SILAC (stable isotope labeling with amino acids in cell culture) are two most commonly used methods for isotopic labeling.

iTRAQ method will produce reporter fragments of varying masses in MS/MS spectrum and it is based on the labeling of the N terminus and side-chain amines of the peptides with tags. While in SILAC method modification in arginine residue and mass shift in the peptides content due to addition of heavy nonradioactive isotope during cell growth stage. There are various label-free methods have been also developed to study the proportional measurement of originator ion flows or spectral counting (Bantscheff *et al.*, 2007).

Use of a particular proteomic technique should be cautiously considered as both gel-based and MS proteomic have various limitations. The use of most advanced proteomic techniques generate a huge amount of non interpretative information. Hence, it is always beneficial to exploit a simpler proteomic technique that is very well focused. A recent development in the advanced technology, particularly bioinformatics, provides a proper platform for easily performing proteomic analyses based on MS and interpretation of the outcomes.

Recently, applications of proteomics technique in meat sciences are speedily increasing. The different aspects of application may vary from determination of meat and meat product quality to detection of frauds in animal origin foods.

3. Proteomics of Importance in Meat Science

Meat products are mainly originated from skeletal muscles. A clear understanding of biochemistry of skeletal muscle is required for understanding the biochemical basis of muscle to meat conversion process. Proteomic studies in muscle biochemistry are rapidly increasing including the use of high resolution 2DGE or liquid chromatography, in association with MS for high-throughput protein identification (Yang *et al.*, 2016). With the use of 2DGE, 129 different proteins involved in cell structure, cell defence, muscular metabolism and other contractile proteins were identified in semi-tendinosus muscle of cattle (Bouley *et al.*, 2004).

Proteomics can be applied to correlate the relationship between meat quality and pre-slaughter conditions of animals. Over-expression of mitochondrial ATPase activity in pig was reported in pigs transported instantly before slaughter as compared to those transported the day before slaughter (Morzel *et al.*, 2004). According to Amid *et al.* (2012) troponin I as well as alpha cardiac muscle actin 1 considered as potent markers for electrically stimulated birds. Another study showed the over expression of beta-enolase, pyruvate kinase and creatine kinase takes place in the breast muscles of broiler chickens slaughtered by neck stabbing after gas stunning in compare to slaughtered by Halal method without stunning (Salwani *et al.*, 2015). A study on proteomic profiles of muscles from surgical-castration and immune-castration studs showed expression of 50 different proteins between them. In addition, proteomics may also be used to explore the dietary effects on broiler growth and final meat quality (Paredi *et al.*, 2012).

Tenderness is considered as one of the most important attribute for meat quality. It is generally determined by the extent of proteolysis of important myofibrillar and cytoskeletal proteins within muscle and subsequent alterations of muscle structure. The proteolysis of carcass is mainly governed by endogenous calcium-dependent calpain system. This system is further regulated by the calpastatin inhibitor. Certain proteomic approaches were capable to recognize ATP synthase subunit alpha and alpha actinin 3, which may be concerned with the regulation of the calpains in muscle (Brule *et al.*, 2010). Proteomics study of distorted postmortem revealed identification of more than 100 different protein spots as a result of proteolysis (Lametsch and Bendixen, 2001). Morzel *et al.* (2004) studied pork muscle during 72 hours ageing and identified changes in troponin T, actin, α-crystallin, myokinase, creatine kinase and mitochondrial ATPase, as well as cypher and myozenin proteins. Using the proteomic study various compounds were proposed as biomarkers related to meat tenderness including heat shock protein (Hsp70-1B) and myosin heavy chain isoforms or lactate dehydrogenase chain B. The proteomic study to determine the heat treatment effect on protein modifications in lamb meat was also performed and showed that the extractability of collagens type I and type III proteins increased after the heat treatment (Yu *et al.*, 2016). It is challenging to analyze the proteins in processed meat products however; with the help of proteomics the quality of processed meat products can be determined along with identification of meat quality biomarkers.

4. Meat Authentication Challenges

The adulteration or substitution of meat is a malicious practice of public concern. The consumers must be protected from meat adulteration or substitution by rapid, specific, and precise meat animal species identification. Several methods are reported for meat speciation *viz.*, anatomical, microscopic, histological, chemical or immunological, DNA-based methods and organoleptic but none of them are suitable for all kind of meat and meat products. For instance, DNA-based methods may not be a suitable tool for analysis of cooked meat. Similarly, antibody detection methods may not be specific and sensitive as there are chances of cross reaction or epitopes thermal modification.

Adulteration in meat can involve both species and tissue replacement. Analysis of DNA and protein are the common practices in meat species identification. Conventionally, protein detection has been reported as the most suitable method for animal species identification. The proteomic based methods are found most suitable for identification of species-specific peptide biomarkers in cooked or processed meat as protein primary structures are not degraded. A multiple reaction monitoring mass spectrometry (MRM-MS) method has been developed by Von Bargen *et al.* (2013) which can detect low concentrations of adulterated pork and horse meat. Protein profiling by 2DE can differentiate the meat of various animal species from both raw and processed meat products of cattle, pig, chicken, duck, turkey and goose (Montowska and Pospiech, 2013). High end mass spectrometry methods for instance rapid evaporative ionization mass spectrometry and rapid ambient mass spectrometry also reported to identify meat species targeting myofibrillar and sarcoplasmic proteins (Montowska *et al.*, 2015).

Besides adulteration of meat of one species with other species, collagen and offal may be used for adulteration of tissue. The collagen is rich in 4-hydroxyproline and used to determine collagen content in meat. The amounts of 4-hydroxyproline present in various part of meat can be used to assess the content of collagen in meat and meat products (Anon, 2008). The LC-MS/MS is found very much suitable for quantification of 4-hydroxyproline (Colgrave *et al.*, 2008). Differentiation of offal from muscle tissue has been successfully performed using mid-infrared spectroscopy (Al Jowder *et al.*, 1999).

In meat products, adulteration of animal fat with vegetable fat might be possible. As vegetable fat have phytosterols *viz.*, stigmasterol and β-sitosterol that are deficient in animal fat. Thus, the presence of these two compounds in meat products indicate adulteration of animal fat with vegetable fat. Several techniques like HPLC (Nair *et al.*, 2006), GC-MS (Szucs *et al.*, 2006), and atmospheric pressure photo-ionization (APPI) LC-MS/MS (Lembcke *et al.*, 2005) are most commonly reported to identify phytosterols in a variety of food matrices. In meat products, adulteration or substitution with interspecies animal fat is the most common practice and this malpractice can be detected by quantification of species-specific fatty acids using Gas chromatographic-flame ionization detection (GC-FID) methods (Precht, 1992).

The animal protein may be substituted by cheap and readily available vegetable proteins of soya, pea and wheat. These substituted proteins can be detected with the help of ELISA methods. Even low-cost animal protein of milk, whey, casein, cheese, eggs, and meat can be fraudulently used to substitute more expensive animal protein. Normally, ELISA tests do not differentiate between different proteins from animals. It is therefore difficult to recognize the substitution of animal protein in meat products with animal origin constituents other than meat. Detection of substituted protein is not only restricted to ELISA. Proteins profiling by Liquid chromatographic (LC) techniques have been successfully performed for qualitative detection of different meats from raw and cooked products of beef, pork and chicken (Chou *et al.*, 2007). Furthermore, LC analysis has been also reported for the quantitative analysis of protein profiles from raw chicken and turkey meat mixtures (Ashoor and Osman, 1988).

Melamine and urea are also reported for adulteration of meat protein because of their high level of nitrogen content (Frick *et al.*, 2009). In compare to meat proteins that contain 16 per cent nitrogen by mass, melamine contains 67 per cent however; urea contains 43 per cent nitrogen (Benedict, 1987). Nitrogen content of melamine and urea can be determined by Kjeldahl analysis. The chromatographic techniques along with tandem mass spectrometry have been reported for estimation of melamine and cyanuric acid (Varelis and Jeskelis, 2008).

The organic compounds including colourants, aromas, preservatives and other reducing chemicals have been used as additives to enhance the appearance meat as fresh. The two techniques namely HPLC and GC methods have been used for appropriate quantification of these additives however, the properties of chemical compounds decide the use of suitable analytical technique. The binding agents such as blood clotting enzymes are the other group of additives. These enzymes can be applied to minced meat or cuts to give preferred shape and mass. The use of LC-MS/MS technique is sufficient to detect fibrinopeptides A and B a fibrinogen product produced by thrombin during the binding process (Grundy *et al.*, 2008).

5. Meat Quality Investigation

Meat is a raw material of high nutritive value and the quality of its end-products are mainly dependent on the quality of the raw materials. The different characteristics of meat such as compositional, sensory and nutritional quality are responsible for meat quality and acceptability by consumers. Meat quality can be affected by rearing of food animals at farm, farm to slaughter house transportation, pre and post slaughter handling as well as processing and storage conditions of carcass. Proteomics or proteome-based markers for meat quality examination can serve as a promising approach to examine different meat quality traits.

The consumer acceptance of meat mainly depends on various sensory qualities like colour, odour and texture. The myoglobin content of muscles regulates colour and odour of meat which may get distorted by microbial load and further redox potential. Meat colour also depends on rate of phosphorylation of muscle proteins, intramuscular fat content and meat tenderness. Various compounds like aldose reductase, creatine kinase, and β-enolase are directly correlated with redness values of meat. The other compounds *viz.*, peroxiredoxin- 2, dihydropteridine reductase, and heat shock protein-27 kDa are also reported to show a positive correlation with the stability of surface colour. The significant proteomic analysis of these compounds can be performed using 2-DE and tandem MS (Suman and Joseph, 2011). Tenderness in meat is a complex trait, affected by genetic factors as well as pre and post slaughter handling of animals and storage conditions of carcass. Different proteome analysis has been performed by targeting the characterization of protein markers for tenderization of carcass at the time of post-mortem storage (Lametsch *et al.*, 2003; Jia *et al.*, 2009).

Other proteome studies indicated the correlation between inferior quality pork and genetic variation in swine species due to mutation in PRKAG gene. It leads to reduction in pork quality due to glycogen over-accumulation in muscles (Milan *et al.*, 1996). Genetic studies using proteomic tools also suggested a correlation between development of boar taint, and androstenone and skatole hormones expression in uncastrated male pigs (Duijvesteijn *et al.*, 2010). Now days, proteomics is being used to further understand the quality of products from pig meats such as dry-cured hams and Bayonne hams (Theron *et al.*, 2009). Systematic identification of naturally generated small peptides from myofibrillar proteins like myosin light chain I, titin and actin enlightens the knowledge on the different proteinases affecting food properties.

6. Muscle to Meat Conversion: Proteomic View

The conversions of "muscle to meat" or "post-mortem changes" are natural biochemical events that take place within the muscle after slaughter/death of food animals and have significant role on meat quality and meat processing. The biochemical events of post-mortem changes mainly depend on both intrinsic as well as extrinsic factors related to carcass. After slaughter, depletion of circulating blood that supplies oxygen resulting into lactic acid accumulation and decrease in muscle pH. Subsequently, reduced water holding capacity (WHC) and calcium release leads to formation of cross-linkages among myosin and actin filaments. Furthermore, there is decrease in the glycogen level in the muscle and energy available to keep the muscle in a relaxed state. The combinations of these events result in beginning of rigor mortis and alterations in energy metabolism with further decrease in pH and muscle flexibility. Disturbance in energy metabolism and fall in pH may lead to lower meat quality as in conditions like pale, soft and exudative (PSE) pork and DFD (Dark Firm and Dry) meat.

The muscle to meat conversion may also be affected by other factors related to the animal husbandry and production system. Man-made factors, predominantly stress arises due to improper transportation, management and slaughter practices at the slaughter house also influence the process of tenderization of meat and ultimately the meat quality. It is well established fact that meat tenderizes during post-mortem storage and postmortem degradation of the myofibrillar proteins *viz.*, desmin, titin and nebulin have a role in improvement in meat tenderness. Proteomics helps to understand the various events occurring in muscle to meat conversion and simultaneously the meat quality. It is a prevailing tool for investigation of postmortem protein degradation. More than 100 proteins were identified in postmortem changes, probably as an outcome of protein degradation (Hwang *et al.*, 2005). Mass spectroscopy analysis of the heavy chain fragment of myosin revealed that degradation of this fragment leads to disruption of the myosin-actin interaction resulting more tender meat (Lametsch *et al.*, 2002). Several other structural or structurally related proteins, for example MLC, troponin T, desmin, capping protein α1 subunit, F-actin capping protein, CapZ, titin and cofilin 2 were also degraded during post-mortem changes (Hwang, 2004).

Proteomic studies also help to understand the WHC of the meat and drip loss. High levels of drip loss can occur in pig meat leads to economical loss, and may affect the quality of both the fresh and processed meat. Creatine phosphokinase (CPK) and desmin have been identified as two potent biomarkers for determination of more drip loss (Van de Wiel and Zhang, 2007). In high drip loss muscle there is over-expression of CPK resulting in rapid fall of pH with altered muscle contraction rate. The chief mechanism involved in drip loss is presence of high level of desmin tend to more myofibril lateral shrinkage and more drip loss (Kristensen and Purslow, 2001). Additionally, the over-expression of HSP70 in muscle tends to decrease in drip loss. Due to over-expression of HSP70, denaturation and aggregation of protein gets reduced with decrease in drip loss. The proteomic study related with PSE showed that a reduced rate of proteolysis among troponin T, light chain of myosin and α-crystallin protein, and HAL gene mutation are concerned with occurrence of PSE (Silveira *et al.*, 2011).

7. Future Perspectives in Meat Quality

Meat of food animals regarded as a major protein source for humans, while meat science proteomics is still lagging behind. Meat science proteomics have been focused basically in the domain of meat quality and infectious diseases research. Meat quality is frequently determined by complex interactions of biological and environmental factors, while often classified according to the appearance of the raw material. Various studies revealed that the tenderness of meat may be different due to difference in postmortem glycolytic rates among different carcasses. Moreover, various meat processing techniques like drying, freezing, heating, high-pressure treatment, marination, fermentation *etc.* affects the processed meat quality by modifying the proteins. The principles and mechanisms behind these processes are inadequately understood; hence proteome studies will be of great significance in optimization of meat processing technologies. Application of proteomics in meat science shows its importance to unravel the dependency on molecular techniques for determination of origin and quality of meat products. Recent developments in application based proteomic techniques provide evidence of its future use in food quality certification. As proteome technologies have been optimized to differentiate complex protein mixtures, these technologies will certainly be a valuable tool for characterization of proteins from complex food products of mixed tissue or species in different ratio. Proteomics will also be a valuable tool for analytic assays development for authentication of meat and meat products.

8. Conclusion

Meat quality is mainly linked to the structure and function of constituent proteins. Heterogeneity of muscle proteins can be studied by the use of proteomics. Proteome studies mainly depend upon four major elements regarded as separation, identification, characterization and quantification of protein. At present meat proteomics considered as innovative and emerging technologies that may impart an important role in regulation of animal product quality. It has been widely used in numerous aspects related to the meat industry. Among various proteomic methods two-dimensional gel electrophoresis, mass spectrometry and chromatographic methods in combination with bioinformatics tools are the most acceptable approach to characterize the proteins related with meat quality. The scope of meat science proteomics is mainly concerned with study of meat speciation, genetic variations in meat quality, effect of pre-slaughter conditions, post-mortem changes and other meat quality traits.

REFERENCES

Al Jowder, O., Defernez, M., Kemsley, E. K., and Wilson, R. H. (1999). Mid-Infrared spectroscopy and chemometrics for the authentication of meat products. Journal of Agricultural and Food Chemistry, 47(8), 3210-3218.

Amid, A., Samah, N. A., and Yusof, F. (2012). Identification of troponin I and actin, alpha cardiac muscle 1 as potential biomarkers for hearts of electrically stimulated chickens. Proteome Science, 10 (1), 1-8.

Anon (2008). Commission Regulation (EC) No 903/2008 of 17 September 2008 on special conditions for granting export refunds on certain pigmeat products. L249. Official Journal of the European Union, 3–8.

Ashoor, S. H., and Osman, M. A. (1988). Liquid chromatographic quantitation of chicken and turkey in unheated chicken–turkey mixtures. Journal-Association of Official Analytical Chemists, 71(2), 403–405.

Bantscheff, M., Schirle, M., Sweetman, G., Rick, J., and Kuster B. (2007). Quantitative mass spectrometry in proteomics: A critical review. Analytical and Bioanalytical Chemistry, 389 (4), 1017 1031.

Bauchart, C., Remond, D., Chambon, C., Patureau Mirand, P., Savary-Auzeloux, I., Reynes, C., and Morzel, M. (2006). Small peptides (<5 kDa) found in ready-to-eat beef meat, Meat Science, 74(4), 658–666.

Benedict, R. C. (1987). Determination of nitrogen and protein content of meat and meat products. Journal-Association of Official Analytical Chemists, 70(1), 69–74.

Bouley, J., Chambon, C., and Picard, B. (2004). Mapping of bovine skeletal muscle proteins using two-dimensional gel electrophoresis and mass spectrometry. Proteomics, 4(6), 1811-1824.

Brule, C., Dargelos, E., Diallo, R., Listrat, A., Bechet, D., Cot-tin, P., and Poussard, S. (2010). Proteomic study of calpain interacting proteins during skeletal muscle aging. Biochimie 92(12), 1923–1933.

Chou, C. C., Lin, S. P., Lee, K. M., Hsu, C. T., Vickroy, T. W., and Zen, J. M. (2007). Fast differentiation of meats from fifteen animal species by liquid chromatography with electrochemical detection using copper nanoparticle plated electrodes. Journal of Chromatography B, 846(1–2), 230–239.

Colgrave, M. L., Allingham, P. G., and Jones, A. (2008). Hydroxyproline quantification for the estimation of collagen in tissue using multiple reaction monitoring mass spectrometry. Journal of Chromatography A, 1212(1–2), 150–153.

Duijvesteijn, N., Knol, E. F., Merks, J. W., Crooijmans, R. P., Groenen, M. A., Bovenhuis, H., and Harlizius, B. (2010). A genome-wide association study on androstenone levels in pigs reveals a cluster of candidate genes on chromosome 6. BMC Genetics, 11, 42.

Frick, G., Dubois, S., Chaubert, C., and Ampuero, S. (2009). Identification by microscopy and MS-based electronic nose of a fraudulent addition to maize gluten. Biotechnology, Agronomy, Society and Environment, 13(S), 45–50.

Grundy, H. H., Reece, P., Sykes,M. D., Clough, J. A., Audsley, N., and Stones, R. (2008). Method to screen for the addition of porcine blood-based binding products to foods using liquid chromatography/triple quadrupole mass spectrometry. Rapid Communications in Mass Spectrometry, 22(12), 2006–2008.

Hollung, K., Veseith, E. Jia, X. Færgestad, E.M. and Hildrum, K.I. (2007). Application of proteomics to understand the molecular mechanisms behind meat quality, Meat Science, 77(1), 97–104.

Hwang, I. H. (2004). Application of gel-based proteome analysis techniques to studying post-mortem proteolysis in meat. Asian-Australasian Journal of Animal Science, 17(9), 1296–1302.

Hwang, I. H., Park, K. S., Kim, S. H., Cho, S. Y., and Lee. K. (2005). Assessment of postmortem proteolysis by gel-based proteome analysis and its relationship to meat quality traits in pig longissimus. Meat Science, 69(1), 79–91.

Jensen, O. N. (2004). Modification-specific proteomics: Characterization of post-translational modifications by mass spectrometry. Current Opinion in Chemical Biology, 8(1), 33–41.

Jia, X., Veiseth-Kent, E., Grove, H., Kuziora, P., Aass, L., Hildrum, K. I., and Hollung, K. (2009). Peroxiredoxin-6- a potential protein marker for meat tenderness in bovine longissimus thoracis muscle. Journal of Animal Science, 87(7), 2391–2399.

Kristensen, L., and Purslow, P. P. (2001). The effect of ageing on the water-holding capacity of pork: role of cytoskeletal proteins. Meat Science, 58(1), 17–23.

Lametsch R., and Bendixen E., 2001. Proteome analysis applied to meat science: Characterizing postmortem changes in porcine muscle. Journal of Agricultural and Food Chemistry, 49(10): 4531–4537.

Lametsch, R., Karlsson, A., Rosenvold, K., Andersen, H. J., Roepstorff, P., and Bendixen, E. (2003). Postmortem proteome changes of porcine muscle related to tenderness. Journal of Agricultural and Food Chemistry, 51(24), 6992–6997.

Lametsch, R., Roepstorff, P. and Bendixen., E. (2002). Identification of protein degradation during post-mortem storage of pig meat. Journal of Agricultural and Food Chemistry, 50(20), 5508–5512.

Lembcke, J., Ceglarek, U., Fiedler, G. M., Baumann, S., Leichtle, A., and Thiery, J. (2005). Rapid quantification of free and esterified phytosterols in human serum using APPILC- MS/MS. Journal of Lipid Research, 46(1), 21–26.

Milan, D., Woloszyn, N., Yerle, M., Le, R. P., Bonnet, M., Riquet, J., Lahbib-Mansais, Y., Caritez, J. C., Robic, A., Sellier, P., Elsen, J. M., and Gellin, J. (1996). Accurate mapping of the "acid meat" RN gene on genetic and physical maps of pig chromosome 15. Mammalian Genome, 7(1), 47–51.

Montowska, M., and Pospiech E. (2013). Species-specific expression of various proteins in meat tissue: proteomic analysis of raw and cooked meat and meat products made from beef, pork and selected poultry species, Food Chemistry, 136, 1461–1469.

Montowska, M., Alexander, M. R., Tucker, G. A., and Barrett, D. A. (2015). Authentication of processed meat products by peptidomic analysis using rapid ambient mass spectrometry. Food Chemistry. 187, 297–304.

Morzel, M., Chambon, C., Hamelin, M., Sante-Lhoutellier, V., Sayd, T., and Monin, G. (2004). Proteome changes during pork meat ageing following use of two different pre-slaughter handling procedures. Meat Science, 67(4), 689-696.

Nair, V. D., Kanfer, I., and Hoogmartens, J. (2006). Determination of stigmasterol, betasitosterol and stigmastanol in oral dosage forms using high performance liquid chromatography with evaporative light scattering detection. Journal of Pharmaceutical and Biomedical Analysis, 41(3), 731–737.

Pan, S., Aebersold, R., Chen, R., Rush, J., Goodlett, D. R., McIntosh, M. W., Zhang, J., and Brentnall, T. A. (2009). Mass spectrometry based targeted protein quantification: methods and applications. Journal of Proteome Research, 8(2), 787–797.

Paredi, G., Raboni, S., Bendixen, E., de Almeida, A. M., and Mozza- relli, A. (2012). Muscle to meat: molecular events and techno- logical transformations: the proteomics insight. Journal of Proteomics, 75(14), 4275-4289.

Precht, D. (1992). Detection of foreign fat in milk-fat, qualitative detection by triacylglycerol formulas. Zeitschrift für Lebensmittel-Untersuchung und – Forschung, 194(1), 1–8.

Salwani, M. S., Adeyemi, K. D., Sarah, S. A., Vejayan, J., Zulkifli, I., and Sazili, A. Q. (2015). Skeletal muscle proteome and meat quality of broiler chickens subjected to gas stunning prior slaughter or slaughtered without stunning. CyTA Journal of Food, 14(3), 1-7.

Silveira, A. C., Freitas, P. F., Cesar, A. S., Cesar, A. S., Antunes, R. C., Guimaraes, E. C., Batista, D. F. A., and Torido, L. C. (2011). Influence of the halothane gene (HAL) on pork quality in two commercial crossbreeds. Genetics and Molecular Research, 10 (3), 1479–89.

Suman. S. P., and Joseph, P. (2013). Myoglobin chemistry and meat colour. Annual Review of Food Science and Technology, 4, 79–99.

Szucs, S., Sarvary, A., Cain, T., and Adany, R. (2006). Method validation for the simultaneous determination of fecal sterols in surface waters by gas chromatography- mass spectrometry. Journal of Chromatographic Science, 44(2), 70–76.

Theron, L., Chevarin, L., Robert, N., Dutertre, C., and Sante-Lhoutellier, V. (2009). Time course of peptide fingerprints in semimembranosus and biceps femoris muscles during Bayonne ham processing. Meat Science, 82(2), 272–277.

Tonge, R., Shaw, J., Middleton, B., Rowlinson, R., Rayner, S., Young, J., Pognan, F., Hawkins, E., Currie, I., and Davison, M. (2001). Validation and development of fluorescence two-dimensional differential gel electrophoresis proteomics technology. Proteomics, 1(3), 377–396.

Van de Wiel, D. F. M., and Zhang, W. L. (2007). Identification of pork quality parameters by proteomics. Meat Science, 77(1), 46–54.

Varelis, P., and Jeskelis, R. (2008). Preparation of [13C3]-melamine and (13C3)-cyanuric acid and their application to the analysis of melamine and cyanuric acid in meat and pet food using liquid chromatography-tandem mass spectrometry. Food Additives and Contaminants. Part A, Chemistry, Analysis, Control, Exposure and Risk Assessment, 25(10), 1208–1215.

Von Bargen, C., Dojahn, J., Waidelich, D., Humpf, H. U., and Brock- meyer, J. (2013). New sensitive high-performance liquid chromatography tandem mass spectrometry method for the detection of horse and pork in halal beef. Journal of Agricultural and Food Chemistry, 61(49), 11986–11994.

Yang, H., Xu, X. L., Ma, H. M., and Jiang, J. (2016). Integrative analy- sis of transcriptomics and proteomics of skeletal muscles of the Chinese indigenous Shaziling pig compared with the Yorkshire breed. BMC Genetics, 17(1), 80.

Yu, T. Y., Morton, J. D., Clerens, S., and Dyer, J. M. (2016). Proteomic investigation of protein profile changes and amino acid residue level modification in cooked lamb *longissimus thoracis et lumborum*: The effect of roasting. Meat Science, 119(1), 80-88.

Chapter 8

Hazards of Meat Industry and the Role of HACCP in their Control

Samir Das* and Koushik Kakoty

Scientist, Division of Animal Health,
ICAR Research Complex ForNEH Region, Umroi Road,
Umiam – 793 103, Meghalaya
**e-mail: drsamirvph@yahoo.com*

ABSTRACT

Meat as a food is an essential in human diet since stone age and with advancement of civilization the fulfillment of meat so required got shifted from hunting to rearing of farm animals. Now it has grown into an industry with mass production, involving scientific rearing of farm animals to scientific slaughtering, packaging, labeling, long term storage, etc. with an aim to give the consumer a safe and wholesome meat which can not only provide nutrition but also is free from any hazard in term of infectious agent and deleterious substances. To ensure safe and wholesome meat to consumer, the industry had to set up some norms and follow those strictly and there came the role of Hazard Analysis Critical Control Point (HACCP). HACCP principle in meat industry helps to identify the potential hazards which can enter the production chain of meat and thereafter to identify the route to entry of these hazards so that these route or entry point termed as critical point can be checked. The whole process needs monitoring and documentation for running the meat industry with an ultimate aim to provide safe food to consumers.

***Keywords:** HACCP, Hazard, Meat, Industry.*

1. Introduction

Hazard Analysis and Critical Control Point (HACCP) is a safety and quality management tool, developed by the Pillsbury Corporation in the year 1960s (Bennet and Steed,1999) in collaboration with National Aeronautics and Space

Administration, which was used for ensuring food safety for astronauts' food. From the 1980s onwards, HACCP principles were gradually adopted by the food industry (Guzewich, 1985).

2. Hazards and their Types

A hazard in meat industry may be defined as a biological, chemical, or physical property that may cause meat or meat products to be unsafe for human consumption. In simple words there are basically three types of hazards in meat industry *i.e.*, biological, chemical, and physical hazards which if not taken due consideration can lead to injury, illness in consumer.

2.1 Biological Hazard

It may be of bacterial, viral, prion, parasitic or mycotic origin. Mostly bacterial or its toxins are the main hazards which may cause food infection, intoxication or toxico-infection.

Food Borne Infection

Food illness when caused by ingestion of pathogenic microorganism sufficient to establish infection in host and thereafter multiplying in the host system causing illness in them. *e.g.*, majority of gram negative bacteria of family enterobactericiae causes food borne infection like *Salmonella, E.coli*, when they are consumed by human being through unhygienic food, hands *etc.* these are established in the intestinal tract where they multiply in large number and because of the cell wall structure of gram negative bacteria having Lipopolysaccharide (LPS) cause release of water from host cell leading to severe watery diarrhea and many a time dysentery (blood or mucous with diarrhea). Usually the symptoms from food infection start after 2 to 3 days of ingestion of pathogens.

Food Intoxication

It is caused by the ingestion of preformed toxins produced by microorganism. Many gram positive bacteria like *Staphylococcus, Streptococcus* cause food intoxication. The usual symptom of food intoxication is vomition and diarrhoea within 02 to few hours of consumption of food.

Toxico-infection

Is caused by ingestion of micro-organism which enter in the body system and after that they multiply in the host with release of toxins e.g, *Clostridium botulinum*. Most of the microbial loading on a carcass derives from the skinning and evisceration procedure. Some of the major biological hazards of meat industry are discussed herewith

2.1.1 Hazards of Bacterial Origin

a) ***Clostridium perfringens***-Cooking kills the growing *C. perfringen* cells responsible for causing food poisoning, but many often the spores survive which later can grow and populate themselves enough to cause food poisoning. If cooked food is not promptly stored or refrigerated, the spores can grow and produce new cells. The cutting, handling, and wrapping operations may each be responsible for the addition of *C. perfringens* spores and vegetative cells. Meat samples can be expected to be contaminated with *C. perfringens*.

b) ***Clostridium botulinum-*** Ingestion of pre-formed potent botulinum neurotoxin produced by the organism mainly in canned meat can cause illness with high mortality rate.

c) ***Campylobacter jejuni–***Meat especially poultry sold at the counter is a major source of *C. jejuni* and *C. coli,* infection

d) ***Escherichia coli-****E. coli* lives in the intestines of man and animals. Some of the strains of *E.coli* like Shiga toxin-producing *E. coli* (STEC) and serovar like *E. coli O157: H7,* produces Shiga toxin causes blood dysentery. *E. coli* O157: H7 can lead to severe complication casing hemolytic uremic syndrome (HUS).

e) ***Listeria monocytogenes-***The disease is prevalent all over the world as an important foodborne zoonosis caused by pathogenic strains belonging to the genus- *Listeria* having six species, of which only *L. monocytogenes* is an opportunistic pathogen of human and animals, whereas, *L. ivanovii* affects only the animals causing abortion (in ruminants), barring occasional cases in man. Its ability to grow at refrigeration temperatures is also a hurdle for meat industry.

f) ***Salmonella* spp.**- Salmonellosis caused by *Salmonella* sp. is responsible for gastroenteritis in man and animal. Members of genus Salmonella are gram-negative rods consists of 2 species *S. enteric* and *S.bongori* with 2435 serovars. The serovar S. Typhi and S. Paratyphi are common pathogenic serovar of human (non-zoonotic) and can enter to food chain by fecal contamination through human handler and can cause serious illness in human cased as Typhiod fever. The other non-typhoidal serovar can also contaminate meat which can come through food of animal origin, *S.Typhimurium* and *S.Enteritidis* are the most prevalent and pathogenic form of zoonotic salmonellosis. These are zoonotic in nature and can cause dysentery in man and animal both.

g) ***Staphylococcus aureus-****Staphylococcus aureus* is commonly found on the skin, roots of hairs, in the nares and throats of people and animals. *Staphylococcus* can cause food poisoning when a food handler contaminates meat and it can also directly come from farm animals. If not taken care, these bacteria multiply quickly at ambient temperature to produce an exo-toxin that causes food poisoning.

h) ***Yersinia enterocolitica-*** also can enter the meat through farm animals and is responsible for food borne illnesses.

2.1.2 Hazards of Parasitic Origin

a) ***Taenia* spp.**-Taeniasis can be caused by *Taenia* spp. of which the mature form resides in human and the larval stages are seen in the tissue of animals. *T.solium* and *T.saginata* are the mature worm seen in the intestinal tract of human for which *Cysticercus cellulosae* is the larval stage seen in the muscle of pork and *Cysticercus bovis* in beef tissue respectively. The disease is usually apparent in animals. The disease is transmitted by consumption of raw and/or undercooked pork/beef products. Additionally, in case of *T. solium* infection where man is definitive host (mature worm) but if accidentally the eggs of *T. solium* enter in man through auto-infection or from contaminated food mainly raw unwashed salad than it leads to formation of larval form in man also mainly seen in eye (ocular cysticercosis) or brain (neurocysticercosis).

b) ***Trichinella spiralis***-It causes a parasitic disease in human being known as Trichinosis, sometime also spelled as trichinellosis, or trichiniasis. It is caused by eating raw or undercooked pork, wild animals infected with the larvae of *Trichinella spiralis*. The disease is common in the developing countries and where pigs are commonly fed raw garbage.

c) ***Toxoplasma gondii***- Toxoplasmosis is a contagious disease of animals and man caused by protozoon, *Toxoplasma gondii*. It is found most frequently in pigs and sheep.

d) ***Cryptosporidium* spp.**- Cryptosporidiosis is a parasitic infection principally of the intestinal epithelium. The disease is noticed all over the world. The disease is caused by coccidian protozoan parasite *Cryptosporidium* spp. *C. parvum* is the most common species in this genus and is found in calves, man, and numerous other mammals. Transmission: fecal-oral transmission is from animals to humans or humans to humans; waterborne transmission is also reported. In humans the disease is characterized by fever and gastrointestinal symptoms(particularly diarrhea) that can resemble influenza.

e) ***Echinococcus* spp.**-Echinococcosis is a disease caused by tape worm mostly transmitted by contaminated food and water. The adult worm *Echinococcus granulosus* lives in dog. Eggs, passed by infected animals (dog), are ingested by farm animals including pigs or man, and get to localize in different organs (in liver, lung, bones, and sometimes, brain) and develop into hydatid cysts containing many larvae (in form of proto-scolices, commonly known as hydatid sand or *Echinococcus* eggs). Man is a dead end host in case of this infection.

2.1.3 Hazards of Fungal Origin

a) ***Aspergillus* spp.**-Can grow on meat during slaughtering, handling and also during the processing of meat product when other contaminated ingredients are added to meat during its processing.

2.2 Hazards of Chemical Origin

Chemical hazards basically enter in the food chain- unintentionally or sometime intentionally as described below:

1. Unintentionally added chemicals: Pesticides, herbicides, animal drugs, fertilizers sanitizers, oils, lubricants, paints which can enter while handling the carcass/meat near to these chemicals used in the utensil, lubrication of machine, cleaning process of industry, cutting of carcass *etc.* Environmental contaminants: lead, cadmium, mercury, arsenic may enter from the environment like water and surrounding where the carcass is handled. Antibiotics, hormones, toxins, inks, packaging materials *etc.* can also enter during growth of animal in farm or at farmers field and while handling, packaging the carcass/meat in the industry.
2. Intentionally added chemicals: During processing of meat and meat products the direct food additives like preservatives (*e.g.*, nitrite), flavouring agents, colour and other additives are used as per the industry guidelines and they are intentionally added to increase shelf life or flavour of meat.

2.3 Hazards of Physical Origin

A physical hazard in meat and its product can be defined as any physical material (usually visible through naked eyes) which is not a normal constitute of meat or meat products but can cause illness or injury to the person consuming this product. Physical hazards in finished meat products can enter in the meat due to careless handling of carcass by handlers, from faulty instruments and there can be breakage in the carcass handling system in term of broken metal piece, hairs, glass, wood, stones, plastics, electronics, jewels, *etc.* Sometime the common pests enter in the carcass chain like worm, cockroaches, insects *etc.* and they are responsible for public outcry if not taken adequate care.

3. Principles of HACCP and their Salient Features

Hazard analysis and critical control point (HACCP)-HACCP in relation to meat industry can be defined as a system for ensuring meat safety by exercising control to prevent any objectionable contamination, entry and multiplication of microorganisms in the meat production chain. This is a system of food quality control to render supply of safe food for human consumption. Emphasis is laid on controlling meat borne hazards particularly the meat borne infections and intoxications during production itself. Hazard in meat refers to unacceptable contamination, survival and/or growth of hazardous microorganisms. In case of meat industry, Hazard Analysis includes factors affecting possible microbial multiplication in meat production, enter of microbes and chemical hazards in meat and its product during production and processing.

The Hazard Analysis and Critical Control Point (HACCP) system is internationally accepted system for error less production of any commodity which is also applied to food safety management for production of safe and wholesome meat for consumer. It is a preventative approach to food safety based on the following seven principles:

1. Identify the hazards that need to be prevented, eliminated or reduced

2. Identify the critical control points (CCPs) at the steps where control is essential
3. Establish the critical limits at CCPs
4. Establish the procedures to monitor the CCPs
5. Establish corrective actions to be taken if a CCP is not under control
6. Establish procedures to verify whether the above procedures are working effectively
7. Establish documents and records to demonstrate the effective application of the above measures

We would like to explain each steps of HACCP in relation to meat industry as follows:

1. Identify any hazards that must be prevented, eliminated or reduced- We have to document all the possible hazards which can enter in meat starting from farm to fork approach. The organism can enter in the meat through the farm animals itself, through human handler in the processing steps and through the environment, like water, material used in the processing plant, packing material; which are earlier discussed in the meat borne hazards. Say *Salmonella* is one of the hazard of bacterial origin.
2. Identify the critical control points (CCPs) at the steps at which control is essential- After identifying the hazards say *Salmonella* one has to document what are the points through which *Salmonella* can enter in the meat or its products. One has to draw the production line and note down at which points *Salmonella* can enter say it can enter through farm animals itself, through handler, and through water. So in the production line where are the maximum chance of entry of this pathogen, during arrival of disease animal with high fever at ante-mortem check, during evisceration of animal, during handling of carcass by infected or carrier human handler, during cleaning operation with water *etc.*
3. Establish critical limits at CCPs- After establishing the critical point from where the hazard may enter in the meat production chain system, the system needs to establish how in each critical point what should be the limit for allowing the meat to pass on for further processing if detected in these points. Say for *Salmonella*, at ante-mortem examination if suspected animal with high fever and dysentery are seen then whether they should be allowed to enter the post-mortem place or not,and basically for *Salmonella* suspected animal it is unfit for slaughter. If *Salmonella* is detected in the water used for washing the carcass, then upto which limit the water can be used, if during human staff medical examination someone suffering from Salmonellosis, whether this person should be allowed to enter the production line or not. For *Salmonella*, the infected person should not be allowed to handle the carcass of meat itself. After the finished product is examined for bacteriological examination and if the sampled product is found to harbor *Salmonella* bacteria then upto which limit the meat should be allowed to pass on for human consumption or it should be rejected.

The standard "Microbiological specifications for food safety" should be followed and decision should be taken accordingly.

4. Establish procedures to monitor the CCPs- The identified hazards in the identified critical control points should be monitored with established protocols. These protocols should be well defined and universal or can be modified as per local condition upto acceptable level only. For *Salmonella* the procedure for its detection should be established and usually isolation of the organism in culture media from meat is followed as per the universal isolation protocols along with its biochemical identification.
5. Establish corrective actions to be taken if a CCP is not under control- If in any production line hazards is seen and documented by the established procedure and are coming more than the limits to be safe for human consumption then what corrective steps need to be done. Like if Salmonella is detected at carcass level then what corrective measures should be done so that the carcass is free from detectable *Salmonella*, they could be improving ante-morterm examination, taking precaution for avoiding intestinal and other visceral contamination with carcass during evisceration, avoiding exterior of hide from contacting carcass during skinning, antimicrobial rinses of carcass, proper scalding procedures; disinfecting knives, *etc.*
6. Establish procedures to verify whether the above procedures are working effectively- Even if the production line and all the detection procedure for hazards are working fine, then also the established procedure should be rechecked time to time so as by default the hazard is not entering because of some procedural fault which had developed in due course of time. Say for *Salmonlla*, all the media should be checked whether they are working or not may be any of the components got expired without notice or new lot of media are not working. The positive and negative control for Salmonella isolation should be always applied with test samples so as not to miss the actual ground contamination.
7. Establish documents and records to demonstrate the effective application of the above measures- All the documents and records should be maintained correctly so that the inspection or new staff can access the data and can interpret accordingly. Say for *Salmonella*, if documents are maintained perfectly that during the last 06 months out of these many samples analysed in these critical point, they observed the contamination twice and that too in carcass which got corrected by improving water quality of the carcass cleaning operation. Then it would be easy to have first look on water quality used for carcass operation for any further inspection.

4. HACCP for Meat Industry

4.1 Microbiological Specifications for Safety and Thereby Monitoring Food Quality

Microbiological status and the sanitary conditions of processing and storage are two important aspects of HACCP and are also very much important to meat industry. In another chapter of this book we have discussed in length and can be referred there.

4.2 Controls and Critical Limits for Biological, Chemical, and Physical Hazards

When all the significant physical, chemical, and biological, hazards are identified along with their points of occurrence, the next work is to identify various measures to prevent these hazards entering into the finished product so as to make the availability of safe food for consumers.

Preventive measures or controls can be defined as the measures or procedures which are used for limiting or controlling entry of the known established hazards whether physical, chemical, or other that can lead to unacceptability of meat and meat products by consumers. While formulating the preventive or control measures, a limit must need to be established to ensure safety. For example, it is a known fact that heat kills most of the pathogenic and spoilage bacteria but there is need to establish a correct time/temperature combinations so that it kills harmful pathogenic bacteria but also preserving the originality of meat.

Various processing steps which are routinely used in the meat industry need to be observed and analyzed properly so that the critically important points from where there is a chance for entry of hazards in meat industry is known. The various steps start even when the animal is raised in the farm, hence farm details are critical, the movement of animals in the farm, in the region, legal-illegal trades should be known so that only proven farm having good track record should be allowed to get entry in the slaughter house. After which all the crucial steps which includes receiving and holding of animal, the stunning procedure, the method of bleeding, skinning, evisceration, washing, chilling, packaging and storage; everything is well documented and recorded so as any deviation in the quality of meat can be traced to the origin. For *e.g.*, during skinning of the carcass it can get contaminated by micro-contamination from its outside hide surface, from floor, cross-contamination by equipment/utensils, through the employee handling during evisceration step from cross-contamination of broken viscera. This is one of the critical step and need to be observed carefully and proper planning is required in this critical step to ensure that contamination does not get its entry during skinning process.

4.3 Importance of HACCP on a Farm and in a Primary Processing

The use of HACCP for the meat industry must begin at the first step from farm where animals are raised as many hazards cannot be eliminated during the slaughtering process. For example, chemical residues and certain microbial pathogens (*e.g. Salmonella, Campylobacter jejuni, Escherichia coli 0157: H7*) if not controlled during raising of animals, then it eventually will go to the further processing line and to check these hazards the animal need to be raised properly taking care of health and hygiene at farm and for which the farmers need to be educated and involved. Eliminating these zoonotic pathogens from farm animals with minimum use of antibiotics is the key to success. This needs a supervision by expert veterinarian who should try to use the minimum drugs while treating these animals and should advice the farmer to focus more on hygiene, vaccination of important endemic diseases and use of drugs judiciously when is required to treat the animals. Chemicals (*e.g.* pesticides, antibiotics) used by farmers must be properly applied to avoid unacceptable residue levels in meat. The farmer community at large need to be educated regarding raising their animal in clean environment avoiding use of unnecessary chemical and simultaneously reducing the burden of pathogens in their animal so that the ultimate meat products is safe to the consumers. They should know the market potential of their good produce and they should be informed and oriented in such a manner that they understand that raising good quality produce can give a premier price to their product.

Two fundamental concepts should be followed during the slaughtering process for achieving the target of getting safe meat. First and foremost is the use of steps which could minimize the degree of contamination on carcasses during the slaughtering process. This may involve training abattoir workers to use proper knives and equipment which are readily cleanable and which can minimizes contamination, by providing adequate work space and time to perform each function correctly,and providing a plant layout that favors microbial control. The

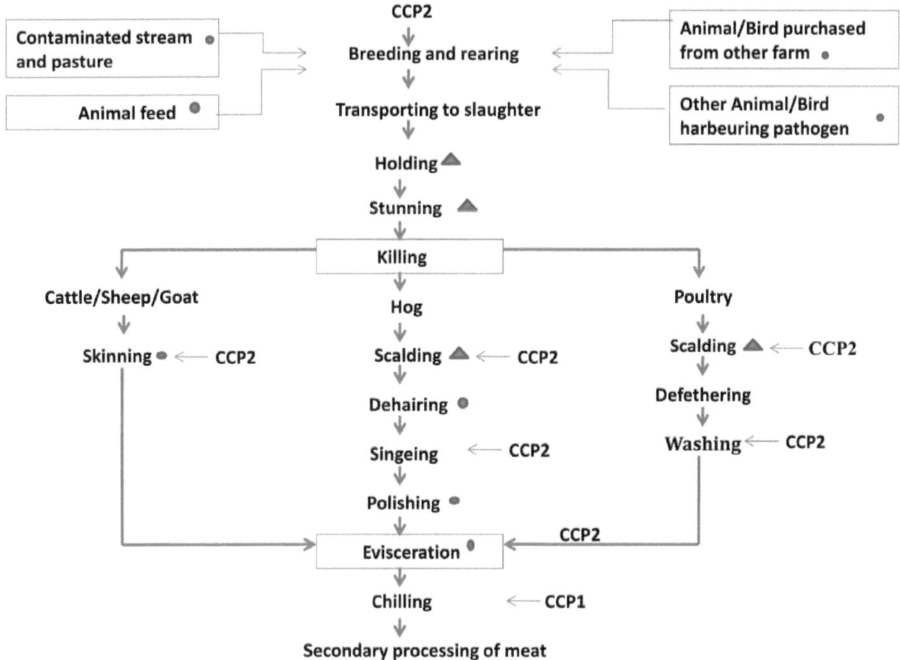

Figure 8.1. Flow Diagram from Farming to Primary Processing Unit. Sites of contamination and control of human pathogens. ◆ = Major contamination; ▲ = Possible contamination; CCP1 = Effective CCP; CCP2 = Not absolute.

second step includes procedures which can remove and/or destroy pathogens which inadvertently contaminate the carcass during slaughtering, *e.g.* spraying with organic acids and this procedure is gaining acceptance over the years by industry and regulatory officials (Dickson and Anderson, 1992). In Serbia, the mandatory HACCP in meat plant showed significant reduction of hygienic indicators on all type of surfaces, meat, equipment, hands of handlers *etc.* showing the positive node for implementing HACCP (Tomasevic *et al.*, 2016).

4.4 Developing HACCP Plan for a Meat Plant

To develop a HACCP plan in meat processing unit the following steps should be followed:

- Development of the meat plant specific HACCP Plan
- Training and retaining the HACCP team
- Describe the meat and the product with its method of distribution.
- Identify the end user and consumers of the meat and its products.
- Develop a flow diagram which describes the process
- Verify this flow diagram
- Then apply the seven principles of HACCP

Description of the Product

This is the first step in the development of the model for the process. It will aid you in describing your meat and meat product(s) so that one can progress through the remainder of model development. Process Flow Diagram: This form should be completed for your process following the completion of the product(s) description. This step includes the course of the process as the product(s) moves from receiving to finished product shipping. It is helpful to complete this portion of your plan while actually walking through your plant and following the production steps involved in the particular product or process.

Hazard Analysis

The Hazard Analysis is a critical step in the development of a plant specific HACCP plan. This portion of plan development must take into consideration the risk or likelihood of occurrence, and the severity of each hazard. Identify the processing steps that present significant hazards and any preventive measures on the Hazard Analysis/Preventive Measures Form.

Critical Control Point (CCP) Determination

Identification and description of the CCP for each identified hazard is the next step in developing the plan. The CCP determination and the information and data that are recorded on the Hazard Analysis/Preventive Measures form will be required for completion of this portion of the plan.

HACCP Plan Development

This portion of the plan development will be used to designate the specific activities, frequencies, critical limits, and corrective actions that ensure that the whole process is under control and adequate to produce a safe product. This part will include all the information gathered to this point in your plan development process steps. In addition, the HACCP plan will include specification of critical limits. These limits will be specified after the identification of the CCP's for the process and will be listed in the HACCP Plan. The critical limit must include at a minimum the regulatory requirement for that specific process step or an equivalent process proven to render the product unadulterated. HACCP is a system of process control for the plant and not an inspection system (BouRached *et al.*, 2004).

4.5 Effective Implementation of Plan

All team members should receive at least a basic introduction to HACCP. All information related to the product should be written in process information form and Process Flow Diagram should be there. The Hazard Analysis/Preventive Measures Form is used to review the steps listed in the Process Flow Diagram and identify where significant hazards could occur and describe the preventive measures, if they exist. CCP Determination Form -The Critical Control Point (CCP) Determination form is used to identify the critical control points in the process. A critical control point is defined as a point, step, or procedure at which control can be applied and a food safety hazard can be prevented, eliminated, or reduced to an acceptable level. All potential hazards so identified in the hazard analysis must be addressed properly so to assure end product safety. Identification of each CCP can be facilitated by the use of a CCP Decision Tree. Record Keeping and verification is one of the crucial steps in effective implementation of plan, it helps in realization of the strength and weakness of plan and the future worker can also get appropriate idea of the plan.

4.6 Impact of Failure and its Control

The Failure Mode and Effect Analysis (FMEA) model has been applied for the risk assessment of meat manufacturing. The Preliminary Hazard Analysis and the Fault Tree Analysis were used to analyze and predict the actual happening of failure modes in a meat production system, which is based on those functions, uniqueness, and/or interactions of the ingredients or the processes, upon which the system depends. Critical control points have been identified and implemented in the cause and effect diagram.

5. Future Perspectives in Meat Quality

HACCP has become an international standard in food safety assurance. Recommended or mandatory use of HACCP is being encouraged by several countries, and industries. The consumers are showing growing acceptance of the system and due to increased awareness are preferring products following such practices. In Australia incorporation of HACCP in meat industry had shown in reduction of customer complaints, availability of better hygiene products in the market, improved morale, a sense of safety and an increase in overseas markets (Khatri and Collins, 2007).

Continuous improvements are integral part of the HACCP programme as new finding and growing experiences can help address new challenges and ultimately result a safer product. Implementation of organic acid decontamination in American abattoirs was also adopted to meet specified performance standards for pathogen reduction as a part for overall HACCP program (Smuldersa and Greerb, 1998).

6. Conclusion

The HACCP concept, which focuses on food safety, is a systematic approach to hazard identification, assessment, and control. The system offers a rational approach to the control of physical, biological, and chemical hazards in foods; it avoids many weaknesses which are present in the traditional end product inspection approach. The focus of the system is to direct attention to the control of key factors that affect the safety of the food. HACCP is applicable to all parts of the food chain, from the production through all the processing steps ultimately to be used in the home.

REFERENCES

Bennet, W.L. and Steed, L.L. (1999).An integrated approach to food safety. Quality Progress,32, 37-42.

BouRached, L., Ascanio, N and Hernández, P. (2004).Design of a Hazard Analysis and Critical Control Points (HACCP) plan to assure the safety of a bologna product produced by a meat processing plant. Archives of Latinoam Nutrition, 54(1),72-80.

Dickson, J.S. and Anderson, M.E. (1992). Microbiological decontamination of food animal carcasses by washing and sanitizing systems: a review. Journal of Food Protection,55,133-140.

Guzewich, J. (1985). HACCP for food service. Journal of Food Protection, 48, 915-916.

Khatri Y. and Collins R. (2007).Impact and status of HACCP in the Australian meat industry. British Food Journal, 109 (5), 343-354.

Smuldersa, F.J.M. and Greerb, G.G. (1998).Integrating microbial decontamination with organic acids in HACCP programmes for muscle foods: prospects and controversies. International Journal of Food Microbiology, 44, 149-169.

Tomasevic I., Kuzmanoviæ J., Anðelkoviæ A., Saraèeviæ M., Stojanoviæ M.M. and Djekic I. (2016). The effects of mandatory HACCP implementation on microbiological indicators of process hygiene in meat processing and retail establishments in Serbia. Meat Science, 114,54-57. doi: 10.1016/j.meatsci.2015.12.008.

Chapter 9

Drugs, Pesticides and Heavy Metals Residues in Meat and their Safety Measures

Ranvijay Singh[1]*, K. Shrman[2] and Sachin Kumar Jain[2]

[1]Associate Professor, Department of Veterinary Public Health,
[2]Assistant Professor, Department of Veterinary Pharmacology and Toxicology,
College of Veterinary Science and Animal Husbandry, Jabalpur, Madhya Pradesh
**e-mail: rvsvet24oct@gmail.com*

ABSTRACT

Consumers prefer safe meat with a willingness to pay accordingly leading for a demand of meat, free from all unwanted residues of antibiotics, drugs, pesticides, heavy metals etc. Not only because of consumer demand for safe meat, it is the responsibility of the state and the meat industry that a safe and wholesome meat is available to population. The unwanted residues in meat is creating health issues in terms of development of antibiotic microbial resistance, and many of these residues are carcinogens, mutagens, and teratogens. Hence, there is a need by the meat industry to follow norms which could lead to production of meat, free from such residue and can be safely available to consumers.

Keywords: *Residue, Meat, Antibiotics, Drugs, Heavy metal.*

1. Introduction

Meat and meat products are important for human diet in many parts of the world because they provide proteins, minerals, vitamins and trace elements required for growth and maintenance of the body. Thus, the meat and its products are helpful in solving the global food problem.

An improvement in the food production and processing technology has increased the chances of contamination of food with various environmental pollutants especially drugs, pesticides and heavy metals. On ingestion, the chemicals or their metabolites are deposited in the body tissues of animals and are consumed by humans through their food *viz.* meat, milk, egg or fish. These elements can be very harmful even at low concentration when ingested over a long time period due to their ability to accumulate in human and animal body. They produce mutagenicity, carcinogenicity, genotoxicity, teratogenicity, hypersensitivity and allergy. These chemicals are neurotoxic, disturb hormone function, body metabolism and failure of vital organs such as liver and kidney. Severe acute toxicity of these drugs or chemicals precipitate death (Gracey *et al.*, 1999). Such toxicity can be reduced by maintaining the withdrawal periods of the drugs or pesticides *viz.* the time required after administration of a drug to an animal to assure that drug residues in the marketable product is below a determined maximum residue limit (MRL). The withdrawal period is variable for different chemicals and depends on the route of application and the product (meat, milk, fish, eggs *etc.*) used for consumption. The relaxation in enforcing the regulations and lack of awareness allowed entry of chemicals in food chain. The withdrawal periods are ranging from 0 to 60 days for animal products. Cooking/heating of meat or milk destroy pathogens, but not residues and sometimes these chemicals may be broken down into components that are more harmful to consumers. There is no permissible limit for the banned drugs in the food.

2. Drugs

2.1 Antimicrobials and Antiparasitic Drug

Antimicrobials are the chemicals which inhibit the growth or kill the bacteria. The antimicrobials obtained partially or wholly by microorganisms (fungus or bacterium) are known as antibiotics. These drugs are used for therapeutic purpose in animals to prevent and control infectious diseases. The antibiotics were commonly used in food animals since the 1940s with the start of intensification of rearing of animals for food purpose. These drugs are commonly used as feed additive to protect the animals from different pathogens and thus, help to gain weight early. It enhances feed conversion and promoting growth. The antimicrobials are generally given to calves, lamb, pig and poultry as growth promoter. About 80 per cent amounts of antimicrobials are used as feed additive all over the world. The excessive and indiscriminate application of antimicrobials in growing animals and poor follow up of withdrawal period in the animals by the farmers/authorities causes the entry of drugs or their residues in food chain. The entry of drugs or their residues in animal body or food chain is associated with bacterial resistance against the antimicrobials and causes emergence of new antibacterial resistance strains (Phillips *et al.*, 2004). In 1988, the Great Britain banned the use of bacitracin, spiramycin,tylomycin and virginiamycin as feed additive; while, European union banned the antimicrobial use in feed additive by 2006 (https: //en.wikipedia.org/wiki/Antibiotic.). In India, many reports from different parts of the country showed the presence of antimicrobials in poultry meat above the recommended level.

Several parasites like flatworms, tapeworms, roundworms and protozoans are affecting the growth and production of animals and are acting as silent killers of animals. To protect the animals from these parasites, anthelmentic and antiprotozoal drugs are commonly given to the animals. The commonly used anthelmentics are oxyclozanide, closantel, rafoxanide, nitoxynil, levamisole, ivermectin, monensin and bendamizadole group *i.e.* albendazole, mebendazole and fenbendazole. The extensive use as prophylactic measures as a growth promoter imparts resistance in parasites against anthelmentics. Such type of resistance is also observed in new generation of anthelmentics.

2.1.1 Health Hazards

Antimicrobial or their residues in food of animal origin produces potential threat to human (cancers, allergic reactions *etc.*). The low levels of antibiotic exposure results in alteration of microflora, and the possible development of resistance which cause failure of antibiotic therapy in clinical situations. A workshop of experts co-sponsored by the World Health Organization (WHO), Food and Agricultural Organization (FAO) and World Animal Health Organization (OIE) in 2003, reported that "that there is clear evidence of adverse human health consequences due to resistant organisms resulting from non-human usage of antimicrobials. Further, the experts were of the view that indiscriminate use of drugs was responsible for increased frequency of treatment failures (in some cases death), increased severity of infections and increased occurrence of disease. Several studies have shown that routine use of antibiotics on the farm promotes drug-resistant superbugs and these pathogens move from the farm to the kitchen, via uncooked meat and poultry. Studies have shown that many strains of *Salmonella, Campylobacter, E. coli, S.aureus* have been reported resistant to multiple drugs. These pathogens are known to cause diseases in both humans and animals and their many strains are of zoonotic importance. CDC estimates that every year, over 400,000 people in the United States are sickened with resistant *Salmonella* or *Campylobacter* (https: //www.cdc.gov/foodsafety.). These strains are difficult to treat and increased the hospital cost. This problem makes more vulnerable to patients undergoing chemotherapy for cancer, dialysis for renal failure, and surgery, especially organ transplantation, for which the ability to treat secondary infections is crucial. Besides this, it enhances the morbidity and mortality. Few antimicrobials are known to be associated with cancer. The drugs like penicillin or streptomycin are sensitive to few persons and lead to hypersensitivity or life threatening allergic reactions, serious nerve damage, severe colon inflammation, swelling of lips, tongue, or face, bleeding and diarrhea. Flunixin is also associated with gastrointestinal erosions, ulcers, renal necrosis and blood in feces.

Irrational use of many anthelmentics causes teratogenicity and congenital defects in animals. Ivermectin causes neurotoxicity *viz.* altering normal activity of the nervous system which can eventually disrupt or even kill neurons. The withdrawal period is variable for different antimicrobials but generally 3-4 days must be given for withdrawl period against antimicrobials and similar or more time periods in case of anthelmentics. The maximum permissible limit of various antimicrobial and anthelmentics are shown in Tables 9.1 and 9.2.

Table 9.1. Maximum Residue Limit (MRL, ug/kg) of Antimicrobials (Codex Alimentarius, 2018)

Sl.No.	Antimicrobials	Animals/ Species/Product	Tissues	Maximum Residue Limit (MRL,ug/kg)
1	Avilamycin	Pigs/Chicken/ Turkey/Rabbits	Muscle	200
			Liver	300
			Kidney	200
			Fat orSkin	200
2	Amoxicillin	Cattle/Sheep/ Pigs	Muscle/Liver/Kidney	50
3	Benzyl Penicillin/Procaine Benzyl Penicillin	Cattle/Chicken/ Pigs	Muscle/Liver/Kidney	50
4	Ceftiofur	Cattle/Pigs	Muscle	1000/
			Liver	2000/
			Kidney	60000/
			Fat	20000
5	Chlortetracycline/Oxytetracycline/Tetracycline	Cattle/Sheep/ Pigs/Poultry	Muscle	200
			Liver	600
			Kidney	1200
		Fish	Muscle	200
		Eggs		400
6	Colistin	Cattle/Sheep/ Pigs/Goat/ Poultry/Turkey/ Rabbit	Muscle	150
			Liver	150
			Kidney	200
			Fat	150
7	Danofloxacin	Cattle/Pigs/ Poultry	Muscle	200
			Liver	400
			Kidney	400
			Fat	100
8	Dihydrostreptomycin/Streptomycin	Cattle/Sheep/ Pigs/Poultry	Muscle	600
			Liver	600
			Kidney	1000
			Fat	600
9	Erythromycin	Poultry/Turkey	Muscle/Liver/Kidney/ Fat	100
		Eggs		50
10	Flumequine	Cattle/Sheep/ Pigs//Poultry	Muscle	500
			Liver	500
			Kidney	3000
			Fat	1000

Sl.No.	Antimicrobials	Animals/ Species/Product	Tissues	Maximum Residue Limit (MRL,ug/kg)
11	Gentamicin	Cattle/Pigs	Muscle	100
			Liver	2000
			Kidney	5000
			Fat	100
12	Lincomycin	Pig/Poultry	Muscle	200
			Liver	500
			Kidney	500
			Fat	100
13	Narasin	Cattle/Pigs// Poultry	Muscle	15
			Liver	50
			Kidney	15
			Fat	50
14	Neomycin	Cattle/Sheep/ Goat/Pigs/ Poultry	Muscle	500
			Liver	500
			Kidney	10000
			Fat	500
15	Spectinomycin	Cattle/Sheep/ Pigs/Poultry	Muscle	500
			Liver	2000
			Kidney	5000
			Fat	2000
16	Spiramycin	Cattle/Sheep/ Poultry	Muscle	200
			Liver	600
			Kidney	300
			Fat	300
17	Sulfadimidine	Not specified	Muscle/Liver/Kidney/ Fat	100
18	Tilmicosin	Cattle/Sheep	Muscle	100
			Liver	1000
			Kidney	300
			Fat	100
		Pig	Muscle	100
			Liver	1500
			Kidney	1000
			Fat	100
		Poultry	Muscle	150
			Liver	2400
			Kidney	600
			Fat	250

Table 9.2. Maximum Residue Limit (MRL, ug/kg) of Anthelmentics (Codex Alimentarius, 2018)

Sl.No.	*Anthelmentics*	*Animals/ Species/ Product*	*Tissues*	*Maximum Residue Limit (MRL,ug/kg)*
1	Cypermethrin and alpha-cypermethrin	Cattle/Sheep	Muscles/Liver/Kidney	50
			Fat	1000
2	Abamectin	Cattle	Liver/Fat	100
			Kidney	50
3	Ivermectin	Cattle	Liver	100
			Fat	40
		Sheep/Pig	Liver	15
			Fat	20
4	Thiabendazole	Cattle/Sheep/ Goat/Pig	Muscles/Liver/Kidney/ Fat	100
5	Albendazole	Not specified	Muscles/Fat	100
			Liver/Kidney	5000
6	Febantel/fenbendazole/ oxfendazole	Cattle/Sheep/ Goat/Horse	Muscles/Kidney/Fat	100
			Liver	500
7	Flubendazole	Pigs/Poultry	Muscles/Liver	10
		Poultry	Muscles	200
			Liver	500
		Eggs		400
8	Closantel	Cattle	Muscles/Liver/	1000
			Kidney/Fat	3000
		Sheep	Muscles/Liver	1500
			Kidney	5000
			Fat	2000
9	Diclazuril	Sheep/Poultry/ Rabbit	Muscles	500
			Liver	3000
			Kidney	2000
			Fat	1000
10	Diminazene (trypanocide)	Cattle	Muscles	500
			Liver	12000
			Kidney	6000
11	Isometamidium (trypanocide)	Cattle	Muscles/Fat	100
			Liver	500
			Kidney	1000
12	Doramectin	Cattle/Pig	Muscles	5-10
			Liver	100
			Kidney	30
			Fat	150

Sl.No.	Anthelmentics	Animals/ Species/ Product	Tissues	Maximum Residue Limit (MRL,ug/kg)
13	Emamectin benzoate	Salmon/Trout	Muscles/Fillet	100
14	Eprinomectin	Cattle	Muscles	100
			Liver	2000
			Kidney	300
15	Imidocarb (antiprotozoal agent)	Cattle	Muscles	300
			Liver	1500
			Kidney	2000
			Fat	50
16	Levamisole	Cattle/Sheep/ Pig/Poultry	Muscles/Kidney/Fat	10
			Liver	100
17	Monensin	Cattle	Muscles/Kidney/Fat	10
			Liver	100
		Sheep/Goat	Muscles/Kidney	10
			Liver	20
			Fat	100
		Poultry	Muscles/Liver/Kidney	10
			Fat	100
18	Moxidectin	Cattle/Sheep	Muscles	20
			Liver	100
			Kidney	50
			Fat	500
19	Nicarbazin (antiprotozoal agent)	Chicken	Muscles/Liver/Kidney/ Fat	200

2.2 Hormones

Hormones are commonly used as growth promoter and therapeutic purpose in animals. The steroidal hormones commonly used are testosterone, estrogen, progesterone, oestradiol, trenbolone, zeranol *etc.* The permission of these steroids are given under restricted conditions such as application at correct site, observation of withholding period *etc.*

2.2.1 Health Hazard

The European Union has banned the use of diethylstiboestrol due to its association with cervical adenocarcinomas in girls (EC, 2001). These hormones are also associated with gynecomastia.

3. Pesticides

According to WHO, "Pesticides are chemical compounds that are used to destroy pests, including insects (insecticides), fungus (fungicides), rodents (rodenticides), algae (algicides) and unwanted plants or weeds (herbicides).To cater the need of growing population and to enhance the agriculture production and preventing spoilage of large quantity of agriculture produce in field and storage, the pesticides are excessively and rampantly used by the farmers all over the world. Over 98 per cent of sprayed insecticides and 95 per cent of herbicides reached destination other than their target species, because they are sprayed or spread across entire agricultural fields. This leads to contamination of environment *viz.* soil and water. The chemicals get entry in the body of animals through consumption of crops, grains, drinking of water or licking. Use of pesticide by Indian farmer in agriculture is less than 350 gm/hectare as against the world average of 500 gm/hectare. Still, 51 per cent of India's food commodities are contaminated with pesticide residues, out of which 20 per cent has pesticide residues above the permissible level.

3.1 Health Hazard

Organochlorines (insecticides) have bioaccumulation properties which enhances its magnification in successive food order. These compounds are extremely durable and persistent in nature for longer periods therefore many chlorinated compounds are banned worldwide. Organophosphates are second generation of insecticides and are highly toxic but have less persistency in nature. These compounds are easily metabolized therefore, produces less residues. Chronic exposures of the pesticides are increasingly linked to immune suppression, hormone disruption, diminished intelligence, reproductive abnormalities, neurotoxicity and cancer. Most of the study from all over the world revealed presence of pesticides from meat, poultry, fish and egg. In India, 26 harmful insecticides have been banned and many more will be banned soon in future because of their harmful effect.

Table 9.3. Maximum Residue Limit (MRL, ug/kg) of Pesticides (Codex Alimentarius, 2018)

Sl.No.	*Pesticides*	*Animals/ Species/ Product*	*Tissues*	*Maximum Residue Limit (MRL,ug/kg)*
1	Deltamethrin	Cattle/Sheep/ Chicken	Muscles	30
			Liver/Kidney	50
			Fat	500
2	Cyhalothrin	Cattle/Sheep/Pig	Muscles/Liver/Kidney	20
			Fat	400
3	Dicyclanil (insecticide)	Sheep	Muscles	150
			Liver/Kidney	125
			Fat	200
4	Fluazuron (insecticide	Cattle	Muscles	200
			Liver/Kidney	500/500
			Fat	7000

4. Heavy Metals

During last decades, human population has faced changes in life and food style, which have led to an increase in demand for processed foods. The rise of food production and development of processing technology have increased the chances of food contamination with various environmental pollutants, especially heavy metals. Some heavy metals also cause toxicity-related mutagenesis and carcinogenesis. Inconveniently, the sign and symptoms of contamination by heavy metals do not appear early or easily, especially in humans. The presence of heavy metal particularly lead and cadmium in meat products have been shown by numerous studies in various parts of world. If residual levels exceed the prescribed standards, due to the cumulative effect in the human body consecutive to repetitive and persistent ingestion, they will induce negative effects on human health, hence the duty of all farmers and meat processors to minimize the possibilities and probabilities of contamination.

The food intoxication of heavy metals is due to their natural presence in the environment or anthropogenic activity like industrialization, urbanization, use of fertilizers, automobiles and pesticides. These metals stay permanently because they cannot be degraded in the environment and enter food chain and thus make their passage into the tissue. They are major threats to human health due to poor metabolization and thus, accumulate in tissues especially the metals like lead, cadmium and mercury. Some heavy metals are deposited as residues in food, during processing. Metals, such as iron, copper, zinc and manganese, are essential metals since they play important role in biological systems, whereas mercury, lead and cadmium are toxic, even in trace amounts. The essential metals can also produce toxic effects at high concentrations. Only a few metals with proven hazardous nature like lead, cadmium and mercury are completely excluded in food for human consumption by European Union (Hoha *et al.*, 2014; Mahmoud *et al.*, 2015; Wafaa *et al.*, 2015).

4.1 Lead

Lead accumulation in animal tissues occur due to grazing on areas near lead industries or ingestion of substances containing high contents of lead. The high content of lead above the recommended level have been found in tissues particularly kidney and liver of ruminants, pigs and poultry. The high levels of lead in adults body causes heart diseases, cancer and infertility. In children, it leads to CNS disorder *viz.* antisocial behavior, low intelligence or hyperactivity due to crossing of blood brain barrier. Pb poisons create physiological and morphological responses in microalgae.

4.2 Arsenic

It is highly poisonous compound and its use in different chemicals have been reduced. It enters into food chain due to its wide application as herbicides, insecticides and rodenticides in farm practices. It has also been used for parasitic control and control of dysentery in pig. Chronic toxicity is due to its use in various compounds and is mainly accumulated in liver, kidney and bones. Arsenic excretes slowly in milk, sweats and feces. In chronic intoxication withholding period of 40 days are required. Polluted water fishes have higher concentration of arsenic. An organic arsenic product named 3-Nitro is commonly used in poultry feed to enhance the colour and brightness was banned in USA as it was associated with cancer in humans (www.fda.gov/downloads/AnimalVeterinary.).

4.3 Cadmium

Cadmium is a toxic element to every animal species and humans. It is almost absent in the human and animal body at birth, however it accumulates with age and can persist for 30-40 years mainly in kidney. The chemical has been reported from various products of meat above the recommended level. The toxicity in animals is due to grazing on pasture irrigated with sewage sludge. Cadmium is well-known to cause damage to kidneys and bones as well as DNA damage and mRNA transcriptional changes in the gills of *Mytilus galloprovincialis*.

4.4 Copper

Copper is essential for good health but very high intakes can cause health problems such as liver and kidney damage. The maximum copper concentration for meat and meat products has been proposed as 0.90–30 mg d^{-1} person. Copper supplemented feeds are prepared for pigs but no cases in humans have been reported due to this source. Animals with copper toxicity reveals enlarged yellow liver, jaundice and hemoglobinurea on postmortem.

4.5 Mercury

The organic and inorganic form of mercury is used in agriculture, horticulture and veterinary practice. Due to high toxicity, the use of the metal in different compounds is minimized, so there is less likelihood of toxicity. The concentration in meat is found due to grazing on sewage treated land and industrial area. The metal is excreted out slowly from sweat, feces, saliva, milk *etc.* The fishes from polluted water can be source of food poisoning.

Table 9.4. Maximum Residue Limit (MRL, ug/kg) of Heavy Metals (FSSAI, 2015)

Sl.No.	*Heavy Metals*	*Animals/Species/Product*	*Maximum Residue Limit (MRL,ug/kg)*
1	Lead	Cattle, Sheep, Pig, Poultry, Fat from meat	0.1ppm or 0.1mg/Kg
		Fish	0.3ppm
2	Arsenic	Fat	0.1ppm
		Fish	76ppm
3	Mercury	Fish	0.5ppm (0.5mg/kgof fresh weight).
4	Cadmium	Fish	0.3ppm
5	Chromium	Fish	12ppm

5. Method of Detection and Determination of Residues

The widespread presence of antimicrobials, pesticides, hormones and heavy metal compounds in animal products, stringent food safety legislations and consumer concerns demand the availability of rapid, sensitive, easy-to-use, reliable, cost effective and broad-spectrum screening methods. The techniques can be readily implemented in survey, surveillance and compliance monitoring schemes for residue detection. Because of the diverse physico-chemical properties of compounds, a variety of analytical techniques are commonly employed to screen the chemicals and their residues in foods of animal origin. Detection and/or determination falls into two categories: Qualitative/Screening methods and quantitative and/or confirmatory methods include immunoassay such as ELISA, gas chromatography with electron capture, flame ionization, or mass spectrometry detection, as well as liquid chromatography (LC) with ultraviolet (UV), fluorometric or electrochemical detection or mass spectrometry (MS).

Microbial inhibition assays (MIA) are used commonly for routine detection/ screening of antimicrobials. The MIAs are sensitive to compounds that inhibit or disturb the growth of a test microorganism. It offers the advantage of detecting the total biological activity associated with unknown residues (non-targeted analysis). A negative secondary screening result indicates the presence of one (or more) class of antimicrobial compound in the sample. Positive samples resulting from the secondary screening assays may be rapidly directed to the appropriate quantitative/ confirmatory further analysis such as,ELISA and liquid chromatography coupled to a tandem mass spectrometer (LC-MS/MS). Enzyme-linked immunosorbent Assay (ELISA) is the most common immunoassay because of its sensitivity and specificity. Test kits are also made available by some laboratories to identify and quantify the antimicrobials and hormones. Radioimmunoassay was also used previously for detection of hormones.

Among biosensors, surface plasmon resonance (SPR) biosensor technology is applied for the detection of antibiotic residues in foods of animal origin. The biosensor responds to an analyte(s), and can interpret concentration as an electrical signal via a combination of a biological recognition element (BRE) and an electrochemical transducer. For quantitative estimation of chemicals/drugs, chromatography technique is in vogue now a day. Different forms of chromatographic techniques are gas chromatography, thin layer liquid chromatography, liquid chromatography and mass spectrometry (LC-MS).Among the chromatographic methods, single residue method (SRM) and multi residue methods (MRM) are applied for the analysis. As the name indicates, multi residue methods (MRM) are designed to identify and quantify a number of pesticides and their toxicologically significant metabolites simultaneously in a range of foods and other substances. These techniques are sensitive, precise and economical enough to be useful for regulatory purposes and greatly acceptable to the scientific community. Liquid chromatography is the most versatile technique for the detection of antibiotics and other chemicals in the food and other products. The most common technique is high performance liquid chromatography (HPLC), is used for qualitative and quantitative determination of antibiotics and other chemicals in food. With use of different detectors and addition of mass spectrometry (MS),the sensitivity and use of liquid chromatography can be increased to manifold *viz.* in parts per billion (ppb). Atomic absorption spectrometry (AAS) is widely used technique for the detection of metals. In this, the free atoms in the form of gas absorb radiation at specific frequency/wavelength and thus the concentration of metal is measured. However, for simultaneous detection of many elements advanced instrument like inductive couple plasma spectroscope (ICP) can be used after proper processing. The detection limit of AAS or ICP is up to ppb.

6. Prevention and Control

6.1 Organic Farming

Organic farming is an integrated system that strives for sustainability in the agricultural system. It maintains soil fertility and biological diversity. It prohibits the use of synthetic pesticides, antibiotics, synthetic fertilizers, genetically modified organisms, and growth hormones and thus, the entry of such chemicals are prevented in crops, foods, animals and environment. In India, Sikkim is first state in horticulture to follow complete organic farming.

6.2 Alternative Therapy

This kind of therapy includes the use of bacteriophage, ayurvedic, homeopathic medicines *etc.* to prevent and control diseases. These medicines do not have residual effect.

6.3 Integrated Pest Management (IPM)

It is economic method of pest control and utilizes minimum amount of insecticides and pesticides to control insects. As per FAO "It is an economically justified method that employs careful consideration of all available pest control measures and techniques and its subsequent integration to discourage the development of pest populations." It reduces or minimizes risks to human health and the environment. IPM emphasizes the least possible disruption to agro-ecosystems during growth of healthy crop.

6.4 Biological Controls (Biocontrol)

It is a method of controlling of insects, mites, weeds and plant diseases using other organism through predation, parasitism and competition mechanisms. It is an important component of integrated pest management. The three basic strategies for pest control includes classical (importation), where a natural enemy of a pest is introduced in the hope of achieving control; inductive (augmentation), in which a large population of natural enemies are administered for quick pest control; and inoculative (conservation), in which measures are taken to maintain natural enemies through regular reestablishment. The side-effects of biocontrol is that it attacks on non-target species and thus has an impact on biodiversity. So, there should be thorough understanding of the possible consequences before introducing any biological method.

6.5 Biotechnology Tools

Biotechnological tools can be boon in prevention and control of residual problem through various ways: strains can be genetically modified (GM) to increase their resistance to pests, use of transgenic in sterile production of insects, vaccines production against various diseases *etc.*

6.6 Diagnostic Methods

Rapid, sensitive, specific and reproducible will help in early detection and quantification of diseases and residues present in animals and foods, respectively. These methods should be widespread and easily accessible.

6.7 Good Managemental Practices

Food animals should be provided with clean and ventilated housing, good nutrition, vaccination and prevention of animal over-crowding to reduce diseases and thereby reduction in antibiotic residues.

6.8 Legislation

Stringent rules and regulation should be made and it should be enforced and implemented throughout the state and country.

6.9 Awareness

People should have knowledge regarding the safety, regulations, proper application technologies and integrated pest management. They should also be aware of abuses and consequences of these drugs.

REFERENCES

Codex Alimentarius (2018). Maximum Residue Limits (Mrls) and Risk Management Recommendations (Rmrs) for residues of veterinary drugs in foods.

E.C. (2001). Commission Regulation No. 466/2001 of 8 March 2001, Official Journal of European Communities 1.77/1.

Food Safety and Standards (Contaminants, Toxins and Residues) (Amendment) Regulations, 2015.

Gracey, J. F., Collins, D.S. and Huey, R. J. (1999).Meat Hygiene (10^{th}Edn.),Chemical Residues in Meat (Ch.13), WB Saunders Company Ltd.

Hoha, G.V., Costãchescu, E., Leahu, A., and Pãsãrin B. (2014). Heavy Metals Contamination Levels in processed meat marketed in Romania. Environmental Engineering and Management Journal, 13 (9),2411-2415.

https: //en.wikipedia.org/wiki/Antibiotic_use_in_livestock

https: //www.cdc.gov/foodsafety/pdfs/cdc-and-food-safety.pdf

Mahmoud, M.A.M., Abdel-Mohsein, H.S. (2015). Health Risk Assessment of Heavy Metals for Egyptian Population via Consumption of Poultry Edibles. Adv. Anim. Vet. Sci, 3(1),58-70.

Phillips, I. (2004). Does the use of antibiotics in food animals pose a risk to human health? A critical review of published data. Journal of Antimicrobial Chemotherapy, 53(1), 28–52.

Wafaa Sh. Al-Zuhairi, Mohammed A. Farhan and Maryam A.Ahemd. (2015). Determination of heavy metals in the heart, kidney and meat of beef,mutton and chicken from baquba and howaydir market in baquba,diyala province, Iraq. International Journal of Recent Scientific Research,6(8),5965-5967.

www.fda.gov/downloads/AnimalVeterinary/GuidanceComplianceEnforcement/GuidanceforIndustry/ucm052519.pdf.

An Overview of Various Methods Involved in the Detection of Pathogenic Microbes in Meat and Meat Products

A. Arun Prince Milton[1]*, G. Bhuvana Priya[2] and Samir Das[1]

[1]*Division of Animal Health, ICAR Research Complex for NEH Region, Umiam, Meghalaya*

[2]*College of Agriculture, Central Agricultural University, Kyrdemkulai, Imphal, Meghalaya*

**e-mail: vetmilton@gmail.com*

ABSTRACT

Detection of pathogenic microbes in meat and meat products is essential to ensure safety. It is also important to detect early to reduce untoward effects on consumer health and food business. Apart from minimizing the risk of outbreaks it also provides product assurance. Timely identification and precise detection of food borne pathogens will decrease the costs of holding food products in cold storage and minimizes product recalls. The important challenges in arriving reliable and rapid detection methods are low contamination of food, stressed or sub-lethally injured bacteria, slow growing microbes (require enrichment), dead microbes (may give false-positive results), viable but non-culturable bacteria (may give false-negative results), the requirement of costly or sophisticated equipments, trained personnel, portability, etc., This chapter presents an overview of different routinely used and novel emerging detection methods. Discussed detection methods include culture based methods, immunological methods, nucleic acid based methods and biosensor based methods. Many novel methods are under continuous development focusing reduction of detection time, improvement of sensitivity and the limit of detection and more importantly attention is being given to develop culture independent methods.

However, we suggest that focus should be given towards the development of more equipment free simple but reliable techniques which are instantly required for many resource limited food testing laboratories located in most of the developing and underdeveloped countries.

1. Introduction

Meat is regarded as a food of high nutritive value and gaining popularity and preference among consumers throughout the world as it can offer high-quality proteins, minerals, vitamins, and other nutrients to improve and sustain human health. Further consumer concerns about food have moved from food security to safety. So as to meet the higher necessities of people and keep the market competitive, the meat industry should produce not only quality stuff but also safer one. As far as meat foods are concerned, there are lot of potential hazard namely microbial pathogens, preformed toxins, parasites, food additives, antibiotics and chemical residues, *etc.* (Knowles *et al.*, 2007). Meat safety is becoming an increasingly global issue due to global meat and poultry trade and increase of consumption, exposing a greater number of people to potential hazards (Viegas *et al.*, 2012). Meat safety is a complex issue facing a number of challenges like traceability issues, regulatory issues, pathogen and residue detection hitches, antibiotic resistance, other consumer concerns, *etc.* The most serious and important meat safety issues causing consumer health problems and product recalls are associated with microbial pathogens (Sofos, 2008). Estimates of economic losses particularly due to meat borne diseases are not available. However, recent estimates of economic costs of food borne illnesses only in low- and middle-income countries have amounted to 110 billion US dollars (Jaffee *et al.*, 2019). Therefore detection of pathogenic microbes in meat and meat products is essential to ensure safety. It is also important to detect early to reduce untoward effects on consumer health and food business. Apart from minimizing the risk of outbreaks, it also provides product assurance. Timely identification and precise detection of food borne pathogens will decrease the costs of holding food products in cold storage and minimizes product recalls (Jadhav *et al.*, 2018).

Another important entity of food safety is attribution of food borne pathogens to a particular source or vehicle responsible for the illness or outbreak. Attribution can be accomplished only with the help of a reliable and rapid detection method which have to be deployed in surveillance and microbial source tracking activities. The information obtained through attribution may be used in controlling the pathogenic microbes and risk assessments (Doyle and Erickson, 2006; Sofos, 2008). Continuous and considerable efforts have been given in developing newer reliable and rapid detection methods and to fine tuning existing methods to detect potential food borne pathogens (Ghatak *et al.*, 2020). Advances in such fields may help in tracing pathogen sources, evaluating control measures and developing, evaluating and verifying different critical control points and critical limits through HACCP programs to control the pathogens (Sofos, 2008). The important challenges in arriving reliable and rapid detection methods are low contamination of food, stressed or sub-lethally injured bacteria, slow growing microbes (require enrichment), dead microbes (may give false-positive results), viable but non-culturable bacteria (may give false-negative results), the requirement of costly or sophisticated equipments, trained personnel, portability, *etc.* (Bhunia, 2014; Wang and Salazar, 2016). In the present discourse, the focus has been given in the comprehensive discussion of various detection methods to detect microbial pathogens in meat and meat products, its advantages and drawbacks.

2. Common Food (Meat and Meat Products) Borne Illnesses

In each year, in the USA alone, 48 million illnesses, 128,000 hospitalizations and 3000 deaths have been attributed to food borne illnesses, resulting in US$78 billion economic expenses. Among them, 9.4 million illnesses have been identified to be due to bacteria (64 per cent), virus (12 per cent) and parasites (24 per cent). Remaining illnesses (38.6 million) are attributed to emerging or unknown pathogens. Further, meat (beef, pork, and poultry and game) commodities were assessed to attribute to 22 per cent and 29 per cent illnesses and deaths, respectively. Particularly, poultry was associated in 19 per cent deaths, with most deaths being attributed to pathogenic bacteria *Salmonella* and *L. monocytogenes* (Scallan *et al.*, 2011; Scharff, 2012; Painter *et al.*, 2013; Bhunia, 2014). Stringent regulatory standards have been formed for some bacterial pathogens, of which *Salmonella*, Shiga-toxigenic *Escherichia coli* (STEC) and *Listeria monocytogenes*, are prohibited in certain foods. Whereas for some bacteria like *Campylobacter* spp., *Yersinia* spp., *Vibrio* spp., *Bacillus* spp., *Shigella* spp., or for viral pathogens, such as Hepatitis A virus and norovirus, the regulatory standards are relaxed, however their occurrence in ready-to-eat (RTE) foods is still undesirable (Rodriguez-Lazaro *et al.*, 2012; Bhunia, 2014).

Outbreaks of food borne enterohaemorrhagic *E. coli* (EHEC) infections have been classically associated with raw, RTE and undercooked beef products (Riley *et al.*, 1983; Tilden *et al.*, 1996). Ground beef and sausages (fermented) are often attributed to other non-O157 STEC illnesses (Mathusa *et al.*, 2010). Sporadic and epidemic listeriosis caused by *L. monocytogenes* has been routinely associated with RTE meat products like delicatessen meats, frankfurters and fermented meat products (Schuchat *et al.*, 1992; Lianou and Sofos, 2007; Skandamis and Nychas, 2015). Food borne non-typhoidal salmonellosis caused by *Salmonella enterica* serotypes has a high association with raw meat of domestic food animals like cattle, pig and poultry (Rhoades *et al.*, 2009; EFSA-ECDC, 2015). *Campylobacter jejuni* seems to predominate in broiler chickens, turkeys and cattle, while *Campylobacter coli* is commonly linked with pigs. Broiler meat is the single important source of human campylobacteriosis (Blackburn and McClure, 2009; EFSA-ECDC, 2015; Kaakoush *et al.*, 2015). Vehicles of foodborne illness caused by *Clostridium perfringens* are typically meat and poultry foodstuff (Labbe and Juneja, 2017). As *Yersinia enterocolitica* is commonly associated with pigs, pork and pork products have been attributed as a potential vehicle of food borne human yersiniosis (Lianou *et al.*, 2017). Food borne illnesses caused by *Clostridium botulinum* have often been attributed to meat roll, reheated chicken, pork sausage and home cured ham (Peck, 2010; Lianou *et al.*, 2017). Among viruses; hepatitis A virus (HAV), hepatitis E virus (HEV) and norovirus (NoV) are identified as significant food borne pathogenic viruses. Most of the epidemiological data support that deli meats have been often associated with food borne viral outbreaks. Food borne hepatitis E illnesses have been often linked with the consumption of raw pork, pig liver sausage, and other pork products (Pavio *et al.*, 2015). Attributions of HEA and NoV infections have been chiefly linked to the consumption of crustaceans, shellfish, and molluscs (Papafragkou *et al.*, 2006; Mattison *et al.*, 2009; EFSA-ECDC, 2015). However, in 3.9 per cent outbreaks caused by NoV in the European Union, meat was the implicated vehicle (EFSA-ECDC, 2015).

3. Detection of Pathogens

Detection and identification of the pathogen is integral to develop an intervention plan to reduce or eliminate the pathogen. Therefore rapid, improved and accurate technologies are needed for timely detection. To evaluate a detection method, two performance criteria are often used *i.e.*, assay accuracy (lower limit of detection as low as one target organism per 25 to 325 g of sample) and time to result (food sample to the final result). A tool that gives more false-negative results is more detrimental than the one which gives more false-positive results, as the former is the biggest concern for outbreaks due to distribution of pathogen contaminated foods. The time taken for giving final results may vary from hours to days. A technology taking lesser time with good accuracy rate is need of the hour for the food industry as the test product could be promptly released in time (Velusamy *et al.*, 2010; Bhunia, 2014; Ghatak, 2020).Since most of the test food products are turning to be negative, it would be wise to deploy a detection method that finds out negative samples rapidly, ideally in hours, so as to execute lengthy analysis (may take 2-3 days) of positive samples to isolate and type the pathogens (Bhunia, 2014). Another issue is the food matrix, as assay performs differently with different food matrices. The complexity of various food matrices like the meat of different species, salt and fat contents, preservatives (processed meat), cooked foods, background microbiota (raw meat) may also stipulate the selection of detection methods (Bhunia, 2014).

3.1 Conventional Culture Based Methods

Culture based methods to detect food borne pathogens employing classical microbiological techniques are widely accepted, and gold standard methods. As they are sensitive, accurate, dependable, and reproducible, they have legal standing throughout the world. A wide range of specific growth media has been developed to selectively isolate the target pathogen, although general sanitation assessment methods (total coliform count, aerobic plate count) are also widely used for quantitative assessment of overall sanitary feature of meat and meat products (Priyanka *et al.*, 2016; Ghatak, 2020). A common procedure of cultural isolation involves homogenizing or stomaching a food (meat) sample, enriching in non-selective or selective enrichment media and then streaking on a selective solid agar medium followed by some biochemical tests and subtyping to ensure the presence and identify the targets(Sharma and Mutharasan, 2013a; Wang and Salazar, 2016). The enrichment is done as the foods usually contain a low concentration of pathogens and also enrichment resuscitates the target pathogen that is stressed or injured due to acid, cold, heat or osmotic shock. Enrichment increases the level of target organism and enables the detachment of target cells intimately adhered to the food matrices and selective enrichment subdues the other competitive background microbiota, thereby facilitating the successful isolation of the target pathogen

(Gracias and McKillip, 2004; Dwivedi and Jaykus, 2011; Bhunia, 2014). For these reasons, short or brief enrichment steps are often combined with some molecular tests like PCR, LAMP *etc.* (Momin *et al.*, 2020). Other advancements like usage of different chromogenic media, growth supplements, specialized substrates and selective antimicrobial agents have improved the classical culture based methods by hastening the process and sometimes visual colonies may be obtained within 24-48 hours. And the presumptive colonies could be further confirmed by molecular or biosensor or immunological based methods (Velusamy *et al.*, 2010; Bhunia, 2014; Singh *et al.*, 2014). The main drawback of cultural methods is its time consuming nature, taking up to 3-7 days to give negative results and for positive results, an additional 2-3 days or more may be needed (Gracias and McKillip 2004; Bhunia, 2014). This time taking nature of the tests may lead to fatalities, distribution of contaminated meat products and unnecessary accumulation of products. Even if cultural methods are laborious, they were considered as cost effective, but taking into account of the rising cost of trained manpower, they are becoming cost-ineffective in present days (Ghatak, 2020). Another important disadvantage is the inability to detect VBNC (viable but non-culturable) state of bacteria, which are prevalent among food patho-gens than hitherto thought (Nicolo *et al.*, 2011; Dinu and Bach, 2013; Li *et al.*, 2014). Some of the meat borne pathogens that fall under VBNC type include *C. jejuni*, *E. coli* (including EHEC), *Salmonella* spp., *Shigella* spp. (Priyanka *et al.*, 2016; Foddai and Grant, 2020). From the food safety perspective, these traditional methods are becoming less attractive as the modern food industry desires more rapid detection methods.

3.2 Immunological Methods

By exploiting the ability of antigen-antibody interaction, various detection and concentration methods were developed for rapid use in the food industry. Cost-effectiveness, simple way of execution and rapidity have made the immunological assays popular in the food industry. The critical factor of the immunological assays is the antibodies as they determine the specificity and sensitivity of any assay (Mangal *et al.*, 2016; Priyanka *et al.*, 2016). Various reports are available on the development of antibodies for food pathogens (Sunwoo *et al.*, 2006; Charlermroj *et al.*, 2012; Meyer *et al.*, 2012). Antibodies are produced in the small laboratory animals and with the arrival of hybridoma technology; monoclonal antibodies that react only with a specific pathogen can be developed (Mangal *et al.*, 2016). Further successful binding of the antibody with antigen depends on the constant expression of the antigens in a food borne pathogen, which are usually influenced by acids, salts, preservatives, temperature, and other chemicals and additives found in the food (Mangal *et al.*, 2016). The limit of detection for immunological assays is roughly 10^4 –10^5 CFU/g of food (Jasson *et al.*, 2010). Immuno-magnetic separation method is a sample concentration method which uses the antibodies coupled with the magnetic particles to capture specific antigen (bacteria) from the food samples. The pathogen concentrated samples can then be detected by any of the immunological or molecular detection methods (Mangal *et al.*, 2016). The three important categories of immunological methods deployed to detect food borne pathogens and commercially popular are enzyme-linked immunosorbent assay

(ELISA), lateral flow immunoassay (LFI) and the latex agglutination test (LAT) (Bhunia *et al.*, 2014; Ghatak, 2020).

3.2.1 Enzyme-Linked Immunosorbent Assay (ELISA)

ELISA is perhaps largely exploited and popular immunological detection method of meat borne pathogens. The important feature of the assay is the amplification of antigen-antibody interaction signal by enzyme-linked secondary antibody and substrate reaction, hence increasing the sensitivity of the assay. Other advantages are quantitative nature and high throughput with the 96-well format (Priyanka *et al.*, 2016; Ghatak, 2020). Sandwich ELISAs employing double antibody are commonly used in commercial kits where antibody coated 96 well plates (primary or capture antibody) are used for antigen capture (pathogen or toxin) from meat or other food samples and a secondary (detecting) antibody coupled with enzyme (horseradish peroxidase or alkaline phosphatase) is added to form antibody-antigen-antibody conjugate (sandwich). The substrate is then added and the enzyme converts the substrate into a detectable form which can be interpreted visually or with an ELISA reader (Mangal *et al.*, 2016). The limit of detection of ELISA is about 10^5 CFU per ml whole bacterial cells or a few ng/mL for toxins (Feng, 1997). Recent developments in ELISA are coupling ELISA with immunomagnetic separation (IMS) and other concentration and capture methods; newer antibody development and efforts are made towards increasing antibody density in the food matrices for more sensitive detection (Wang and Salazar, 2016). Nanomaterials have also been used in antibodies as they provide enhanced antibody density with their large surface-to-volume ratio (Chunglok *et al.*, 2011; Cho and Irudayaraj, 2013). ELISA assay using gold nanoparticle based antibodies and coupled with IMS have detected 15 CFU/mL of *S.* Typhimurium and 3 CFU/mL of *E. coli* O157: H7 in spiked food samples in 2 hour time (Cho and Irudayaraj, 2013). With the advent of monoclonal antibodies, Scotcher *et al.* (2010) have developed a sandwich ELISA to detect botulinum neurotoxin serotypes B (BoNT/B) from food with better sensitivity than mouse bioassay. The disadvantage of ELISA is the requirement of pure and specific antibodies, the lower limit of detection, perishable nature of reagents, need for investment costs like specialized ELISA readers (Law *et al.*, 2014; Alahi and Mukhopadhyay, 2017; Ghatak, 2020).

3.2.2 Lateral Flow Assays (LFI)

In lateral flow assays (LFI), the capture antibody is immobilized on a nitrocellulose membrane and the detection antibody coupled with gold or latex particles is positioned near the sample application pad. A blotting paper located at the other end of the sample application pad enables fluid movement as it absorbs through the membrane. When a homogenized sample is applied on the pad, the antigen binds to the latex-or gold- particles-coupled detection antibody, and the antigen-detection antibody complex flows laterally on the nitrocellulose membrane by capillary action and reaches the capturing zones, one specific for target organism and another specific for unbound antibodies conjugated with latex or gold (control line). The development of two lines indicates a positive result, whereas one line (control) shows a negative result (Bhunia, 2014; Wang and Salazar, 2016). The main advantages of the LFI are rapidity (10-15 min), cost-effectiveness, long-term stability, simple and minimum skill required. These advantages make the assay more suitable for on-site or field testing with high-throughput nature (Shim *et al.*, 2007; Park *et al.*, 2008). Some of the recent advancements in LFI are multiplex detection and development of novel labels, formats and supporting materials (Shan *et al.*, 2015). A multiplex LFI was recently developed for differential detection of BoNT/A and/B in spiked foods (Ching *et al.*, 2012). The main limitation of LFI is that they require high pathogen concentration (in a range of 10^7–10^9) for a positive reaction. Nevertheless, use of magnetic nanoparticles in place of gold particles has improved the assay sensitivity manifold and the signal could be collected with an automated reader (Anfossi *et al.*, 2013; Joung *et al.*, 2014).

3.2.3 Latex Agglutination Assay (LAT)

In latex agglutination assay (LAT), the meat borne pathogens are detected using antibody coated latex beads which agglutinate with a specific antigen (pathogen or toxin) and produce visible clumps against a dark background. The assay is very simple and cost-effective and commercial kits for several food borne pathogens are available (D'Aoust *et al.*, 1991; Ghatak, 2020). The main limitation of LAT is lower sensitivity or limit of detection, which requires 10^7 bacterial cells per reaction approximately (Feng, 1997). Other drawbacks are the short shelf life of beads and qualitative nature (Miller *et al.*, 2008; Mangal *et al.*, 2016).

3.3 Nucleic Acid Based Methods

Nucleic acid based detection methods function by targeting specific DNA/RNA sequences of the targeted food borne pathogens. Species, serotype or strain specific DNA or RNA are detected employing synthetic oligonucleotides that are complementary to the sequences of the target pathogens(Zhao *et al.*, 2014; Salazar *et al.*, 2015). Certain food borne pathogens like *C. botulinum*, *S. aureus*, *etc.*, produce preformed toxins capable of producing food borne intoxication. Genes encoding such toxins can also be detected using nucleic acid based methods (Zhao *et al.*, 2014). The advantages of nucleic-acid based methods are their higher specificity and sensitivity, less labor intensive, time-effective, less human errors and easy gel or visual based or real-time interpretation. These methods are also faster than the immunological and culture based methods (Mandal *et al.*, 2011; Lee *et al.*, 2014). Various formats of nucleic acid based methods are polymerase chain reaction (PCR), multiplex PCR, viability PCR, realtime PCR, reverse transcriptase (RT) PCR, isothermal based methods like LAMP, NASBA, PSR *etc.*, DNA microarray and hybridization methods.

3.3.1 PCR Based Methods

Today, PCR is one of the most extensively deployed nucleic acid based methods in food testing with the availability of numerous commercial assays. Many advances have also happened on the described original protocol (Priyanka *et al.*, 2016, Ghatak, 2020). It involves amplification of the portion of a suspect genetic material thereby enabling visualization of the amplified product through agarose gel electrophoresis. They require a cocktail of reagents including polymerase enzyme, primer, *etc.*, and a thermocycler. The specificity of the assay depends upon the complementary binding of a primer (short single stranded DNA) with the suspected pathogen sequence (Law *et al.*, 2014; Mangal *et al.*, 2016; Souii *et al.*, 2016; Umesha and Manukumar, 2018; Ghatak, 2020).The important drawbacks of the techniques are its qualitative nature, the requirement of nucleic acid extraction and post PCR processing like gel electrophoresis and lack of ability for selective detection of live or viable pathogens leading to false positive results (Priyanka *et al.*, 2016; Momin *et al.*, 2020). Although the operational cost of the method is cheap, it requires high initial investment to setup a facility which require sophisticated instruments like thermocycler, gel electrophoresis unit, gel documentation system, *etc.* (Momin *et al.*, 2020). As PCR assays are also vulnerable to interferences by various food matrices, a brief cultural enrichment is required to improve assay performance and also to differentiate viable from dead cells (Rossen *et al.*, 1992; Momin *et al.*, 2020). Preprocessing concentration or separation with bead based or centrifugation methods may also solve the issue (Rossen *et al.*, 1992; Lantz *et al.*, 1998; Yang *et al.*, 2007).

To achieve selective detection of viable cells, certain dyes like ethidium monoazide (EMA) or propidium monoazide (PMA) have been used with PCR, which is termed as viability PCR (Nogva *et al.*, 2003; Pan and Breidt, 2007). These dyes can only enter punctured cell membranes thus binding to the DNA and cause irreversible damage to them resulting in inhibition of PCR amplification. Therefore, only live bacteria with intact cell membrane get amplified (Trevors, 2012; Emerson *et al.*, 2017). This approach has been extensively explored for the detection of food borne pathogens by blending with PCR, real-time PCR and LAMP (Priyanka *et al.*, 2016; Wang and Salazar, 2016). A limitation of the viability PCR is that the cell membrane integrity is not always a dependable indicator of the cell viability as some evidence suggest that sometimes cell membrane may remain intact even in the absence of the metabolic activity or sometimes the bacteria may have perforated cell wall during some point of their growth leading to false positive and false negative results, respectively (Stiefel *et al.*, 2015; Ayrapetyan and Oliver, 2016).Another approach for selective detection of live or viable cells is to target messenger RNA (mRNA) through reverse transcription-PCR (RT-PCR) as mRNAs are present only in metabolically active bacterial cells (Sheridan *et al.*, 1998; Lleo *et al.*, 2000). RT-PCR requires reverse transcriptase enzyme to convert mRNA into cDNA (complementaryDNA) and cDNAs are used as a template for endpoint or real-time PCR amplification (Wang and Salazar, 2016).However, to detect bacterial pathogens in food, RT-PCR is least preferred as it is too laborious and the RNA in the samples get degraded rapidly that may lead to false negative reports (Xiao *et al.*, 2012).

Multiplex PCR is a variant of conventional PCR that detects multiple target pathogens using a multiple set of primers in a single reaction tube. Additional information can be obtained from a single assay that otherwise would need more run, more time and more reagents (Vandenvelde *et al.*, 1990; Mangal *et al.*, 2016). Multiplex PCR based assays are more suitable to food industries as the limited volume samples and cost are the key consideration. For the standardization of multiplex PCR assays, it is important that the amplicon sizes of various targets must be different and proper optimization of annealing temperatures for each primer sets is highly important (Mangal *et al.*, 2016). Multiplex PCR based detection methods have been successfully deployed for detection of meat and other food borne pathogens (Ching *et al.*, 2012; Lee *et al.*, 2014; Xu *et al.*, 2019).

3.3.2 Real Time PCR Based Methods

Real-time or quantitative PCR (qPCR) enables a real-time or continuous monitoring of the amplification by measuring the surge in fluorescence activity of either a specialized dye-tagged probe or DNA binding dye. Quantification is done by calculating the volume of fluorescence and comparing with the experimental control (Dwivedi and Jaykus, 2011; Zhao *et al.*, 2014). The quantitation is either based on the absolute number of copies or relative quantities when normalized to input copy number or additional reference or normalizing genes. Two categories of real-time PCR methods are (i) nonspecific fluorescent (SYBR green I or EtBr) dye based;(ii) sequence specific DNA probes containing oligonucleotides labeled with fluorescent reporter such as TaqMan, Scorpions, molecular beacons *etc.*, that allows detection after hybridization of the probe with its target complementary DNA (Priyanka *et al.*, 2016; Ghatak, 2020). Probe based methods provide higher sensitivity with more reproducibility than SYBR Green based assays. Real-time PCR has been routinely deployed for detecting a number of food borne pathogens and several real-time PCR based commercial kits are available in the market (Mangal *et al.*, 2016). The advantages of the real-time PCR that have made it as a very attractive method are increased sensitivity and rapidity. As the amplification is monitored this method obviates the need for gel electrophoresis. The main limitations of this technique are costlier instrumentation and requirement of expertise as the process is complex (Ghatak, 2020). Another format of qPCR is the quantitative reverse transcription PCR (RT-qPCR), which uses cDNA as the template and enables quantitation of mRNA levels of a target pathogen (Bustin 2000; Wong and Medrano 2005). RT-qPCR is the preferred method of choice for the detection of food borne viruses in food (Morillo *et al.*, 2012; Szabo *et al.*, 2015; Terio *et al.*, 2017).

3.3.3 Isothermal Amplification Based Methods

Recent developments in molecular biology have demonstrated the possibility of amplification of DNA in isothermal conditions without the use of thermocyclers (Gill and Ghaemi, 2008). Many such techniques have attracted food industry and were harnessed for the detection of food borne pathogens from meat and meat products. As most of these detection techniques can be visually interpreted, they obviate the necessity of gel electrophoresis thus making them rapid and simple (Hara-Kudo *et al.*, 2005; Momin *et al.*, 2020) one of the widely used techniques is Loop-mediated isothermal amplification (LAMP) assay (Priya *et al.*, 2018; Priya *et al.*, 2020). Notomi *et al.* (2000) developed LAMP which makes use of the hairpin-forming ability of primers and strand displacement activity of certain DNA polymerases. LAMP is an exceptionally specific technique as the amplification starts only after recognizing six regions of the target DNA. Many commercial LAMP based kits are available for detecting meat borne or other food borne pathogens like *Salmonella, Campylobacter, Listeria,* verotoxin-producing *Escherichia coli, etc.* (Mangal *et al.*, 2016). And the addition of reverse transcriptase has made possible to amplify DNA from

RNA (RT LAMP). Real time LAMP is also possible with a portable instrument to monitor the amplification real-time (Mashooq *et al.*, 2016). Foodborne viruses can be detected rapidly with real-time RT LAMP (Fukuda *et al.*, 2006; Yoneyama *et al.*, 2007). The limitations of LAMP based techniques are multiple primer requirement and carry-over or leftover contamination leading to false-positive results (Momin *et al.*, 2020). There are many other potential isothermal based techniques developed and deployed for detection of meat borne pathogens such as Nucleic acid sequence based amplification (NASBA) (Zhai *et al.*, 2019); recombinase polymerase amplification (RPA) (Kim and Lee, 2016); Single primer isothermal amplification (SPIA) (Yang *et al.*, 2020); polymerase spiral reaction (PSR) (Momin *et al.*, 2020), *etc.* The advantages of these techniques are equipment free nature, cost-effectiveness, rapidity and sensitivity even comparable with real-time PCR (Momin *et al.*, 2020).

3.3.4 DNA Microarray

DNA microarray is a small chip-like or glass device consisting of immobilized short (25 to 80 bp) hybridization probes that are complementary to the genes or markers of multiple target pathogens, which on reaction with sample DNA (dye labeled) bind and emit fluorescence. An important feature of the DNA microarray is the hybridization of the dye labeled target DNA fragments with the immobilized probes of the array (Rasooly and Herold, 2008; McLoughlin, 2011; Severgnini *et al.*, 2011; Lopez-Campos *et al.*, 2012). The signal intensity of the fluorescence is directly proportional to the target pathogen concentration (Lauri and Mariani, 2009). This technique has enabled high throughput and simultaneous detection of multiple targets. One of the first studies to deploy microarray technique in food borne pathogens has demonstrated simultaneous detection of 22 pathogens in broth culture (Wang *et al.*, 2007). However, very few commercial kits are available for applications in the food industry (Rasooly and Herold, 2008). The main limitation of this technique is that they demand greater expertise for chip fabrication, probe design and assay use. Moreover, costs involved in instrumentations are huge with the requirement of hybridization unit, chip reader and specialized software for interpretation (Rasooly and Herold, 2008; Lopez-Campos *et al.*, 2012).

3.4 Biosensor Based Methods

Biosensors are analytical devices which convert a biological response into an electrical signal. There are two major components of any biosensors: a bio-receptor to recognize or sense the target analyte and a transducer to convert recognition into a quantifiable electrical signal. Hence, classification of biosensors can be done based on the kind of bio-component used, mode or mechanism of signal transduction, *etc.* (Velusamy *et al.*, 2010; Lavecchia *et al.*, 2010; Singh *et al.*, 2016). The transduction elements are generally optical, piezoelectric, electrochemical and magnetoelastic and the recognition elements that are responsible for specificity can be enzymes, cell receptors, antigens, antibodies, aptamers, bacteriophages, nucleic acid probes and antimicrobial peptides (Bhunia, 2014). Biosensors are sensitive, rapid, cost effective, require minimal sample preparation and suitable for on-site detection. Different sensing or recognizing elements, transducing technologies and biosensor configurations have been reported for the detection of food borne pathogenic organisms (Singh *et al.*, 2013; Singh *et al.*, 2016). Biosensors that are categorized into optical, electrical, optochemical, electrochemical, piezoelectric, magnetoelastic and mass based sensors demonstrate utility in the food industry (Bhunia, 2014; Lv *et al.*, 2018).

Optical biosensors which work on the principles of absorbance, surface plasmon resonance, fluorescence, chemiluminescence, light polarization, total internal reflection and rotation are widely studied for the detection of the food borne pathogens and some are now even commercialized (Terry *et al.*, 2005).Among the optical based sensors, evanescent wave sensors, surface plasmon resonance (SPR) sensors, light-scattering sensors, Raman spectros-copy, Fourier transformed infrared (FTIR) sensors and hyperspectral imaging sensors are good in producing real-time or near real-time results (Bhunia, 2014). BARDOT (bacterial rapid detection using optical scattering technology), based on light scattering sensor system was developed to identify bacterial colonies on agar plates by passing a low power red diode (1 mW, 635 nm) laser beam through the bacterial colony resulting in unique scatter signatures for different bacterial groups (Bae *et al.*, 2007; Banada *et al.*, 2007). Evanescent-wave fluoroimmunosensors also called as fiberoptic biosensors produce evanescent waves after interaction of inward light with fluorophore-bound target elements on the surface waveguides (Huff *et al.*, 2012). The evanescent waves after undergoing total internal reflection inside the waveguide get real-time detected by a fluorometer. The strength of the signal is proportionate to the quantity of the analyte bound. Two such commercial systems namely RAPTOR and Analyte 2000 are available. RAPTOR is fully automated, compact and portable with a cartridge containing four waveguides precoated with specific antibody probes. It has also been used to detect food borne pathogens and toxins with a sensitivity range of 10^2–10^4 cells and nanogram amounts of toxins (Sharma and Mutharasan, 2013a; Ohk and Bhunia, 2013). Fibreoptic immunosensors have been used for the detection of *E. coli* O157: H7, *L. monocytogenes*, staphylococcal enterotoxins and *Salmonella* spp. from different food matrices. A combination of brief enrichment step or concentration using paramagnetic beads can improve the sensitivity and can provide results in 4 to 6 hours (Bhunia, 2014). In SPR biosensors, capture molecules (ligand) are halted on a gold film fixed on a prism. After interaction of light with the analyte-ligand complex, an alteration of the refractive index takes place at the gold-dielectric surface which generates an SPR response and as a result, a shift in the light wavelength occurs, which is then detected by a camera or a photodiode. SPR biosensors are more suitable for the detection of small molecules like bacteria, protein, chemicals, toxins, *etc.* (Bhunia, 2014). Commercial SPR systems like Spreeta and Biacore are presently available. SPR biosensors have been reported to detect *S.* Enteritidis and *C. jejuni* in chicken with a sensitivity of 10^6 CFU/mL and 10^3 CFU/mL, respectively (Wei *et al.*, 2007; Waswa *et al.*, 2007).

Raman spectroscopy is a nondestructive imaging technique that uses diode lasers to record rotational and vibrational properties. When the laser interacts with bacteria or other microbes, it produces Raman scattering, referred to as inelastic scattering. As Raman scatter signals are generally weak, amplification of signal strength is done using silver or gold nanoparticles conjugated to biosensing molecules like antibodies and it is termed as surface-enhanced Raman spectroscopy (Craig *et al.*, 2013). Raman spectroscopy has detected food borne pathogens like *Salmonella* spp., *E. coli* O157. H7 and *L. monocytogenes* with a limit of detection of 10^2–10^4 cells (Golightly *et al.*, 2009). Hyperspectral imaging (HSI) and HSI based sensors gather the spatial intensity data and generate a real time spatiospectral map of the item using a small-wavelength bandwidth from the visible to infrared ranges (Gowen *et al.*, 2007; Yoon *et al.*, 2011). It does not require any probes or labelling reagents, thus becoming suitable for real-time testing of food products. It has been used for testing beef, poultry and seafood (Elmasry *et al.*, 2012; Feng and Sun, 2012). HSI has also been used for testing food borne pathogens in different food matrices (Tao and Peng, 2014; Cheng and Sun, 2015; Lasch *et al.*, 2018).

Electrochemical biosensors are classified into amperometric or voltammetric (alteration in the quantified current), potentiometric (change in quantified voltage between the electrodes), and conductometric/impedimetric (when a shift in the capacity to transport charge) based on the parameters measured (Sharma and Mutharasan 2013a; Singh *et al.*, 2016). Many studies have applied electrochemical biosensors to detect food borne bacterial pathogens such as *Salmonella* spp., *Listeria* spp., *E. coli* O157: H7, *S.* Typhimurium *etc.*, in different food matrices (Gehring and Tu, 2005; Maalouf *et al.*, 2007; Tang *et al.*, 2010; Li *et al.*, 2011; Tolba *et al.*, 2012; Dong *et al.*, 2013; Joung *et al.*, 2013; Wang *et al.*, 2013). Piezoelectric quartz crystal microbalance (QCM) is one of the most important types of piezoelectric biosensor that has the advantage of label free detection, biocompatible electrodes for ligand immobilization, ease of use, real-time monitoring, *etc.* Piezoelectric based sensors have also been applied for detection of pathogens in various food matrices (Maraldo and Mutharasan 2007a; Maraldo and Mutharasan 2007b; Sharma and Mutharasan, 2013b). Magnetoelastic biosensors are made of ferromagnetic alloys and it oscillates in the time changing magnetic field at a fundamental frequency. These oscillations produce a magnetic flux or resonance that can be detected by a non-contacting pickup coil. When a target comes in contact with the sensor surface, the additional mass causes a shift in resonance frequency and the signal can be detected remotely by the pickup coil. Hence, magnetoelastic sensors are wireless and remote devices that may be very useful in remote monitoring of food borne pathogens in foods (Ruan *et al.*, 2003). Several studies have reported the use of magnetoelastic sensors for detection of food borne pathogens with no or minimal sample preparation (Guntupalli *et al.*, 2007; Chai *et al.*, 2012).

Matrix-assisted laser desorption ionization-time of flight mass spectrometry (MALDI-TOF-MS), produces spectral signatures for specific pathogens according to its protein profiles (Sandrin *et al.*, 2013). This method works by laser bombardment of an analyte entrenched and crystallized in an appropriate matrix resulting in ionization and vaporization of the analyte and flight of the ions in an electric field. A peptide mass fingerprint (PMF) is arrived by computing the ratio of the mass to charge of the vaporized charged particles and its time of flight. The identification of the analyte (pathogen) is done by relating the PMF with a standard database (Pavlovic *et al.*, 2013; Singhal *et al.*, 2015; Sauget *et al.*, 2017). The detection method is accurate, suitable for high-throughput screening and cost effective although inceptive investments are high. This technology demands a high degree of expertise and as this system needs purified isolates from samples, it may take days to deliver results (Pavlovic *et al.*, 2013; Ghatak, 2020). MALDI-TOF has been successfully applied for the detection of several food borne pathogens like *Salmonella, Campylobacter, Staphylococcus, etc.* (Dieckmann and Malorny, 2011). Nanotechnology has been coupled with biosensors as it has special properties that can facilitate detection. However further investigation is required in the areas of a reduced likelihood of nano-sized transducers to act with an analyte, toxicity of nanomaterials and manipulation of specific nano-sized substances. Developments in new recognition methods, new transducing materials, multiplex detection configurations, wireless devices, integrated biosensing systems and microfluidic techniques would uphold biosensors to turn out to be a mature and method of choice for application in the growing food industry.

4. Conclusions

In addition to the narrated detection methods, some other platforms are also available including next-generation sequencing (NGS) based platforms, automated microbial identification systems like VITEK, Phoenix 100, Sensititre *etc.* Many other novel methods are under development targeting reduction of detection time, improvement of sensitivity and limit of detection and more importantly attention is being given to develop culture independent methods. However, we suggest that focus should be given towards the development of more equipment free simple but reliable techniques which are instantly required for many resource-limited food testing laboratories located in most of the developing and underdeveloped countries.

REFERENCES

Alahi, M.E.E., and Mukhopadhyay, S.C. (2017). Detection methodologies for pathogen and toxins: a review. Sensors, 17, 1–20.

Anfossi, L., Di Nardo, F., Giovannoli, C., Passini, C., and Baggiani, C. (2013). Increased sensitivity of lateral flow immunoassay for ochratoxin A through silver enhancement. Anal. Bioanal. Chem. 405(30), 9859–9867.

Ayrapetyan, M., and Oliver, J.D. (2016) The viable but non-culturable state and its relevance in food safety. Curr Opin Food Sci., 8, 27–133. https: //doi. org/10.1016/j.cofs.2016.04.010.

Bae, E., Banada, P.P., Huff, K., Bhunia, A.K., Robinson, J.P., and Hirleman, E.D. (2007). Biophysical modeling of forward scattering from bacterial colonies using scalar diffraction theory. Appl. Opt., 46(17), 3639–3648.

Banada, P.P., Guo, S., Bayraktar, B., Bae, E., Rajwa, B., Robinson, J.P., Hirleman, E.D., and Bhunia, A.K. (2007). Optical forward-scattering for detection of *Listeria monocytogenes* and other *Listeria* species. Biosens. Bioelectron., 22(8), 1664–1671.

Bhunia AK. (2014). One day to one hour: how quickly can food borne pathogens be detected? Future Microbiol, 9, 935–46.

Blackburn, C.de W., and McClure, P.J. (2009). Campylobacter and Arcobacter. In: C.de W., Blackburn, and P.J. McClure (Eds.), Foodborne Pathogens: Hazards,

Risk Analysis and Control, second ed. Woodhead Publishing Ltd., Cambridge, UK, pp. 718–762.

Bustin, S.A. (2000). Absolute quantification of mRNA using real-time reverse transcription polymerase chain reaction assays. J. Mol. Endocrinol., 25,169-193.

Chai, Y., Li, S., Horikawa, S., Park, M.K., Vodyanoy, V., and Chin, B.A. (2012) Rapid and sensitive detection of *Salmonella typhimurium* on eggshells by using wireless biosensors. J Food Prot., 75, 631–636.

Charlermroj, R., Oplatowska, M., Kumpoosiri, M., Himananto, O., Gajanandana, O., Elliott, C.T., and Karoonuthaisiri, N. (2012). Comparison of techniques to screen and characterize bacteria-specific hybridomas for high-quality monoclonal antibodies selection. Anal Biochem., 421, 26–36.

Cheng, J.H., and Sun, D.W. (2015). Rapid quantification analysis and visualization of *Escherichia coli* loads in grass carp fish flesh by hyperspectral imaging method. Food Bioprocess Technol.,8, 951–959.

Ching, K.H., Lin, A., McGarvey, J.A., Stanker, L.H., and Hnasko, R. (2012). Rapid and selective detection of botulinum neurotoxin serotype-A and -B with a single immunochromatographic test strip. J. Immunol Methods, 380,23–29.

Cho, I., and Irudayaraj, J. (2013). In-situ immuno-gold nanoparticle network ELISA biosensors for pathogen detection. Int J Food Microbiol., 164,70–75.

Chunglok, W., Wuragil, D.K., Oaew, S., Somasundrum, M., and Surareungchai, W. (2011). Immunoassay based on carbon nanotubes-enhanced ELISA for *Salmonella enterica* serovar Typhimurium. Biosens Bioelectron, 26,3584–3589.

Craig, A.P., Franca, A.S., and Irudayaraj, J. (2013). Surface-enhanced Raman spectroscopy applied to food safety. Annu. Rev. Food Sci. Technol., 4, 369–380.

D'Aoust, J.Y., Sewell, A.M., and Greco, P. (1991). Commercial latex agglutination kits for the detection of food borne Salmonella. J. Food Prot., 54, 725–730.

Dieckmann, R., and Malorny, B. (2011). Rapid screening of epidemiologically important Salmonella enterica subsp.enterica serovars by whole-cell matrix-assisted laser desorption ionization-time of flight mass spectrometry. Appl. Environ. Microbiol., 77, 4136-4146.

Dinu, L.D., and Bach, S. (2013). Detection of viable but non-culturable *Escherichia coli* O157: H7 from vegetable samples using quantitative PCR with propidium monoazide and immunological assays. Food Control,31(2), 268–273.

Dong, J., Zhao, H., Xu, M., Ma, Q., Ai, and S. (2013). A label-free electrochemical impedance immunosensor based on AuNPs/PAMAM-MWCNT-Chi nanocomposite modified glassy carbon electrode for detection of *Salmonella* Typhimurium in milk. Food Chem., 141, 1980–1986.

Doyle, M. P., and Erickson, M. C. (2006). Emerging microbiological food safety issues related to meat. Meat Science, 74, 98–112.

Dwivedi, H.P., and Jaykus, L. (2011). Detection of pathogens in foods: the current state-of-the-art and future directions. Crit Rev Microbiol., 37,40–63.

EFSA-ECDC (European Food Safety Authority-European Centre for Disease Prevention and Control), (2015). The European Union summary report on trends and sources of zoonoses, zoonotic agents, and food-borne outbreaks in 2014. EFSA Journal, 13, 4329, pp.191.

Elmasry, G., Kamruzzaman, M., Sun, D.W., and Allen, P. (2012). Principles and applications of hyperspectral imaging in quality evaluation of agro-food products: a review. Crit. Rev. Food Sci. Nutr., 52(11), 999–1023.

Emerson, J.B., Adams, R.I., Román, C.M.B., Brooks, B., Coil, D.A., Dahlhausen, K., Ganz, H.H., Hartmann, E.M., Hsu, T., Justice, N.B., Paulino-Lima, N.B., Luongo, J.C., Lymperopoulou, D.S., Gomez-Silvan, C., Rothschild- Mancinelli, B., Balk, M., Huttenhower, C., Nocker, A., Vaishmpayan, P., and Rothschild, L.J. (2017) Schrödinger's microbes: tools for distinguishing the living from the dead in microbial ecosystems. Microbiome, 5,86. https: //doi.org/10.1186/ s40168-017-0285-3.

Feng, Y.Z., and Sun, D.W. (2012). Application of hyperspectral imaging in food safety inspection and control: a review. Crit. Rev. Food Sci. Nutr., 52(11), 1039–1058.

Feng, P. (1997). Impact of molecular biology on the detection of foodborne pathogens. Mol Biotechnol., 7(3), 267-278.

Foddai, C.G.A. and Grant, I.R. (2020). Methods for detection of viable food borne pathogens: current state-of-art and future prospects. Applied Microbiology and Biotechnology, 104, 4281–4288.

Fukuda, S., Takao, S., Kuwayama, M., Shimazu, Y., and Miyazaki, K. (2006). Rapid detection of norovirus from fecal specimens by real-time reverse transcription-loop-mediated isothermal amplification assay. J. Clin. Microbiol., 44, 1376-1381.

Gehring, A., and Tu, S. (2005). Enzyme-linked immunomagnetic electrochemical detection of live *Escherichia coli* O157: H7 in apple juice. J. Food Prot., 68, 146–149.

Ghatak S. (2020). Strategies for elimination of food borne pathogens, their influensive detection techniques and drawbacks. In: Meat Quality Analysis Advanced Evaluation Methods, Techniques, and Technologies Pages 267-286 https: // doi.org/10.1016/B978-0-12-819233-7.00015-X

Gill, P. and Ghaemi, A. (2008). Nucleic Acid Isothermal Amplification Technologies–A Review, Nucleosides, Nucleotides and Nucleic Acids, 27,3, 224-243. DOI: 10.1080/15257770701845204.

Golightly, R.S., Doering, W.E., and Natan, M.J. (2009). Surface-enhanced Raman spectroscopy and homeland security: a perfect match? ACS Nano, 3(10), 2859–2869.

Gowen, A.A., O'Donnell, C.P., Cullen, P.J., Downey, G., and Frias, J.M. (2007). Hyperspectral imaging – an emerging process analytical tool for food quality and safety control. Trends Food Sci. Technol., 18(12), 590–598.

Gracias, K., and McKillip, J. (2004). A review of conventional detection and enumeration methods for pathogenic bacteria in food. Can J. Microbiol., 50,883–890.

Guntupalli, R., Hu, J., Lakshmanan, R.S., Huang, T.S., Barbaree, J.M., and Chin, B.A. (2007).A magnetoelastic resonance biosensor immobilized with polyclonal antibody for the detection of *Salmonella typhimurium*. Biosens Bioelectron., 22 (7), 1474-1479.

Hara-Kudo, Y., Yoshino, M., Kojima, T., and Ikedo, M. (2005). Loop-mediated isothermal amplification for the rapid detection of *Salmonella*. FEMS Microbiol Lett., 253,155–161.

Huff, K., Aroonnual, A., Littlejohn, A.E.F., Rajwa, B., Bae, E., Banada, P.P., Patsekin, V., Hirleman, E.D., Robinson, J.P., Richards, G.P., and Bhunia A.K. (2012). Light-scattering sensor for real-time identification of *Vibrio parahaemolyticus, Vibrio vulnificus* and *Vibrio cholerae* colonies on solid agar plate. Microb. Biotechnol., 5(5), 607–620.

Jadhav, S.R., Shah, R.M., Karpe, A.V., Morrison, P.D., Kouremenos, K., Beale, D.J. and Palombo, E.A. (2018) Detection of Foodborne Pathogens Using Proteomics and Metabolomics-Based Approaches. Front. Microbiol., 9,3132. doi: 10.3389/ fmicb.2018.03132.

Jaffee, S., Henson, S., Unnevehr, L., Grace, D., and Cassou, E. (2019). The Safe Food Imperative: Accelerating Progress in Low- and Middle-Income Countries. World Bank, Washington, DC. https: //doi.org/10.1596/978-1-4648-1345-0.

Jasson, V., Jacxsens, L., Luning, P., Rajkovic, A., and Uyttendaele, M. (2010). Alternative microbial methods: An overview and selection criteria. Food Microbiol., 27, 710–730.

Joung, C., Kim, H., Lim, M., Jeon, T., Kim, H., and Kim, Y. (2013). A nanoporous membrane-based impedimetric immunosensor for label-free detection of pathogenic bacteria in whole milk. Biosens Bioelectron., 44, 210–215.

Joung, H.A., Oh, Y.K., and Kim, M.G. (2014).An automatic enzyme immunoassay based on a chemiluminescent lateral flow immunosensor. Biosens. Bioelectron., 53, 330–335.

Kaakoush, N.O., Castan˜o-Rodrý´guez, N., Mitchell, H.M., and Man, S.M. (2015). Global epidemiology of Campylobacter infection. Clinical Microbiology Reviews, 28, 687–720.

Kim, J.Y., and Lee, J. (2016). Rapid Detection of Salmonella Enterica Serovar Enteritidis from Eggs and Chicken Meat by RealTime Recombinase Polymerase Amplification in Comparison with the Two Step Real Time PCR. Journal of Food Safety. https: //doi.org/10.1111/jfs.12261

Knowles, T., Moody, R., and McEachern, M.G. (2007) European food scares and their impact on EU food policy, British Food Journal, 109(1), 43–67.

Labbe, R., and Juneja, V. (2017). *Clostridium perfringens*, p 235–242. In; C., Dodd, T.,Aldsworth, R.,Stein, D.,Cliver, and H. Riemann (Ed), Foodborne Diseases, 3rd ed. Elsevier, London, United Kingdom.

Lantz, P.G., Knutsson, R., Blixt, Y., Al Soud, W.A., Borch, E., and Radstrom P. (1998). Detection of pathogenic *Yersinia enterocolitica* in enrichment media and pork by a multiplex PCR: a study of sample preparation and PCR-inhibitory components. Int J Food Microbiol., 45,93–105.

Lasch, P., Stammler, M., Zhang, M., Baranska, M., Bosch, A., Majzner, K. (2018). FT-IR hyperspectral imaging and artificial neural network analysis for identification of pathogenic bacteria. Anal Chem., 90, 8896–8904.

Lauri, A., and Mariani, P.O. (2009). Potentials and limitations of molecular diagnostic methods in food safety. Genes Nutr., 4,1–12.

Lavecchia, T., Tibuzzi, A., and Giardi, M.T. (2010). Biosensors for Functional Food Safety and Analysis. In; Bio-Farms for Nutraceuticals: Functional Food and Safety Control by Biosensors edtd. M T. Giardi, G. Rea and B. Berra. chapter 20 Landes Bioscience and Springer Science + Business Media.

Law, J.W.F., Ab Mutalib, N.S., Chan, K.G., and Lee, L.H. (2014). Rapid methods for the detection of food borne bacterial pathogens: principles, applications, advantages and limitations. Front. Microbiol., 5, 1–19.

Lee, N., Kwon, K.Y., Oh, S.K., Chang, H.J., Chun, H.S., and Choi, S.W. (2014). A multiplex PCR assay for simultaneous detection of *Escherichia coli* O157: H7, *Bacillus cereus, Vibrio parahaemolyticus, Salmonella* spp., *Listeria monocytogenes,* and *Staphylococcus aureus* in Korean ready-to-eat food. Food borne Pathog Dis., 11, 574–580.

Li, D., Feng, Y., Zhou, L., Ye, Z., Wang, J., Ying, Y., Ruan, C., Wang, R., and Li, Y. (2011). Label-free capacitive immunosensor based on quartz crystal Au electrode for rapid and sensitive detection of *Escherichia coli* O157: H7. Anal Chim Acta., 687, 89–96.

Li, L., Mendis, N., Trigui, H., Oliver, J.D., and Faucher, S.P. (2014). The importance of the viable but non-culturable state in human bacterial pathogens. Front Microbiol., 5,258, 1–20.

Lianou, A., Panagou, E. Z., and Nychas, G.J. E. (2017). Meat safety–I foodborne pathogens and other biological issues. In F. Toldrá (Ed.). Lawrie's Meat Science, 8th ed., pp. 521–552. Springer. https: //doi.org/10.1016/B978-0-08-100694-8.00017-0.

Lianou, A., and Sofos, J.N. (2007). A review of the incidence and transmission of Listeria monocytogenes in ready-to-eat products in retail and food service environments. Journal of Food Protection, 70, 2172–2198.

Lleò, M.M., Pierobon, S., Tafi, M.C., Signoretto, C., and Canepari, P. (2000). mRNA detection by reverse transcription-PCR for monitoring viability over time in an *Enterococcus faecalis* viable but non culturable population maintained in a laboratory microcosm. Appl Environ Microbiol., 66, 4564–4567. https: //doi. org/10.1128/AEM.66.10.4564-4567.2000

Lopez-Campos, G., Martý´nez-Sua´rez, J.V., Aguado-Urda, M., and Lo´pez-Alonso, V. (2012). Microarray detection and characterization of bacterial foodborne pathogens. Food, Health, and Nutrition. Publisher; Springer, pp. 13-33.

Lv, M., Liu, Y., Geng, J., Kou, X., Xin, Z., and Yang, D. (2018). Engineering nanomaterials-based biosensors for food safety detection. Biosens Bioelectron., 106,122128. doi: 10.1016/j.bios.2018.01.049

Maalouf, R., Fournier-Wirth, C., Coste, J., Chebib, H., Saýkali, Y., Vittori, O., Errachid, A., Cloarec, J., Martelet, C., and Jaffrezic-Renault, N. (2007). Label-free detection of bacteria by ectrochemical impedance spectroscopy: comparison to surface plasmon resonance. Anal Chem., 79, 4879–4886.

Mandal, P.K., Biswas, A.K., Choi, K., and Pal, U.K. (2011). Methods for rapid detection of food borne pathogens: an overview. Am. J. Food Technol., 6, 87–102.

Mangal, M., Bansal, S., Sharma, S.K., and Gupta, R.K. (2016). Molecular detection of food borne pathogens: a rapid and accurate answer to food safety. Crit. Rev. Food Sci. Nutr., 56, 1568-1584.

Maraldo, D., and Mutharasan, R. (2007a). 10-minute assay for detecting *Escherichia coli* O157: H7 in ground beef samples using piezoelectric-excited millimeter-size cantilever sensors. J Food Prot., 70, 1670–1677.

Maraldo, D., and Mutharasan, R. (2007b). Detection and confirmation of staphylococcal enterotoxin B in apple juice and milk using piezoelectric-excited millimeter-sized cantilever sensors at 2.5 fg/mL. Anal Chem., 79,7636–7643.

Mashooq, M., Kumar, D., Niranjan, A. K., Agarwal, R. K., and Rathore, R. (2016). Development and evaluation of probe based real time loop mediated isothermal amplification for Salmonella: a new tool for DNA quantification. *J Microbiol Methods.*, 126, 24–29.

Mathusa, E.C., Chen, Y., Enache, E., and Hontz, L. (2010). Non-O157 Shiga toxin-producing *Escherichia coli* in foods. Journal of Food Protection, 73, 1721–1736.

Mattison, K., Bidawid, S., and Farber, J. (2009). Hepatitis viruses and emerging viruses. In: C.de W., Blackburn, and P.J., McClure (Eds.), Food borne Pathogens: Hazards, Risk Analysis and Control, second ed. Wood head Publishing Ltd., Cambridge, UK, pp. 891–929.

McLoughlin, K.S. (2011). Microarrays for pathogen detection and analysis. Brief Funct. Genomics, 10,342–353.

Meyer, T., Schirrmann, T., Frenzel, A., Miethe, S., Stratmann-Selke, J., Gerlach, G., Strutzberg-Minder, K., D¨ubel, S., and Hust, M. (2012). Identification of immunogenic proteins and generation of antibodies against *Salmonella* Typhimurium using phage display. BMC Biotechnol., 12,1–15.

Miller, R.S., Speegle, L., Oyarzabal, O.A., and Lastovica, A.J. (2008). Evaluation of three commercial latex agglutination tests for identification of *Campylobacter spp*. J. Clin. Microbiol., 46, 3546–3547.

Momin, K. M., Milton, A. A. P, Ghatak, S., Thomas, S. C., Priya, G. B., Das, S., Shakuntala,I., Sanjukta, R., Puro, K., and Sen, A. (2020). Development of a novel and rapid polymerase spiral reaction (PSR) assay to detect Salmonella in pork and pork products. Molecular and Cellular Probes, 50, 101510. https: //doi.org/10.1016/j.mcp.2020.101510.

Morillo, S.G., Luchs, A., Cilli, A., Timenetsky, M.D.S.T. (2012). Rapid detection of norovirus in naturally contaminated food: food borne gastroenteritis outbreak on a cruise ship in Brazil. Food Environ. Virol., 4,124–129. https: //doi. org/10.1007/s12560-012-9085-x

Nicolo, M.S., Gioffre, A., Carnazza, S., Platania, G., Silvestro, I.D., and Guglielmino, S.P.P. (2011). Viable but nonculturable state of foodborne pathogens in grapefruit juice: a study of laboratory. Food Borne Pathog. Dis., 8(1), 11–17.

Nogva, H.K., Drømtorp, S.M., Nissen, H., and Rudi, K. (2003) Ethidium monoazide for DNA-based differentiation of viable and dead bacteria by 52-nuclease PCR. Biotechniques, 34, 804–813. https: //doi.org/10.2144/03344rr02

Notomi, T., Okayama, H., Masubuchi, H., Yonekawa, T., Watanabe, K., Amino, N., and Hase, T. (2000). Loop-mediated isothermal amplification of DNA. Nucleic Acids Res, 28(12), E63. doi: 10.1093/nar/28.12.e63.

Ohk, S.H., and Bhunia, A.K. (2013). Multiplex fiber optic biosensor for detection of *Listeria monocytogenes, Escherichia coli* O157: H7 and *Salmonella enterica* from ready-to-eat meat samples. Food Microbiol., 33(2), 166–171.

Painter, J.A., Hoekstra, R.M., Ayers, T., Tauxe, R.V., Braden, C.R., Angulo, F.J., and Griffin, P.M. (2013). Attribution of foodborne illnesses, hospitalizations, and deaths to food commodities by using outbreak data, United States, 1998-2008. Emerging Infectious Diseases, 19, 407–415.

Pan, Y., and Breidt, F. Jr. (2007). Enumeration of viable Listeria monocytogenes cells by real-time PCR with propidium monoazide and ethidium monoazide in the presence of dead cells. Appl. Environ. Microbiol., 73, 8028–8031. https: //doi. org/10.1128/AEM.01198-07.

Papafragkou, E., D'Souza, D.H., and Jaykus, L. (2006). Foodborne viruses: prevention and control. In: S.M.,Goyal (Ed.), Viruses in Foods. Springer Science, Business Media, LLC, New York, NY, pp. 289–330.

Park, S., Kim, H., Paek, S., Hong, J.W., and Kim, Y. (2008). Enzyme-linked immuno-strip biosensor to detect *Escherichia coli* O157: H7. Ultramicroscopy, 108,1348–1351.

Pavio, N., Meng, X.-J., and Doceul, V. (2015). Zoonotic origin of hepatitis E. Current Opinion in Virology, 10, 34-41.

Pavlovic, M., Huber, I., Konrad, R., and Busch, U. (2013). Application of MALDI-TOF MS for the identification of food borne bacteria. Open Microbiol. J., 7, 135-141.

Peck, M.W. (2010). Clostridium botulinum. In: V.K., Juneja, and J.N., Sofos (Eds.), Pathogens and Toxins in Foods: Challenges and Interventions. ASM Press, Washington, pp. 31–52.

Priya, G. B., Agrawal, R. K., Milton, A. A. P., Mishra, M., Mendiratta, S. K., Luke, A., Inbaraj, S., Singh, B.R., Kumar, D., Ravi Kumar G.V.P.P.S., and Rajkhowa S. (2020). Rapid and visual detection of Salmonella in meat using invasin A (invA) genebased loop-mediated isothermal amplification assay. LWT - Food Science and Technology, 126, 109262.

Priya, G. B., Agrawal, R. K., Milton, A. A. P., Mishra, M., Mendiratta, S. K., Agarwal, R. K., Luke,A., Singh, B.R., and Kumar, D. (2018). Development and evaluation of isothermal amplification assay for the rapid and sensitive detection of *Clostridium perfringens* from chevon. Anaerobe, 54, 178–187. doi: 10.1016/j.anaerobe.2018.09.005

Priyanka, B., Patil, R.K., and Dwarakanath, S. (2016). A review on detection methods used for food borne pathogens. Indian J. Med. Res., 144, 327–338.

Rasooly, A., and Herold, K.E. (2008). Food microbial pathogen detection and analysis using DNA microarray technologies. Food borne Pathog. Dis., 5, 531-550.

Rhoades, J.R., Duffy, G., and Koutsoumanis, K. (2009). Prevalence and concentration of verotoxigenic *Escherichia coli, Salmonella* and *Listeria monocytogenes* in the beef production chain: a review. Food Microbiology, 26, 357–376.

Riley, L.W., Remis, R.S., Helgerson, S.D., McGee, H.B., Wells, J.G., Davis, B.R., Hebert, R.J., Olcott, E.S., Johnson, L.M., Hargrett, N.T., Blake, P.A., Cohen, and M.L. (1983). Hemorrhagic colitis associated with a rare Escherichia coli serotype. New England Journal of Medicine, 308,681–685.

Rodriguez-Lazaro, D., Cook, N., Ruggeri, F.M., Sellwood, J., Nasser,A., Nascimento, M.S.J., D'Agostino, M., Santos,R., Saiz,J.C., Rze¿utka, A., Bosch, A., Gironés,R. Carducci, A., Muscillo,M., Kovaè, K., Diez-Valcarce, M., Vantarakis, A., von Bonsdorff, C., Husman,A.M.R., Hernández,M., and van der Poel, W.H.M. (2012). Virus hazards from food, water and other contaminated environments. FEMS Microbiol. Rev., 36(4), 786–814.

Rossen, L., Norskov, P., Holmstrom, K., and Rasmussen, O.F. (1992). Inhibition of PCR by components of food samples, microbial diagnostic assays and DNA-extraction solutions. Int J Food Microbiol., 17, 37-45.

Ruan, C., Zeng, K., Varghese, O.K., and Grimes, C.A. (2003). Magnetoelastic immunosensors: amplified mass immunosorbent assay for detection of *Escherichia coli* O157: H7. Anal Chem., 75, 6494–6498.

Salazar, J.K., Wang, Y., Yu, S., Wang, H., and Zhang, W. (2015). Polymerase chain reaction-based serotyping of pathogenic bacteria in food. J Microbiol Methods, 110, 18–26.

Sandrin, T.R., Goldstein, J.E., and Schumaker, S. (2013). MALDI TOF MS profiling of bacteria at the strain level: a review. Mass Spec. Rev. 32(3), 188–217.

Sauget, M., Valot, B., Bertrand, X., and Hocquet, D. (2017). Can MALDI-TOF mass spectrometry reasonably type bacteria? Trends Microbiol., 25, 447-455.

Scallan, E., Hoekstra, R.M., Angulo, F.J., Tauxe, R. V. , Widdowson, M., Roy,S.L., Jones,J.L., and Griffin, P.M. (2011). Foodborne illness acquired in the United States – major pathogens. Emerg. Infect. Dis. 17(1), 7–15.

Scharff, R. (2012). Economic burden from health losses due to foodborne illness in the United States. J. Food Prot., 75(1), 123–131.

Schuchat, A., Deaver, K.A., Wenger, J.D., Plikaytis, B.D., Mascola, L., Pinner, R.W., Reingold, A.L., and Broome, C.V. (1992). Role of foods in sporadic listeriosis I. Case-control study of dietary risk factors. JAMA, 267, 2041–2045.

Scotcher, M.C., Cheng, L.W., and Stanker, L.H. (2010). Detection of botulinum neurotoxin serotype B at sub mouse LD50 levels by a sandwich immunoassay and its application to toxin detection in milk. PLoS ONE, 5, e11047.

Severgnini, M., Cremonesi, P., Consolandi, C., De Bellis, G., and Castiglioni, B. (2011). Advances in DNA microarray technology for the detection of foodborne pathogens. Food and Bioprocess Tech., 4,936–953.

Shan, S., Lai, W., Xiong, Y., Wei, H., and Xu, H. (2015). Novel strategies to enhance lateral flow immunoassay sensitivity for detecting food borne pathogens. J. Agric Food Chem., 63,745–753.

Sharma, H., and Mutharasan, R. (2013a). Review of biosensors for foodborne pathogens and toxins. Sensor Actuat B-Chem., 183,535–549.

Sharma, H., and Mutharasan, R. (2013b). Rapid and sensitive immunodetection of *Listeria monocytogenes* in milk using a novel piezoelectric cantilever sensor. Biosens Bioelectron., 45,158–162.

Sheridan, G.E.C., Masters, C.I., Shallcross, J.A., Mackey, B.M. (1998) Detection of mRNA by reverse transcription-PCR as an indicator of viability in *Escherichia coli* cells. Appl Environ Microbiol., 64, 1313–1318.

Shim, W., Choi, J., Kim, J., Yang, Z., Lee, K., Kim, M., Ha, S., Kim, K., Kim, K., and Kim, C. (2007). Production of monoclonal antibody against *Listeria monocytogenes* and its application to immunochromatography strip test. J Microbiol Biotechn., 17,1152–1161.

Singh, A., Poshtiban, S., Evoy, S. (2013). Recent advances in bacteriophage based biosensors for food-borne pathogen detection. Sensors, 13,1763–1786.

Singh, A.K., Bettasso, A.M., Bae, E., Rajwa, B., Dundar, M.M., Forster, M.D., Liu, L., Barrett, B., Lovchik, J., Robinson, J.P., Hirleman, E.D., and Bhunia, A.K. (2014). Laser optical sensor, a label-free on-plate *Salmonella enterica* colony detection tool. *MBio* 5(1), e01019-13.

Singh, P. K., Jairath, G., Ahlawat, S. S., Pathera, A., and Singh, P. (2016). Biosensor: An emerging safety tool for meat industry. Journal of Food Science and Technology, 53(4), 1759- 1765.

Singhal, N., Kumar, M., Kanaujia, P.K., and Virdi, J.S. (2015). MALDI-TOF mass spectrometry: an emerging technology for microbial identification and diagnosis. Front. Microbiol., 6, Article 791.

Skandamis, P., and Nychas, G.J.E. (2015). Pathogens: risks and control. In: F. Toldra (Ed.), Handbook of Fermented Meat and Poultry, second ed. John Wiley and Sons Ltd., Chichester, UK, pp. 389–412.

Sofos, J.N. (2008). Challenges to meat safety in the 21st century. Meat Science 78, 3–13.

Souii, A., M'hadheb-Gharbi, M.B., and Gharbi, J. (2016). Nucleic acid-based biotechnologies for food-borne pathogen detection using routine time-intensive culture-based methods and fast molecular diagnostics. Food Sci. Biotechnol., 25, 11–20.

Stiefel, P., Schmidt-Emrich, S., Maniura-Weber, K., and Ren, Q. (2015). Critical aspects of using bacterial cell viability assays with the fluorophores SYTO9 and propidium iodide. BMC Microbiol., 15, 36. https: //doi.org/10.1186/s12866-015-0376-x.

Sunwoo, H.H., Wang, W.W., and Sim, J.S. (2006). Detection of *Escherichia coli* O157: H7 using chicken immunoglobulin Y. Immunol Lett., 106, 191–193.

Szabo, K., Trojan, E., Anheyer-Behmenburg, H., Binder, A., Schotte, U., Ellerbroek, L., Klein, G., and Johone, R. (2015) Detection of hepatitis E virus RNA in raw sausages and liver sausages from retail in Germany using an optimized method. J Intern. Microbiol., 2015,149–156. https: //doi.org/10.1016/j.ijfoodmicro.2015.09.013.

Tang, D., Tang, J., Su, B., and Chen, G. (2010). Ultrasensitive electrochemical immunoassay of staphylococcal enterotoxin B in food using enzyme-nanosilica-doped carbon nanotubes for signal amplification. J. Agric. Food Chem., 58,10824–10830.

Tao, F., and Peng, Y. (2014). A method for nondestructive prediction of pork meat quality and safety attributes by hyperspectral imaging technique. J. Food Eng., 2014, 126, 98–106.

Terio, V., Bottaro, M., Pavoni, E., Losio, M.N., Serraino, A., Giacometti, F., Martella, V., Mottola, A., Di Pinto, A., and Tantillo, G. (2017). Occurrence of hepatitis A and E and norovirus GI and GII in ready-to-eat vegetables in Italy. Int. J. Food Microbiol. 249, 61–65. https: //doi.org/10.1016/j.ijfoodmicro.2017.03.008.

Terry, L.A., White, S.F., and Tigwell, L.J. (2005). The application of biosensors to fresh produce and the wider food industry. J. Agric. Food Chem., 53,1309–1316.

Tilden, J., Young, W., McNamara, A.M., Custer, C., Boesel, B., Lambert-Fair, M.A., Majkowski, J., Vugia, D., Werner, S.B., Hollingsworth, J., and Morris Jr., J.G. (1996). A new route of transmission for Escherichia coli O157: H7: infection from dry fermented salami. Public Health Briefs, 86, 1142–1145.

Tolba, M., Ahmed, M.U., Tlili, C., Eichenseher, F., Loessner, M.J., and Zourob, M. (2012). A bacteriophage endolysin-based electrochemical impedance biosensor for the rapid detection of *Listeria* cells. Analyst, 137,5749–5756.

Trevors, J.T. (2012) Can dead bacterial cells be defined and are genes expressed after cell death? J. Microbiol. Methods, 90, 25–28. https: //doi.org/10.1016/j.mimet.2012.04.004

Umesha, S., and Manukumar, H.M. (2018). Advanced molecular diagnostic techniques for detection of food-borne pathogens: current applications and future challenges. Crit. Rev. Food Sci. Nutr., 58, 84-104.

Vandenvelde, C., Vestraeten, M., and Van Beers. (1990). Fast multiplex polymerase chain reaction on boiled clinical samples for rapid viral diagnosis. J Virol Methods, 30, 215-227.

Velusamy, V., Arshak, K., Korostynska, O., Oliwa, K., and Adley, C. (2010) An overview of foodborne pathogen detection: in the perspective of biosensors. Biotechnol Adv., 28, 232–254.

Viegas, I, Santos JML, Barreto A, and Fontes MA. (2012). Meat Safety: A Brief Review on Concerns Common to Science and Consumers, Int. Jrnl. of Soc. of Agr. and Food,19(2), 275-288.

Wang, Y., Ping, J., Ye, Z., Wu, J., and Ying, Y. (2013). Impedimetric immunosensor based on gold nanoparticles modified graphene paper for label-free detection of *Escherichia coli* O157: H7. Biosens Bioelectron 49, 492–498.

Wang, Y., and Salazar, J.K. (2016). Culture-independent rapid detection methods for bacterial pathogens and toxins in food matrices. Comp Rev in Food Sci. and Food Safety, 15(1),183–205.

Wang, X.W., Zhang, L., Jin, L.Q., Jin, M., Shen, Z.Q., An, S., Chao, F.H., and Li, J.W. (2007). Development and application of an oligonucleotide microarray for the detection of food-borne bacterial pathogens. Appl. Microbiol. Biotechnol., 76,225–233.

Waswa, J., Irudayaraj, J., and DebRoy, C. (2007). Direct detection of E. coli O157: H7 in selected food systems by a surface plasmon resonance biosensor. LWT-Food Sci. Technol., 40, 187–192.

Wei, D., Oyarzabal, O.A., Huang, T., Balasubramanian, S., Sista, S., and Simonian, A.L. (2007). Development of a surface plasmon resonance biosensor for the identification of *Campylobacter jejuni*. J. Microbiol. Methods, 69,78–85.

Wong, M. L., and Medrano, J. F. (2005). Real-time PCR for mRNA quantitation. Biotechniques, 39, 1, 75-85.

Xiao, L., Zhang, K., Wang, H.H. (2012). Critical issues in detecting viable Listeria monocytogenes cells by reverse real-time transcriptase PCR. J Food Prot., 75,512–517. https: //doi.org/10.4315/0362-028X.JFP-11-346

Xu, J., Zhang, P., Zhuang, L., Zhang, D., Qi, K., Dou, X., and Gong, J. (2019). Multiplex polymerase chain reaction to detect Salmonella serovars Indiana, Enteritidis, and Typhimurium in raw meat. Journal of Food Safety, 39(5). https: //doi.org/10.1111/jfs.12674

Yang, H., Qu, L., Wimbrow, A.N., Jiang, X., Sun, Y. (2007). Rapid detection of *Listeria monocytogenes* by nanoparticle-based immunomagnetic separation and real-time PCR. Int. J. Food Microbiol., 118,132–138.

Yang, Q., Xu, H., Zhang, Y., Liu, Y., Lu, X., Feng, X., Tan, J., Zhang, S., and Zhang, W. (2020). Single primer isothermal amplification coupled with SYBR Green II: Real-time and rapid visual methods for detection of *Listeria monocytogenes* in raw chicken, LWT - Food Science and Technology, doi: https: //doi.org/10.1016/j.lwt.2020.109453.

Yoneyama, T., Kiyohara, T., Shimasaki, N., Kobayashi, G., Ota, Y., Notomi, T., Totsuka, A., and Wakita, T. (2007).Rapid and real-time detection of hepatitis A virus by reverse transcription loop-mediated isothermal amplification assay. J Virol Methods., 145, 162-168. 10.1016/j.jviromet.2007.05.023.

Yoon, S.C., Park, B., Lawrence, K.C., Windham, W.R., and Heitschmidt, G.W. (2011). Line-scan hyperspectral imaging system for real-time inspection of poultry carcasses with fecal material and ingesta. Computers Electron. Agri., 79(2), 159–168.

Zhai, L., Liu, H., Chen, Q. Lu https: //pubmed.ncbi.nlm.nih.gov/30637640/-affiliation-1, Z., Zhang, C., Lv, F., and Bie, X. (2019). Development of a real-time nucleic acid sequence–based amplification assay for the rapid detection of Salmonella spp. from food. Braz. J. Microbiol. 50, 255–261. https: //doi.org/10.1007/s42770-018-0002-9

Zhao, X., Lin, C.W., Wang, J., and Oh, D.H. (2014). Advances in rapid detection methods for foodborne pathogens. J. Microbiol Biotechnol., 24, 297–312.

Chapter 11

Novel Approaches in Meat Preservation

Swati Gupta

Assistant Professor, Department of Livestock Products Technology, College of Veterinary Science and A.H., Navsari Agricultural University, Navsari, Gujarat

e-mail: gswati24@yahoo.com

ABSTRACT

Meat has been a staple food for human beings since time immemorial. Demand for healthy, convenient and safe meat products with natural flavour and fresh appearance has led to the development of new preservation technologies. Novel meat preservation technologies have the potential to address the consumer demands. Novel preservation technologies such as ultrasound, cold plasma, ohmic heating, pulsed light technology, high pressure processing, irradiation, biopresevation etc. provide new opportunities to develop consumer driven integrated strategies. These novel preservation technologies are safe, easy and environment-friendly, however, there is the need for more focused research to close the gap between laboratory scale interventions and the meat processing industry.

Keywords: *Meat, Novel, Preservation.*

1. Introduction

Man depends on products of plant and animal origin for food. The quality and safety of food products are the most important factors for consumers. Meat is an integral part of the human diet and provides all essential nutrients required for growth, maintenance and health perseverance. Meat is a perishable food and its preservation in a sound and safe condition is a challenge to mankind since long. The diverse nutrients in meat give an ideal environment for the growth of meat spoilage microorganisms and common food-borne pathogens. The meaning of the word "preserve" is to keep safe, retain quality and prevent spoilage. Consumers demand healthy, safe, high quality meat products with natural flavour and fresh appearance. To harmonize all these demands without compromising safety, it is necessary to implement new preservation technologies in the meat industry. The main purpose of meat preservation is to prevent it from microbial spoilage and to minimize oxidation as well as enzymatic spoilage. The basic principles of preservation techniques are:

- Prevention or delay in the growth of microorganisms
  - (i) Avoiding invasion of microorganisms
  - (ii) Inhibiting the growth and activity of microorganisms
  - (iii) Killing or destruction of microorganisms
- Prevention or delay of self decomposition
  - (i) Inactivation of naturally existing inherent enzymes in meat
  - (ii) Prevention or delay of chemical reactions

2. Methods of Meat Preservation

Several techniques have been utilized for preservation of meat since primitive times. Traditionally, meat preservation is based on control by temperature, by moisture and by inhibitory processes. Drying, chilling, freezing, curing, smoking and canning have been used for meat preservation from many years. Drying is one of the oldest methods of meat preservation. Thermal processing is the most widely used method for destroying microorganisms and imparting food with a lasting shelf life. Conventional preservation methods may result in some undesired changes, such as loss of colour, flavour, texture and nutritional value. Increasing consumer demand for food with a high nutritional value and a 'fresh like' taste has led to the development of new mild processes and novel approaches. The main aim of novel preservation techniques is to kill or inactivate undesirable microorganisms and enzymes without damaging nutritional and sensory properties. Novel preservation technologies have the potential to address the consumer demands. Some of the novel preservation technologies for meat and meat products have been discussed here.

2.1 Ultrasound

Ultrasound is a form of energy in a solid or fluid with a frequency greater than the maximum frequency audible to human ear (16-18 kHz). The lowest ultrasonic frequency is 20 kHz and the upper limit is considered to be 5MHz for gases and 500MHz for liquids and solids (Mason *et al.*, 1988).

Ultrasound is divided into two categories *i.e.* low and high power energy (Awad *et al.*, 2012; Jayasooriya *et al.*, 2004). Low power ultrasound involves the use of high frequencies (2-10 MHz) at low power (up to 10 watts) where as high power ultrasound involves the use of low frequencies (20-100 KHz) at high power (hundreds of watts to ten kilowatts) and also known as power ultrasonics.

High power ultrasonic applications generally depend on complex vibration in the propagating media, which produces cavitation in biological tissue. Physical, mechanical or chemical effects of ultrasonic waves alter material properties through immense pressure, shear and temperature gradient in the medium. High pressure, shear and temperature gradient can destroy cell membranes and DNA, thus leading to cell death (Chen *et al.*, 2012). Sometimes, the ultrasound waves can not propagate through the sample, because of the presence of small gas bubbles in a sample.

Efficiency of ultrasound varies due to differences in ultrasound parameters such as experimental design, frequency, time and temperature of application, treatment protocols, type of meat, species and strains of microorganisms. Furthermore, the effect of ultrasound depends on the size, shape of the microorganisms, the type of cells as well as their physiological states. Bigger cells are more sensitive than smaller ones. Coccal shaped are more resistant than rod shaped bacteria. Younger cells are more sensitive than older ones (Piyasena *et al.*, 2003). Ultrasound has potential to inactivate pathogenic microorganisms such as *Salmonella* spp., *Escherichia coli, Staphylococcus aureus, Listeria monocytogenes* (Rastogi, 2011). Kordowska-Wiater and Stasiak (2011) reported reduction of Gram-negative bacteria (*Salmonella anatum, E. coli, Proteus* spp. and *Pseudomonas fluorescens*) from the surface of chicken skin after ultrasound treatment. Herceg *et al.* (2013) studied the effect of high-intensity ultrasound on the inactivation of suspensions containing *E. coli, Staphylococcus aureus, Salmonella* spp., *Listeria monocytogenes and Bacillus cereus* and found more inactivation after longer periods of treatment. Ultrasound and steam treatment to contaminated chicken carcasses in a processing line resulted in reduction of *Campylobacter* spp. (Hanieh *et al.*, 2014).

Ultrasound has a significant effect on the texture and maturation of meat by weakening myofibrillar and connective tissues (Alarcon-Rojo *et al.*, 2015). Chang *et al.* (2012) applied power ultrasound (40 kHz, 1500W) on semitendinosus beef muscle and found no significant effect on colour and heat-insoluble collagen, however, the thermal stability properties of collagen and the texture were affected. Chicken breast muscles treated with ultrasound (24 kHz for 15 s at 12W cm^{-2}) showed reduction in the shear force value (Xiong *et al.*, 2012).

Ultrasound is a precise, non-destructive, non-invasive and nonchemical 'green' technology for the meat industry. Further experiments are needed to design effective ultrasonic systems for different types of meat and meat products.

2.2 Cold Plasma

Cold plasma is a novel preservation technology that uses energetic, reactive gases to inactivate contaminating microbes on meat. The effect of sterilization using plasma was first documented in the year 1960 and the technology was patented in 1968 by Menashi (Afshari and Hosseini, 2014). Plasma produced at an ambient temperature is referred as cold plasma because the temperature in plasma reactor stays near the room temperature (Banu *et al.*, 2012). Cold plasma is a mixture of partially ionized gas that contains different reactive species such as positive and negative ions, free radicals, photons and reactive neutral species such as reactive oxygen and nitrogen species (Stoica *et al.*, 2014). It is also called as the fourth state of matter which starts from solid to liquid state, liquid to gas and finally gas to plasma (Misra, 2011).

Plasma is considered as a distinct state of matter because it does not have a regular shape or volume and it can form filaments or beams under magnetic fields. The properties of plasma generally depend on the power and type of gas. The most common method of generating and sustaining cold plasma is by applying an electric field to a neutral gas. Noble gases *i.e.* helium or argon are used to generate plasma. Cold atmospheric plasma can be generated using direct current (DC) or alternating current (AC) power supplies at low and high gas pressures (Bardos and Barankova, 2010). A range of different frequencies for AC power supplies from low kHz frequencies, radio frequencies to microwave frequencies have been used to generate plasma discharges. Devices that have been used for generation of plasma at atmospheric pressure are the corona discharges, gliding arc discharge, micro hollow cathode discharges, dielectric plasma needle barrier discharge, atmospheric pressure plasma jet. When there is collision between electrons or photons with the neutral atoms and molecules in the feed gas, electrons and ions are produced in the gas phase.

The plasma treatment can effectively kill or inactivate wide range of microorganisms *viz.*, bacteria, yeast, molds and viruses. When the microorganisms are exposed to radical bombardment on the surface of the cell with great intensity, provokes lesions on the surface of the cell leads to destruction of the living cell rapidly. Lesion formation is due to accumulation of electrostatic forces on the exterior surface of the living cell (Dey *et al.*, 2016). Nitrogen and oxygen gas plasma are good sources of reactive oxygen-based and nitrogen-based species such as O, O_2, O_3, OH, NO, NO_2 (Cabiscol *et al.*, 2000; Mai-Prochnow *et al.*, 2014). Reactive species have the most important role in inactivation of microorganisms. They cause oxidative effect on the outer surface of microbial cells. In addition to reactive species, UV photons can modify the DNA of the microorganisms. The exact mechanism of interaction between the microorganisms and species is still unclear. It may be due to direct permeabilization of the cell membrane and cell wall or critical damage of intracellular proteins and nucleic acids from oxidative species damage. Certain reactions such as oxidation and peroxidation occur inside and outside the cell which is mainly catalyzed by the plasma ions (Dobryinin *et al.*, 2009).

The efficiency of cold plasma depends on type of food product, design, voltage, gas pressure, gas composition, distance of microorganisms from the discharge glow, characteristics of microorganisms like load, type and physiological state (Ziuzina *et al.*, 2014; Stratakos and Koidis, 2015).

Kim *et al.* (2011) studied the effect of atmospheric pressure plasma in bacon using helium and a mixture of helium and oxygen. They reported 1.89 and 4.58 log reductions in the number of total aerobic bacteria respectively. The efficacy of gas plasma treatment is greatly affected by surface topography. Noriega *et al.* (2011) examined chicken skin and muscle inoculated with *Listeria innocua* and found more than 3 log reductions on muscle in 4 minutes cold plasma treatment and 1 log reduction on skin in 8 minutes cold plasma treatment under optimal conditions. Rod *et al.* (2012) investigated the application of cold atmospheric pressure plasma for decontamination of inoculated *Listeria innocua* in Bresaola (sliced ready-to-eat meat product) and found reduction of *Listeria innocua.* Ulbin-Figlewicz *et al.* (2015)

reported reduction in the population of psychrotroph bacteria, total microorganisms, yeasts and moulds to 2.7, 2.96, and 3.08 log cfu/cm^2, respectively after helium plasma treatment for 10 minutes. Bae *et al.* (2015) found 99 per cent reduction of Murine norovirus (MNV-1) titer and 90 per cent reduction of hepatitis A virus (strain HM-175) titer following 5 minutes of APP jet treatment without concomitant changes in meat quality.

Cold plasma is a novel, ultra-fast, environmentally friendly technique as increase in product temperature is less than 5 °C and it requires short treatment time. Bactericidal molecules can be generated solely from air and reactive gas species revert back to original gas within minutes to hours after treatment. However, limitation of plasma reactive species is in penetration of foods (Song *et al.*, 2009).

2.3 Ohmic Heating

Ohmic heating has drawn interest to ensure the quality and safety of meat products. Ohmic heating is defined as a process in which heat is internally generated by the passage of alternating electrical current through a body such as a food system due to electrical resistance. In this process, heat is directly dissipated into the food material rather than conduction or convection. The heating occurs in the form of internal energy transformation (from electric to thermal) within the material (Sastry and Barach, 2000). It offers the potential for safer meat products by effectively inhibiting microbial growth through uniform temperature distribution in the product (Ozkan *et al.*, 2004; Sastry and Li, 1996).

Ohmic heating is volumetric and heats both phases simultaneously. Primarily, microbial inactivation in ohmic heating is thermal in nature. The presence of electric field in ohmic heating may cause mild non-thermal lethal effect (Sun *et al.*, 2008). The low frequency (usually 50–60 Hz) allows cell walls to build up charges and form pores (FDA-CFSAN, 2000). Excessive exposure causes cell death due to the leakage of intracellular components through the pores (Lee and Yoon, 1999). The size of pores formed may vary depending on the strength of the electric field. Electroporation occurs because the cell membrane has a specific dielectric strength, which can be exceeded by the electric field. The dielectric strength of a cell membrane is related to the amount of lipid present in the membrane itself.

Ohmic heating is influenced by several factors, such as electrical conductivity, electrodes, field strength, ionic concentration and particle size. The rate of ohmic heating is directly proportional to the square of electric field strength and electrical conductivity (Sastry and Palaniappan, 1992). Electrical conductivity is affected by the temperature, ionic strength, microstructure of the food material and field strength of the experimental setup (Parrott, 1992). The electrical conductivity of meat increases linearly with the temperature during ohmic heating at constant voltage gradient. Muscle fibre direction, fat content and the type of meat affect the electrical conductivity (Sarang *et al.*, 2008; Bozkurt and Icier, 2010). Heterogeneous structure of meat affects the uniform distribution of heat. Compounds with poor conductivity, especially the fat in meat products, do not generate heat at the same rate as muscle thus creating cold spots (Shirsat *et al.*, 2004). Whole meat or processed meat with low fat showed higher electrical conductivity and reduced ohmic heating period.

Greater the ionic concentration, faster is the heating rate. During heating of biological tissue electrical conductivity increases due to increase in ionic mobility, structural changes in the tissue such as cell wall breakdown, expulsion of non-conductive gas bubbles, softening and lowering in aqueous phase viscosity (Sasson and Monselise, 1977). The conductivity increases in the presence of ionic substances while it decreases in the presence of non polar constituents (Sastry and Palaniappan, 1992). Salted meat and minced beef showed the highest electrical conductivity level compared with the intact meat heated in either a parallel or a perpendicular direction (Zell *et al.*, 2009).

In ohmic heating, heating rate is affected by particle size, particle orientation and location (Kim *et al.*, 1996). As the particle size increases, the heating rate decreases. The effect of orientation on the conductivity can be ignored for small particles like emulsions, colloids (less than 5 mm), but it has a significant influence on electrical properties and the relative heating rates for larger particles (15–25 mm) (Mckenna *et al.*, 2006). Ohmic heating has proved successful in heating meat emulsion and meat batters (Piette *et al.*, 2004; Brunton *et al.*, 2006). Piette *et al.* (2004) reported the reduction of 9.06 $\log_{10}$cfu/g *Enterococcus faecalis* with a core temperature of 80°C in bologna sausages. Ohmic cooking is an alternative for continuous sterilization of foods containing particulates like meatballs and other meat products (Yildiz turp *et al.*, 2013).

Better energy efficiency, higher yields, shorter processing times, environment friendly, maintaining the colour, nutritional value and the safety of the meat products are the advantages of ohmic cooking. On the other hand, it is more costly than conventional methods of processing.

2.4 Pulsed Light Technology

Pulse light technology is a nonthermal method of food preservation that opens up new possibilities in meat and meat products preservation. The pulse light technology basically comprises the emittance of short-duration, high-power pulses of light from an inert-gas flash lamp, which causes microbial cell deterioration. The flash lamp converts 45 per cent to 50 per cent of the input electrical energy to pulsed radiant energy. Inert gas such as xenon or krypton is used in flash lamps. Generally, Xenon is preferred gas for the microbial inactivation because of its higher conversion efficiency. FDA has also approved Xenon lamps for the treatment of food (FDA, 1996).

The emitted broad spectrum radiation of xenon lamps distribution is 25 per cent UV, 45 per cent visible light and 30 per cent infrared. The high current discharge through gas filled flash lights results in millisecond flashes of broad spectrum white light which is about 20000 times more intense than sun light. Therefore, a few flashes in short exposure time are sufficient for the pasteurizing or sterilizing treatment. The rate of flashes is 1-20 flashes/second (Abida *et al.*, 2014).

Different mechanisms have been proposed to explain the inactivation effect of pulsed light on microorganisms. The lethal effect of pulsed light may be attributed to its rich broad spectrum ultraviolet content. The photochemical effect causes DNA damage such as clonogenic death: double/single strand DNA breaks, formation of pyrimidone dimers in DNA (Takeshita *et al.*, 2003; Cheigh *et al.*, 2012). The pulsed UV light causes genetic damage to cells such as photolysis, loss of colony-forming ability and destruction of nucleic acid. In absence of UV light, the photothermal effect is the primary cause of destruction. Photothermal mechanism is due to temperature increase determined by heat dissipation of light pulses penetrating the product. Photophysical effect such as changes in ion flow, increased cell membrane permeability, depolarization of cell membrane and structural deformation of cells and spores are also responsible for antimicrobial effects (Ohlsson and Bengtsson, 2002).

The inactivation efficacy of pulsed light depends upon type of microorganism, the distance from the light source, interaction between light and the substrate or between light and the microbial cells, intensity and the number of pulses delivered. Smaller bacteria are more resistant to pulsed light than larger ones (Wekhof, 2000). Some authors reported a significantly higher resistance of Gram-positive in comparison to Gram-negative bacteria in their experiments. Mucoid as well as pigment-forming bacteria exhibit higher resistance to pulse light *e.g. Pseudomonas aeruginosa*. Fungi are more resistant to pulse light than bacteria and spores are more resistant than viable cells (Rowan *et al.*, 1999; Anderson *et al.*, 2000; Henrich *et al.*, 2016)

Shorter wavelengths provide deeper penetration into the food than longer wavelengths (Dagerskog and Osterstrom, 1979). Interaction between light and the substrate or between light and the microorganisms is very important for the efficacy of the pulsed light treatment. The composition of the medium and the wavelength of the incident light decide the reflection, refraction, scattering and absorbing of the light. As the distance from light source and depth of the substrate increases, the absorption and scattering diminishes. The incident beam of light undergoes refraction due to difference in the optical density between the substrate and the surrounding air (Rajkovic *et al.*, 2010). Suitable packaging materials for pulsed light treatments are polyethylene, polypropylene, polyolefines and polyvinyl chloride (Keklik *et al.*, 2010)

Keklik *et al.* (2009) treated chicken frankfurters with pulsed UV light for 5, 15, 30, 45 and 60 sec from the quartz window and reported reduction between 0.1 to 1.9 log cfu/cm^2 in vacuum packaged samples and 0.3 to 1.9 log cfu/cm^2 in unpackaged samples. High-power pulsed light of 1,000 pulses, treatment duration 200 seconds and total ultraviolet light dose 5.4 Joule/cm^2 was found to reduce viability of *Salmonella typhimurium* and *Listeria monocytogenes* inoculated on the surface of chicken by 2-2.4 $\log_{10}$ cfu/ml (Paskeviciute *et al.*, 2011). Organoleptic properties of treated chicken did not detect any changes in raw or cooked chicken meat. Hierro *et al.* (2011) inoculated *Listeria monocytogenes* artificially in vacuum packaged ham and bologna slices and treated with pulsed light with fluences of 0.7, 2.1, 4.2 and 8.4 Joule/cm^2. It was found that the microbial load of ham and bologna slices reduced by 1.78 cfu/cm^2 and 1.11 cfu/cm^2 respectively. The combination of pulse light and vacuum packaging showed shelf-life extension of 30 days in ham compared with only vacuum packaging. Haughton *et al.* (2011) evaluated the efficacy of high intensity light pulse technology (3 Hz, maximum of 505 J/pulse and pulse duration of 360 µs) for the decontamination of raw chicken and found reduction in *Campylobacter jejuni, E. coli* and *Salmonella enteritidis*. Ganan *et al.* (2013) tested efficacy of pulsed light to reduce *Listeria monocytogenes* and *Salmonella enterica* serovar typhimurium on the surface of ready-to-eat dry cured meat products (*salchichon* and loin) and reported reductions between 1.5 and 1.8 log_{10} cfu/cm^2 for both microorganisms.

Pulsed light technology offers several advantages such as process flexibility, minimum space requirement, no adverse effect on nutrient content, safe, easy to use and environment friendly. The limitation of pulse light technology is complex surfaces, where it is difficult to illuminate or reach all parts of the surface to obtain a sterilizing effect (Ohlsson and Bengtsson, 2002). The potential of pulsed light on meat is still under investigation and need for more focused research to close the gap between laboratory scale interventions and the meat processing industry.

2.5 High Pressure Processing (HPP)

High pressure processing is a potential and novel preservation method for the meat industry. High pressure processing is carried out with intense pressure in the range of 100-1000 MPa (Yordanov and Angelova, 2010). It was discovered by Bert H. Hite in 1899 and thoroughly investigated in food since the early 1980s.

The effect of HPP processing on chemical and microbiological changes in food is governed by the Le Chatelier's principle, Pascal's principle, and Microscopic ordering principle. High pressure stimulates phase transitions, chemical reactions and changes in molecular configuration that are accompanied by a decline in volume. The break in ionic bonds due to high pressure leads to decline in volume due to the electrostriction of water. High pressure modifies only noncovalent bonds and does not affect covalent bonds, therefore, natural qualities of the products is maintained.

A high-pressure system consists of a high pressure vessel and its closure, pressure generation system, temperature control device and material handling system. The pressure vessel is the most important component of HPP (Hendrick *et al.*, 2005; Yordanov and Angelova, 2010). Flexible or partially rigid packaging material such as polyethylene, polypropylene, and ethylene vinyl alcohol films are commonly used in high pressure processing. In addition, coextruded films with polymeric barrier layers (Polyvinylidene chloride, ethylene vinyl alcohol, polyvinyl alcohol, polyamide), adhesive laminated films on a polymer base, or inorganic layer such as aluminium foil are also used. Food products in flexible packaging are loaded into trays or containers that are automatically conveyed into the high pressure chamber filled with pressure transmitting fluid. Pressure-transmitting fluids are used to transmit uniform pressure in the products. Some of the factors involved in selecting the medium are ability of the pressure-transmitting fluid to protect inner vessel surface from corrosion, the process temperature range and the viscosity of the fluid under pressure. Most commonly used pressure-transmitting fluids are water, castor oil, ethanol, glycol, silicone oil, sodium benzoate solutions and inert gases (Yaldagard *et al.*, 2008).

High pressure can be generated by direct or indirect compression or by heating the pressure medium. In direct compression, the pressure medium is directly pressurized by a piston, driven at its large-diameter end by a low-pressure pump.

In indirect compression, a high-pressure intensifier is used to pump the pressure medium from the reservoir into the closed vessel until the desired pressure is reached. In heating pressure medium, pressure is generated by temperature induced expansion of pressure medium. It can be used only for those applications where high pressure in combination with elevated temperature is required (Thakur and Nelson, 1998). There is uniform distribution of pressure throughout food material irrespective of size and shape due to isostatic condition and the shape and dimensions of the food material remains same in the container/packet (Han *et al.*, 2011).

High pressure processing is effective in inactivation of pathogenic and spoilage microorganisms. High-hydrostatic pressure treatments alter membrane functionalities and perturb the physicochemical balance of the cell (Perrier-Cornet *et al.*, 1995). In addition, high pressure causes changes in cell morphology and biochemical reactions, protein denaturation and inhibition of genetic mechanisms *e.g.* DNA replication and transcription. The denaturation of key enzymes and the disruption of ribosomes may also be responsible for microbial inactivation

Applied pressure, temperature, time and product parameters such as pH, a_w, salt content and the presence of other antimicrobials are the critical factors for high pressure processing that determine the lethality of microorganisms in particular food matrix (Rendueles *et al.*, 2011; Bajovic *et al.*, 2012). Generally, Gram-positive bacteria are more resistant to pressure than Gram-negative bacteria, moulds and yeasts. Cells in the exponential phase are more sensitive to pressure treatments than in the log or stationary phases of growth (Balny and Masson, 1993). Bacterial spores are very resistant to high hydrostatic pressure. Hydrostatic pressure above 100 MPa causes rapid inactivation of many vegetative bacteria. High pressure treatment of meat at 200 MPa can prevent trichinellosis and be considered as an effective substitute for heating. High pressure treatment (300-600 MPa) for a short period of time can be used as pasteurization to inactivate the vegetative pathogenic and spoilage microorganisms (Campus, 2010). For sterilization the range over 600 MPa and combination with high temperature is needed (Cheftel, 1995). Pressure of 800 MPa for 60 min at 60°C decreased the spore count of *Bacillus stearothermophilus* from 10^6 to 10^2/ml (Hayakawa *et al.*, 1989).

Koseki *et al.* (2007) spiked *Listeria monocytogenes* at 5 $\log_{10}$cfu/gm in sliced cooked ham and found that the counts were below the detection limit after 500 MPa for 10 min treatment. Jofre *et al.* (2009) reported effective inactivation of pathogenic and spoilage organisms such as *Listeria monocytogenes, Salmonella enterica, Staphylococus aureus, Yersinia enterocolitica and Campylobacter jejuni, E. coli* and yeast *Debaryomyces hansenii* after pressure treatment at 600MPa for 6 min at 31°C in RTE meat products (Cooked ham, dry cured ham and marinated beef loin).

Sensory acceptance of high pressure treated meat products depend on colour, texture, aroma, and taste modifications induced by the process. Problems of sensory acceptance occur with high pressure treated raw meat is mainly due to visible colour changes. Above 200 MPa pressure results in whitening effect due to increase in lightness of raw meat colour. It may be due to protein coagulation with a resulting loss of solubility of sarcoplasmic and/or myofibrillar proteins that affect structure and surface properties, globin denaturation and heme group release or displacement (Carlez *et al.*, 1995; Goutefongea *et al.*, 1995). The extent of change in colour is also influenced by time of high pressure exposure, fat content *etc*. High fat meat products (20 to 25 per cent) show greater colour changes as compared to low fat content (9 per cent) (Carballo *et al.*, 1997; Jimenez-Colmenero *et al.*, 1997; Simonin *et al.*, 2012). The cooked and cured meat products colour is less affected by pressure than raw meat (Karlowski *et al.*, 2002). It has been reported that pre-rigor treatment for a few minutes at 100-200 MPa induces meat tenderization (Ohmori *et al.*, 1991). Rubio *et al.* (2007) evaluated the effect of HPP (500 MPa for 5 min at 18 °C) on the quality of vacuum packaged dry cured beef (Cecina de Leon) product and observed no changes in sensory characteristics for 210 days during refrigerated storage (6 °C). However, few authors have reported increase in thiobarbituric acid values in high pressure treated meat (Cheah and Ledward, 1997), chicken breast muscle (Orlien *et al.*, 2000) and deboned turkey meat (Tuboly *et al.*, 2003). Pressure treatment of vacuum-packed beef and chicken below 600 MPa did not induce significant changes in the raw aroma profile during 14 days of chilled storage in comparison with fresh untreated meat (Schindler *et al.*, 2010).

High pressure processing is environment friendly technology since it requires only electric energy and there are no waste products however high capital investment, resistance of bacterial spores to pressure, enzymatic and oxidative degradation are some limitations of this technology. High pressure processing has been approved for ready-to-eat (RTE) meat and poultry products by United States Department of Agriculture-Food Safety and Inspection Services (USDA-FSIS).

2.6 Irradiation

Irradiation of meat and meat products is done by controlled application of energy of ionizing radiations such as γ-rays, x-rays and electron beams. The approved radionuclotides for food irradiation include Cobalt-60 and Caesium-137 with a half life of 5.27 years and 30.19 years respectively. Cobalt-60 emits two gamma rays of 1.17 and 1.33 million electron volt (MeV) where as Caesium-137 emits gamma rays of 0.66 MeV. Generally, Co60 is used because it produces stronger gamma rays and it is not soluble in water. Gamma irradiation uses high-energy gamma rays with high penetration power. γ-rays can pass through living tissues without interacting with them because they have high energy and so small. Electron beam (E-beam) irradiation uses a stream of high-energy electrons, known as beta rays. Electron beams are produced by commercial electron accelerators such as linear accelerator or Van de Garaff accelerator. X-rays are produced by bombardment of a metal plate with high power electron beams and have intermediate penetration. Electron beams up to 10 MeV and X-rays upto 5 MeV are permitted.

Irradiation causes damage to the genetic material of microorganisms directly or indirectly which affect their multiplication or normal cell function. Radiolysis of water results in formation of hydroxyl radicals, hydrogen atoms, hydrogen peroxide and hydrated protons. Exposure of DNA to hydroxyl radicals causes both single and double strand break. Hydrogen atoms react by abstracting hydrogen from C-H bonds. The sensitivity depends on species of microorganisms, size of the

DNA, the rate of repair of damaged DNA *etc.* In general, viruses are more resistant than bacterial spores; vegetative organisms, moulds, parasites and insects follow descending order in resistance.

The quantity of radiation absorbed by the product being irradiated is expressed in rads. 1 rad is equivalent to 100 ergs of energy absorbed per gram of product. 1 gray represents an absorption of 1 Joule per kilogram of product and is equivalent to 100 rad. Dosimeters are placed with the meat or meat products being irradiated to measure the radiation to which it is exposed. Dose of irradiation varies with the objective of irradiation. Depending on the objective and dose, the following terms are commonly used:

- ☆ Radurization - Irradiation used to cause substantial reduction in numbers of viable specific spoilage microbes. Common dose level employed for meat is 0.75-2.5 kGy.
- ☆ Radicidation - Irradiation used to eliminate or reduce viable specific non-spore forming pathogens other than viruses so that there is no potential health hazard. Dose level employed is 2.5-10 kGy.
- ☆ Radappertization - Irradiation used is considered equivalent of "commercial sterility" as understood in the canning industry. The objective is to kill both vegetative cells and spores. Dose levels used is 10-50 kGy.

Irradiation facility consists of irradiation chamber, source of irradiation, conveyor system to move the products inside and out. Irradiation facilities must be designed and constructed in order to ensure the control of the radiological hazard for the personal as well as the environment. The irradiation chamber should be constructed such that there is no leakage of radiation outside the chamber. The source of irradiation (isotope or electron beam) must be placed within a biological shield - a building of concrete which completely surrounds the irradiation unit with walls of such thickness that there is no possibility of radiation exposure outside the shield. The isotope radiation source, when not in use, should be stored in a deep tank of water or a dry storage container which absorbs the radiation. Irradiation plants must be licensed by the government agency responsible for the regulation of irradiation applications and installations. A plant that has been licensed to irradiate food should be subjected to regular quality control and quality assurance procedures.

Park *et al.* (2010) reported that use of gamma rays irradiation up to 10 kGy on beef patties can be used for reducing bacterial populations with no adverse effect on quality and most of sensory characteristics of beef sausage patties. Irradiated meat may show colour changes depending on dose of irradiation, species, muscle and type of packaging. Post irradiation changes may occur in foods due to increase in free amino acid content. Vitamin B_1 among water soluble vitamins and Vitamin E among fat soluble vitamins are the most radio sensitive. Therefore, loss of vitamins due to irradiation depends on the nature and composition of food.

Irradiation can induce formation of isooctane- soluble carbonyl compounds in the lipid fraction and low molecular weight, acid-soluble carbonyls in the protein fraction of meat. Among the volatile components, 1-heptene and 1-nonene are influenced by irradiation dose and aldehydes (propanal, pentanal, hexanal) are influenced by packaging type. Sulfur-containing volatiles formed from sulfur-containing compounds (Dimethyltrisulfide, bismethylthiomethane) also contribute to fishy/irradiation odour. Reduction in the temperature during the irradiation process decreases the effects on odour/flavour because free radical generation and dispersion are reduced (Brewer, 2009). Radiation-induced hydrocarbons (1-tetradecene, pentadecane, hexadecane, 1,7-hexadecadiene, heptadecane, 8-heptadecene, eicosane), dimethyl disulfide, and methane can be used as markers for irradiated sausages (Nam *et al.*, 2011). Innovative packaging methods (Vacuum packaging, MAP, double packaging) along with antioxidants may be used to decrease the detrimental effects of irradiation.

Food irradiation is approved by more than 50 countries and endorsed by governmental and international organizations such as FAO, World Health Organisation (WHO), Codex Alimentarius and FDA. The Food and Agricultural Organization (FAO) and World Health Organization (WHO), declared that irradiation of any food upto a dose of 10 kGy (1 Mrad) causes no health hazard. Irradiation of fresh and frozen poultry was approved in 1992; red meat irradiation was approved in 1997 (Food Safety and Inspection Service, 1999). US Food and Drug Administration (USFDA) and the European Commission approved the use of X-ray technology with a maximum energy of 5 MeV, and then the FDA amended the maximum level to 7.5 MeV in 2004.

Irradiated product must bear the international logo of irradiation "radura" in green colour and must either have the word "Irradiated" in the product name, or the pack must be labelled "Treated with irradiation". Labeling requirements vary from country to country. In Australia, New Zealand and the EU require the labeling of any food that contains an irradiated ingredient whereas in the United States, labeling applies only where the whole food item is treated.

Table 11.1 Dose Limit given by FSSAI for Meat and Meat Products in India

Food	*Purpose*	*Dose Limit (KGy)*	
Meat and meat products including poultry (fresh and frozen)	Elimination of pathogenic micro organisms	1.0	7.0
	Shelf -life extension	1.0	3.0
	Control of human parasites	0.3	1.0

Irradiation is an effective technology that can be used in meat and meat products to reduce or eliminate the pathogens although it requires a high investment and maintenance cost.

2.7 Biopreservation and Natural Antimicrobials

Natural antimicrobial substances can be used to replace chemical preservatives and to obtain 'green label' products. Biopreservation refers to use of natural or controlled microflora and/or their antibacterial products for extending storage life and enhanced safety of foods (Galvez *et al.*, 2008). Phytochemicals are pleasantly accepted by majority of consumers in comparison with synthetic preservatives.

The biopreservatives to be used in meat products should be compatible, effective, safe, stable and economical.

Various spices and essential oils have preservative properties and have been used to extend the storage life of meat products. Aqueous extract of garlic inhibited the growth of microbial contaminants including facultative aerobic, mesophilic and faecal coliforms on the surface of poultry carcasses in refrigerated condition (Oliveira *et al.*, 2005). Allicin is the main ingredient of garlic that has antimicrobial activity against both gram-positive and gram-negative bacteria such as *Staphylococcus aureus, Salmonella typhi, E. coli* and *Listeria monocytogenes*. Essential oils are volatile liquids that are extracted from plants. They have demonstrated antimicrobial, antiviral, antimycotic, antitoxigenic, antiparasitic and insecticidal properties (Hyldgaard *et al.*, 2012) and have the greatest effect on the Gram-positive bacteria followed by Gram-negative bacteria. Their activity is mainly associated with their major compounds such as thymol, eugenol, carvacrol and trans-cinnamaldehyde (Bullerman *et al.*, 1977; Kim *et al.*, 1995; Lambert *et al.*, 2001). The mechanism of action of essential oils consist of disturbing the cytoplasmic membrane and disrupting the proton motive force, electron flow, active transport and coagulation of cell contents (Burt, 2004). Inhibitory effects of clove and cinnamon oils in ground chicken against *L. monocytogenes* (Hoque *et al.*, 2008), thyme and balm oils in fresh chicken breast (Fratianni *et al.*, 2010) and hop extracts in marinated pork (Kramer *et al.*, 2015) against meat spoilage bacteria have been reported. Spice mix of thyme, bay leaf and garlic has been reported to inhibit the growth of *S. aureus, Bacillus* sp., *Proteus* sp., *A. flavus* and *Mucor* sp. in market beef samples (Olaitan *et al.*, 2010).

The application of antimicrobial peptides from lactic acid bacteria (LAB) that target food spoilage and pathogenic microorganisms without toxic or other adverse effect has received great attention. Lactic acid bacteria includes various major genera: *Lactobacillus, Lactococcus, Carnobacterium, Enterococcus, Leuconostoc, Pediococcus* and *Weissella etc.* Other genera are: *Aerococcus, Microbacterium, Propionibacterium* and *Bifidobacterium* (Carr *et al.*, 2002). The lactic acid bacteria (LAB), generally considered as "Food Grade" organisms by regulatory agencies. The antimicrobial system present in lactic acid bacteria offers scope for development of an effective natural preservation of food either as purified chemical agents or as viable cultures. The lactic acid bacteria produce several antimicrobial substances such as organic acids, diacetyl, hydrogen peroxide, reuterin, reutericyclin, antifungal peptides and bacteriocins. LAB act antagonistically against a wide range of food borne pathogens and spoilage microorganisms like *Staphylococcus, Salmonella, Clostridium, Psuedomonas, Listeria monocytogenes* and *Yersinia enterocolitica* (Jacobson *et al.*, 2003). Antagonstic behaviour of lactic acid bacteria may be due to nutrient and oxygen competition, competition for attachment or adhesion sites and production of a wide range of inhibitory substances.

LAB produce a variety of bacteriocins and most of the bacteriocins are composed of 20-60 amino acid residues. Bacteriocins are ribosomally synthesized antimicrobial peptides or proteins (Jack *et al.*, 1995) that vary in molecular weight, mode of action, spectrum of activity, genetic origin and biochemical properties. The bacteriocins produced by LAB offer several desirable properties that make them suitable for meat preservation such as "generally recognized as safe substances", nontoxic on eukaryotic cells, usually pH and heat tolerant. They have a relatively broad antimicrobial spectrum, against many food-borne pathogenic and spoilage bacteria and having little influence on the gut microbiota as they become inactivated by digestive proteases.

Nisin is a bacteriocin produced by the *Lactococcus lactis subsp. lactis*. Nisin contains 34 amino acids and has a molecular mass of 3510 Da. It exists as dimmer or tetramer and can form ring structures with lathionine and β-methyl lanthionine. Antibacterial activity of nisin against sensitive organisms depends upon the species of organisms, concentration of nisin and pH of medium. It causes disruption of cytoplasmic membrane which results in leakage of essential cellular materials from cells by inactivating sulphhydril groups in the cytoplasmic membrane, depolarization of cytoplasmic membrane with depletion of the proton motive force, efflux of different substrates and cations, inhibition of the respiratory activity, partial efflux of internal ATP and aggregation of cytoplasmic membrane resulting in the formation of water filled pores (Bruno *et al.*, 1992; Abee *et al.*, 1995). Nisin A and Nisin Z are naturally occurring nisin variants that differ in a single amino acid residue but show similar activities.

Nisin shows antibacterial activity against a wide range of Gram-positive bacteria *i.e. Bacillus, Clostridium, Listeria, Staphylococcus* (Fujitha *et al.*, 2007). Application of nisin to the surface of meat resulted in reduction of *L. monocytogenes* (El-Katheib *et al.*, 1993; Fang and Lin, 1994). Davies *et al.* (1999) reported the effectiveness of nisin to control spoilage bacteria in vacuum-packed Bologna type sausage. It is heat sensitive at natural pH and its activity is also reduced by interaction with phospholipid components which may limit its applications in food containing emulsifier.

Pediocin is produced by *Pediococcus acidilactici*, generally recognized as a safe (GRAS) and commonly used in fermented sausage. Pediocin is small (<10 kDa), non-lanthionine Class II LAB bacteriocin containing, membrane active peptides (Derez *et al.*, 2005; Xiraphi *et al.*, 2008). Pediocin PA-1 permeabilizes the cytoplasmic membrane of the receptor bacteria which results in leakage of ions and small molecules. Most pediocins function under a wide range of pH (Rodriguez *et al.*, 2002) and found to be effective against *Listeria monocytogenes, Enterococcus faecalis, Staphylococcus aureus* and *Clostridium perfringens*.

Sakacins are bacteriocins produced by *Lactobacillus sakei*. The best known sakacins are sakacins A, G, K, P and Q. The sakacins have been studied against *Listeria* sausages and cured meat products. They are also used to inhibit bacterial growth that may cause ropiness, sliminess or other product defects.

Many bacteriocins and producer strains are known, however, nisin and pediocin PA-1 produced by LAB are commonly used for biopreservation of foods (Gurakan, 2007). Micocin®, a *Carnobacteria maltaromaticum* bacteriocin has been released by Alberta University. To control Listeria in meat products use of Micocin® is legal in Canada, United States, Mexico, Costa Rica and Colombia (Health Canada, 2011).

3. Conclusion

There is a need of novel technologies for meat preservation due to increasing demand for more natural and minimally processed food. Novel preservation technologies should be developed considering the changes in food habits of consumers as well as globalization of the market. Novel technologies have been in constant development in order to meet consumer demands. Finally, it is worth mentioning the need for thorough investigation in the above fields of novel meat preservation approaches. More research is still needed to be carried out and to design effective system to promote their commercialization.

REFERENCES

Abee, T., Krockel, L. and Hill, C. (1995). Bacteriocins: modes of action and potentials in food preservation and control of food poisoning, International Journal of Food Microbiology, 28, 169-185.

Abida, J., Rayees, B., and Masoodi, F.A. (2014). Pulsed light technology: a novel method for food preservation. International Food Research Journal, 21(3), 839-848.

Afshari, R., and Hosseini, H. (2014). Non-thermal plasma as a new food preservation method, its present and future prospect, Journal of Paramedical Sciences, 5(1), 116-120.

Alarcon-Rojo, A.D., Janacua, H., Rodriguez, J.C., Paniwnyk, L. and Mason, T.J. (2015). Power ultrasound in meat processing. Meat Science, 107, 86–93

Anderson, J.G., Rowan, N.J., MacGregor, S.J., Fouracre, R.A. and Farish, O. (2000). Inactivation of food-borne enteropathogenic bacteria and spoilage fungi using pulsed-light. IEEE Transactions on Plasma Science, 28, 83-88.

Awad, T.S., Moharram, H.A., Shaltout, O.E., Asker, D. and Youssef, M.M. (2012). Applications of ultrasound in analysis, processing and quality control of food: a review. Food Research International, 48, 410–427.

Bae, S.C., Park, S.Y., Choe, W., and Ha, S.D. (2015). Inactivation of murine norovirus-1 and hepatitis A virus on fresh meats by atmospheric pressure plasma jets. Food Research International, 76,342–347.

Bajovic, B., Bolumar, T., and Heinz, V. (2012). Quality considerations with high pressure processing of fresh and value added meat products. Meat Science, 92, 280-289.

Balny, C. and Masson, P. (1993). Effects of high pressure on proteins. Food Review International, 9, 611.

Banu, M.S. (2012). Cold Plasma as a Novel Food Processing Technology. International Journal of Emerging Trends in Engineering and Developments, 4 (2), 803-818.

Bardos L., and Barankova H. (2010). Cold atmospheric plasma: Sources, processes, and applications, Thin Solid Films, 518, 6705–6713.

Bozkurt, H. and Icier, F. (2010). Electrical conductivity changes of minced beef–fat blends during ohmic cooking. Journal of Food Engineering, 96, 86–92.

Brewer, M.S. (2009). Irradiation effects on meat flavour-a review. Meat Science, 81, 1-14.

Bruno, M.E.C., Kaiser, A. and Montville, T.J. (1992). Depletion of proton motive form by nisin in *Listeria monocytogenes* cells. Applied Environmental Microbiology, 58, 2255.

Brunton, N.P., Lyng, J.G., Zhang, L. and Jacquier, J.C. (2006). The use of dielectric properties and other physical analyses for assessing protein denaturation in beef biceps femoris muscle during cooking from 5 to 85°C. Meat Science, 72 (2): 236–244.

Bullerman, L.B., Lieu, F.Y., and Seier, S.A. (1977). Inhibition of growth and aflatoxin production by cinnamon and clove oils - cinnamic aldehyde and eugenol. Journal of Food Science, 42(4), 1107–1109.

Burt, S. (2004). Essential oils: Their antibacterial properties and potential applications in foods - a review. International Journal of Food Microbiology, 94(3), 223–253.

Cabiscol, E., Tamarit, J., and Ros, J. (2000). Oxidative stress in bacteria and protein damage by reactive oxygen Species. International Microbiolgy, 3(1), 3-8.

Campus, M. (2010). High pressure processing of meat, meat products and seafood. In: Food Engineering, Ed. Brendan C. Siegler. Nova Science Publishers, Inc.

Carballo, J., Fernandez, P., Carrascosa A.V., Solas, M.T., and Jim´enez-Colmenero, F. (1997). Characteristics of low- and high-fat beef patties: effect of high hydrostatic pressure. Journal of Food Protection, 60 (1), 48–53.

Carlez A, Veciana-Nogues T, Cheftel J-C. 1995. Changes in colour and myoglobin of minced beef meat due to high pressure processing. LWT-Food Science Technology, 28(5): 528–538.

Carr, F. J., Hill, D., and Maida, N. (2002). The lactic acid bacteria: a literature survey. Critical Review Microbiology, 28, 281-370.

Chang, H.J., Xu, X.L., Zhou, G.H., Li, Ch. B. and Huang, M. (2012). Effects of characteristics changes of collagen on meat physico-chemical properties of beef semitendinosus muscle during ultrasonic processing. Food and Bioprocess Technology, 5, 285–297.

Cheah, P.B., and Ledward, D.A. (1997). Catalytic mechanism of lipid oxidation following high pressure treatment in pork fat and meat. Journal of Food Science, 62, 1135-1139.

Cheftel, J.C. (1995). Review: high pressure, microbial inactivation, and food preservation. Food Science Technolology International, 1, 75-90.

Cheigh, C.I., Park, M.H., Chung, M.S., Shin, J.K. and Park, Y.S. (2012). Comparison of intense pulsed light- and ultraviolet (UVC)-induced cell damage in Listeria monocytogenes and Escherichia Coli O157: H7. Food Control, 25, 654–659.

Chen, J.H., Ren, Y., Seow, J., Liu, T., Bang, W.S. and Yuk, H.G. (2012). Intervention technologies for ensuring microbiological safety of meat: Current and future trends. Comprehensive Reviews in Food Science and Food Safety, 11, 119–132.

Dagerskog, M. and Osterstrom, L. (1979). Infra-red radiation for food processing I: A study of the fundamental properties of infra-red radiation. Lebensmittel-Wissenschaft. and Technologie, 12, 237–242.

Davies, E.A., Minle, C.F., Bevis, H.E., Potter, R.W., Harris, J.M., Williams, G.C., Thomas, L.V., and Broughton, J.D. (1999). Effective use of nisin to control acid bacterial spoilage in vacuum-packed bologna-type sausage. Journal of Food Protection, 62(9), 1004–1010.

Deraz, S.F., Karlsson, E.N., Hedstrom, M., Andersson, M.M. and Mattiasson, B. (2005). Purification and characterization of acidocin D20079, a bacteriocin produced by Lactobacillus acidophilus DSM 20079. Journal of Biotechnology, 117, 343-354.

Dey, A.,Rasane, P. Choudhury, A., Singh, J., Maisnam, D., Rasane, P. (2016).Cold Plasma Processing: A review. Journal of Chemical and Pharmaceutical Sciences, 9 (4), 2980-2984.

Dobryinin, D., Fridman, G. and Fridman, A. (2009). Physical and biological mechanisms of direct plasma interaction with living tissue. New Journal of Physics,11, 115020, DOI: 10.1088/1367-2630/11/11/115020.

El-Katheib, T., Yousef, A.E., and Ockerman, H.W. (1993) Inactivation and attachment of listeria monocytogenes on beef muscle treated with lactic acid and selected bacteriocins. Journal of Food Protection, 56, 29–33.

Fang, T.J. and Lin, L.W. (1994) Growth of Listeria monocytogenes and Pseudomonas fragi on cooked pork in a modified atmosphere packaging/nisin combination system. Journal of Food Protection, 57, 479–485.

FDA (Food and Drug Administration). (1996). Code of Federal Regulations (CFR) Title 21 Part 179. Irradiation in the production, processing and handling of food. Office of the Federal Register, US Government Printing Office, Washington DC. FDA 21CFR179.41.

FDA-CFSAN (Food, Drug Administration-Center for Food Safety, Applied Nutrition) (2000). Kinetics of microbial inactivation for alternative food processing technologies- Ohmic and inductive heating. http: //www.cfsan.fda.gov/~comm/ift-ohm.html

Food Safety and Inspection Service (1999). Irradiation of Meat Food Products. Federal Register 64 (264)72150-72166.

Fratianni, F., De Martino, L., Melone, A., De Feo, V., Coppola, R., and Nazzaro, F. (2010). Preservation of chicken breastmeat treated with thyme and balm essential oils. Journal of Food Science, 75, M528–M535.

Fujita, K., Ichimasa, S., Zendo, T., Koga, S. and Yoneyama, F. (2007). Structural analysis and characterization of lacticin Q, a novel bacteriocin belonging to a new family of unmodified bacteriocins of gram-positive bacteria. Applied Environmental Microbiology, 73, 2871-2877.

Galvez, A., Lucas Lopez, R., and Abriouel, H. (2008). Application of bacteriocins in the control of foodborne pathogenic and spoilage bacteria. Critical Review Biotechnology, 28,125-152.

Ganan M., Hierro E., Hospital X.F., Barroso E., and Fernandez M. (2013). Use of pulsed light to increase the safety of ready-to-eat cured meat products. Food Control, 32(2), 512–517.

Goutefongea, R., Rampon, V., Nicolas, J., Dumont, J.P. (1995). Meat colour changes under high pressure treatment. Proceeding of the 41st International Congress of Meat Science and Technology; San Antonio, Texas, 20–25 August 1995: p 384–385.

Gürakan, G.C. (2007). Biopreservation by lactic acid bacteria. In Metabolism and applications of Lactic acid bacteria, Ed. Barbaros Ozer. Published by Research Signpost, India, 15-31.

Han, Y., Jiang, Y., Xu, X., Sun, X., Xu, B., and Zhou, G. (2011). Effect of high pressure treatment on microbial populations of sliced vacuum-packed cooked ham. Meat Science, 88(4), 682-688.

Hanieh, S., Niels, H., Nonboe, K.U., Corry, J.E.L. and Purnell, G. (2014). Combined steam and ultrasound treatment of broilers at slaughter: A promising intervention to significantly reduce numbers of naturally occurring campylobacters on carcasses. International Journal of Food Microbiology, 176,23–28.

Haughton, P. N., Lyng, J. G., Morgan, D. J., Cronin, D. A., Fanning, S. and Whyte, P. (2011). Efficacy of high-intensity pulsed light for the microbiological decontamination of chicken associated packaging, and contact surfaces. Food borne Pathogen Disease 8(1), 109–117.

Hayakawa, I., Kanno, T. and Fujo, Y. (1989). In "Engineering and Food," (W.E.L. Spiess and H. Schubert, eds.), Elsevier Appl. Sci. Publ., p. 1901.

Health Canada. (2011) Bureau of Microbial Hazards. Policy on Listeria monocytogenes in ready-to-eat foods; FD-FSNP 0071. http: //www.hc-sc.gc.ca/fn-an/legislation/pol/policy listeria monocytogenes_2011-eng.php

Heinrich, V. Zunabovic, M., Varzakas, T., Bergmair, J. and Kneifel, W. (2016) Pulsed light treatment of different food types with a special focus on meat: a critical review. Critical Reviews in Food Science and Nutrition, 56(4), 591-613.

Hendrick, M.*E.g.*, Knorr, D., Loey, A.V., and Heinz, V. (2005). Ultra high pressure treatment of foods, Kluwer Academic Plenum Publishers, New York, 297-309.

Herceg, Z., Markov, K., Šalamon, B.S., Jambrak, A.R., Vukušiæ, T. and Kaliterna, J. (2013). Effect of high intensity ultrasound treatment on the growth of food spoilage bacteria. Food Technology and Biotechnology, 51, 352–359.

Hierro, E., Barroso, E., de la Hoz, L., Ordonez, J.A., Manzano, S., and Fernandez, M. (2011). Efficacy of pulsed light for shelf-life extension and inactivation of *Listeria monocytogenes* on ready-to-eat cooked meat products. Innovative Food Science and Emerging Technologies, 12, 275-281.

Hoque, M.M., Bari, M.L., Juneja, V.K., and Kawamoto, S. (2008). Antimicrobial activity of cloves and cinnamon extracts against food borne pathogens and spoilage bacteria, and inactivation of *Listeria monocytogenes* in ground chicken meat with their essential oils. Report of National Food Research Institute, 72, 9–21.

Hyldgaard, M., Mygind, T., and Meyer, R.L. (2012). Essential oils in food preservation: Mode of action, synergies, and interactions with food matrix components. Frontiers in Microbiology, 3, 12.

Jack, R.W., Tagg, J.R. and Ray, B. (1995). Bacteriocins of gram positive bacteria. Microbiology Reviews, 59, 171-200.

Jacobson, T., Budde B.B., and Koch, A.G. (2003). Application of Leuconostoc carnosus for biopreservation of cooked meat products. Journal of Applied Microbiology, 95(2), 242-249.

Jayasooriya, S.D., Bhandari, B.R., Torley, P., and D'Arey, B.R. (2004). Effect of high power ultrasound waves on properties of meat: A review. International Journal of Food Properties, 7, 301–319.

Jim´enez-Colmenero, F, Carballo, J., Fern´andez, P., Barreto, G., and Solas, M.T. (1997). High-pressure-induced changes in the characteristics of low-fat and high-fat sausages. Journal of the Science, Food and Agriculture, 75(1), 61–66.

Jofre, A., Aymerich, T., Grebol, N., and Garriga, M. (2009). Efficiency of high hydrostatic pressure at 600MPa against food-borne microorganisms by challenge tests on convenience meat products. LWT-Food Science and Technology, 42(5), 924-928.

Karlowski, K., Windyga, B., Fonberg-Broczek, M., Sciezynska, H., Grochowska, A., Gorecka, K., Mroczek, J., Grochalska, D., Barabasz, A., Arabas, J., Szczepek, J., and Porowski, S. (2002). Effects of high pressure treatment on the microbiological quality, texture and colour of vacuum-packed pork meat products. High Press Research, 22(3), 725–732.

Keklik N.M., Demirci A., and Puri V.M. (2009). Inactivation of *Listeria monocytogenes* on unpackaged and vacuum-packaged chicken frankfurters using pulsed UV-light. Journal of Food Science, 74, 8, M431-M439.

Keklik N.M., Demirci A., and Puri V.M. (2010). Decontamination of unpackaged and vacuum-packaged boneless chicken breast with pulsed ultraviolet light. Poultry Science, 89, 3, 570–581.

Kim, B., Yun, H., Jung, S., Jung, Y., Jung, H., Choe, W., and Jo, C. (2011). Effect of atmospheric pressure plasma on inactivation of pathogens inoculated onto bacon using two different gas compositions, Food Microbiology, 28 (1), 9-13.

Kim, H.J., Choi, Y.M., Yang, A.P.P., Yang, T.C.S., Taub, I.A., Giles, J., Ditusa, C., Chall, S. and Zoltal, P. (1996). Microbiological and chemical investigation of ohmic heating of particulate foods using a 5 kW ohmic system. Journal of Food Processing and Preservation, 20: 41–58.

Kim, J., Marshall, M.R., and Wei, C. (1995). Antibacterial activity of some essential oil components against five foodborne pathogens. Journal of Agricultural and Food Chemistry, 43(11), 2839–2845.

Kordowska-Wiater, M. and Stasiak, D.M. (2011). Effect of ultrasound on survival of gram negative bacteria on chicken skin surface. Bull. Vet. Inst. Pulawy 55, 207–210.

Koseki, S., Mizuno, Y., and Yamamoto, K. (2007). Predictive modelling of the recovery of *Listeria monocytogenes* on sliced cooked ham after high pressure processing. International Journal of Food Microbiology, 119(3), 300-307.

Kramer, B., Theilmann, J., Hickisch, A., Muranyi, P., Wunderlich, J., and Hauser, C. (2015). Antimicrobia activity of hop extracts against foodborne pathogens for meat applications. Journal of Applied Microbiology, 118, 648–657.

Lambert, R.J.W., Skandamis, P.N., Coote, P.J., and Nychas, G.J.E. (2001). A study of the minimum inhibitory concentration and mode of action of oregano essential oil, thymol and carvacrol. Journal of Applied Microbiology, 91(3), 453–462.

Lee, C.H. and Yoon, S. (1999). Effect of ohmic heating on the structure and permeability of the cell membrane of *Saccharomyces cerevisiae*. IFT Annual Meeting. Chicago, July 24–28.

Mai-Prochnow, A., Murphy, A.B., McLean, M., Kong, M.G., and Ostrikov, K. (2014). Atmospheric pressure plasmas: Infection control and bacterial responses, International Journal of Antimicrobial Agents, 43 (6), 508–517.

Mason, T.J. and Lorimer, J.P. (1988). Sonochemistry: Theory, Applications and Uses of Ultrasound in Chemistry; Ellis Horwood Limited Publishers: Chichester, p. 1–251.

McKenna, B. M., Lyng, J., Brunton, M., and Shirsat, N. (2006). Advances in radio frequency and ohmic heating of meats. Journal of Food Engineering, 77, 215–229. Mertens, B. and Knorr, D. (1992). Developments of nonthermal processes for food preservation. Food Technology, 46(5),129-132.

Misra, N.N., Tiwari, B.K., Raghavarao, K.S.M.S., and Cullen, P.J. (2011). Nonthermal plasma inactivation of food-borne pathogens. Food Engineering Reviews, 3, 159-170

Nam, K.C., Lee, E.J., Ahn, D.U. and Kwon, J. (2011). Dose-dependent changes of chemical attributes in irradiated sausages. Meat Science, 88 (1), 184-188.

Noriega, E., Shama, G., Laca, A., Diaz, M., and Kong, M.G. (2011). Cold atmospheric gas plasma disinfection of chicken meat and chicken skin contaminated with Listeria innocua. Food Microbiology, 28(7), 1293–1300.

Ohlsson, T. and Bengtsson, N. (2002). Minimal processing technologies in the food industry. In Ohlsson, T. and Bengtsson, N. (Eds). Cambridge, England: Woodhead Publishing, p. 112. CRC Presss.

Ohmori, T., Shigehisa, T., Taji, S., and Hayashi, R. (1991). Effect of high pressure on the protease activities in meat. Agricultural and Biological Chemistry, 55, 357-361.

Olaitan, A. O., Chukwudi, U. S., and Margaret, Y. (2010). Antimicrobial potentials of some spices on beef sold in Gwagwalada market, FCT, Abuja. Academia Arena, **2**(7): 15-17.

Oliveira, K.A.M., Santos-Mendonca, R.C., Gomide, L.A.M., and Vanetti, M.C.D. (2005). Aqueous garlic extract and microbiological quality of refrigerated poultry meat. Journal of Food Processing and Preservation, **29**: 98–108.

Orlien, V., Hansen, E., Skibsted, L.H. (2000). Lipid oxidation in high-pressure processed chicken breast muscle during chill storage: critical working pressure in relation to oxidation mechanism. European Food Research and Technology, 211, 99-104.

Ozkan, N., Ho, I., and Farid, M. (2004). Combined ohmic and plate heating of hamburger patties: Quality of cooked patties. Journal of Food Engineering, 63, 141–145.

Park, Y., Yoon, J.N., Park, I.J.,Han, B.S., Song, J.H., Kim, W.G., Kim, H.J., Hwang, S.B., Han, and Lee, J.W. (2010). Effects of gamma irradiation and electron beam irradiation on quality, sensory, and bacterial populations in beef sausage patties. Meat Science, 85(2), 368-372.

Parrott, D.L. (1992). Use of OH for aseptic processing of food particulates. Food Technology, 45, 68-72.

Paskeviciute, E., Buchovec, I. and Luksiene, Z. (2011). High Power Pulsed light for decontamination of chicken from food pathogens: A study on antimicrobial efficiency and organoleptic properties. Journal of Food Safety, 31, 61-68.

Perrier-Cornet J.P., Marenchal P.A., and Gervais P. (1995). A new design intended to relate high pressure treatment to yeast cell mass transfer. Journal of Biotechnology, 44, 49-58.

Piette, G., Buteau, M.L., De Halleux, D., Chiu, L., Raymond, Y., and Ramaswamy, H. S. (2004). Ohmic cooking of processed meats and its effects on product quality. Journal of Food Science, 69, 71–78.

Piyasena, P., Mohareb, E. and McKellar, R.C. (2003). Inactivation of microbes using ultrasound: A review. International Journal of Food Microbiolology, 87, 207–216.

Rajkovic, A., Tomasevic, I., Smigic, N., Uyttendaele, M., Radovanovic, R. and Devlieghere, F. (2010). Pulsed ultraviolet light as an intervention strategy against *Listeria monocytogenes* and *Escherichia coli* O157: H7 on the surface of a meat slicing knife. Journal of Food Engineering, 100, 446-451.

Rastogi, N.K. (2011). Opportunities and Challenges in Application of Ultrasound in Food Processing. Critical Reviews in Food Science and Nutrition, 51(8), 705-722.

Rendueles, E., Omer, M.K., Alvseike, O., Alonso-Calleja, C., Capita, R., and Prieto, M. (2011). Microbiological food safety assessment of high hydrostatic pressure processing: review. LWT-Food Science and Technology, 44(5), 1251–1260.

Rod, S.K., Hansen, F., Leipold, F., and Knochel, S. (2012). Cold atmospheric pressure plasma treatment of ready-to-eat meat: Inactivation of *Listeria innocua* and changes in product quality. Food Microbiology, 30 (1), 233-238.

Rodriguez, J.M., Martinez, M.I., and Kok, J. (2002). Pediocin PA-1, a wide-spectrum bacteriocin from lactic acid bacteria. Critical Reviews in Food Science and Nutrition, 42, 91-121.

Rowan, N.J., MacGregor, S.J., Anderson, J.G., Fouracre, R.A., Mcllvaney, L. and Farish, O. (1999). Pulsed-light inactivation of food-related microorganisms. Applied and Environmental Microbiology, 65, 1312-1315.

Rubio, B., Martinez, B., Garcia-Cachan, M. D., Rovira, J., and Jaime, I. (2007). Effect of high pressure preservation on the quality of dry cured beef "Cecina de Leon". Innovative Food Science and Emerging Technologies, 8(1), 102-110.

Sarang, S., Sastry, S.K. and Knipe, L. (2008). Electrical conductivity of fruits and meats during ohmic heating. Journal of Food Engineering, 87, 351–356.

Sasson, A., and Monselise, S.P. (1977). Electrical conductivity of Shamouti orange peel during fruit growth and postharvest senescence. Journal of American Soc. Horticul. Science, 102,142–144.

Sastry, S. K., and Li, Q. (1996). Modelling the ohmic heating of foods. Food Technology, 50, 246–248.

Sastry, S.K., and Barach, J.T. (2000). Ohmic and inductive heating. Journal of Food Science Supplement, 65(4), 42-46.

Sastry, S.K., and Palaniappan, S. (1992). Mathematical modeling and experimental studies on ohmic heating of liquid-particle mixtures in a static heater. Journal of Food Engineering, 15, 241-261.

Schindler, S., Krings, U., Berger, R.G., and Orlien V. (2010). Aroma development in high-pressure-treated beef and chicken meat compared to raw and heat treated. Meat Science, 86(2), 317–323.

Shirsat, N., Lyng, J.G., Brunton, N.P. and McKenna, B. (2004). Ohmic processing: Electrical conductivities of pork cuts. Meat Science, 67, 507–514.

Simonin, H., Duranton, F., and Lamballerie, M. (2012). New insights into the high-pressure processing of meat and meat products. Comprehensive Reviews in Food Science and Food Safety, 11, 285-306.

Song, H.P., Kim, B., Choe, J.H., Jung, S., Moon, S.Y., and Choe, W. (2009). Evaluation of atmospheric pressure plasma to improve the safety of sliced cheese and ham inoculated by 3-strain cocktail *Listeria monocytogenes*. Food Microbiology, 26(4), 432-436.

Stoica, M., Alexe, P. and Mihalcea, L. (2014). Atmospheric cold plasma as new strategy for foods processing - An overview. Innovative Romanian Food Biotechnology, 15, 1-8.

Stratakos, A.C., and Koidis, A. (2015). Suitability, efficiency and microbiological safety of novel physical technologies for the processing of ready-to-eat meats and pumpable products. International Journal of Food Science and Technology, 50(6), 1283-1302.

Sun, H., Kawamura, S., Himoto, J., Itoh, K., Wada, T. and Kimura, T. (2008). Effects of ohmic heating on microbial counts and denaturation of proteins in milk. Journal of Food Science and Technology Research, 14, 117–123.

Takeshita, K., Shibato, J., Sameshima, T., Fukunaga, S., Isobe, S., Arihara, K. and Itoh, M. (2003). Damage of yeast cells induced by pulsed light irradiation. International Journal of Food Microbiology, 85,151–158.

Thakur, B.R. and Nelson, P.E. (1998). High-pressure processing and preservation of food. Food Reviews International, 14(4), 427-447.

Tuboly, E., Lebovics, V.K., Gaál, O., Mészáros, L., Farkas, J. (2003). Microbiological and lipid oxidation studies on mechanically deboned turkey meat treated by high hydrostatic pressure. Journal of Food Engineering, 56, 241-244.

Ulbin-Figlewicz, N., Brychcy, E., and Jarmoluk, A. (2015). Effect of low pressure cold plasma on surface microflora of meat and quality attributes. Journal of Food Science Technology 52(2),1228–1232

Wekhof, A. (2000). Disinfection with flash lamps. PDA Journal of Pharmaceutical Science and Technology, 54: 264-276.

Xiong, G.Y., Zhang, L.L., Zhang, W. and Wu, J. (2012). Influence of ultrasound and proteolytic enzyme inhibitors on muscle degradation, tenderness, and cooking loss of hens during aging. Czechoslo Journal of Food Science, 30, 195–205.

Xiraphi, N., Georgalaki, M., Rantsiou, K., Cocolin, L., Tsakalidou, E., and Drosinos, E.H. (2008). Purification and characterization of a bacteriocin produced by *Leuconostoc mesenteroides* E131. Meat Science, 80, 194-203.

Yaldagard, M., Seyed, A.M., and Tabatabaie, F. (2008). The principles of ultra high pressure technology and its application in food processing/preservation: A review of microbiological and quality aspects. African Journal of Biotechnology, 7(16), 2739-2767.

Yildiz turp, G., Sengun, I.Y., Kendirci, P. and Icier, F. (2013). Effect of ohmic treatment on quality characteristic of meat: A review. Meat Science, 9: 441-448.

Yordanov, D. and Angelova, G. (2010). High pressure processing for foods preserving. Biotechnology and Biotechnological Equipment, 24, 1940-1945.

Zell, M., Lyng, J.G., Cronin, D.A. and Morgan, D.J. (2009). Ohmic heating of meats: Electrical conductivities of whole meats and processed meat ingredients. Meat Science, 83,563–570.

Ziuzina, D., Patil, S., Cullen, P.J., Keener, K.M., and Bourke, P. (2014). Atmospheric cold plasma inactivation of *Escherichia coli, Salmonella enterica serovar typhimurium* and *Listeria monocytogenes* inoculated on fresh produce. Food Microbiology, 42, 109-116.

Chapter 12

Packaging Advancements for the Safety of Meat and Meat Products

Subhash Kumar Verma[1]* and Meena Goswami Awasthi[2]

[1]Assistant Professor, Dept. of Livestock Products Technology, College of Veterinary Science and Animal Husbandry, Chhattisgarh Kamdhenu University, Anjora, Durg, C.G.

[2]Assistant Professor, Department of Livestock Products Technology, College of Veterinary Science and Animal Husbandry, UP Pandit Deen Dayal Upadhyaya Pashu Chikitsa Vigyan Vishwavidyalaya Evam Go Anusandhan Sansthan (DUVASU), Mathura, U.P.

**e-mail: subhash90verma@gmail.com*

ABSTRACT

Packaging is an efficient containment, preservation and protection method for meat product with all necessary information required during packing, transport, storage, sale and use, along with the provision of convenience, taking into consideration all legal and environmental issues. Modernization and industrialization have participated for tremendous changes in life styles and packaging industry has to respond to those changes. Package also functions by reducing the output from industrial production to a manageable and desirable size. The essence of a new technology to monitor the food spoilage from farm to fork has emerged to reduce hazards such as food borne diseases. Various methodologies of packaging technology have been developed over the years for food quality monitoring during supply chain. The packaging requirements of a particular product are matched with the functional properties offered by a particular packaging material. New concepts of active packaging, intelligent packaging and nanotechnology offer innovative solutions which play an important role in improving or monitoring food quality and safety and extending shelf-life.

Keywords: *Meat, MAP, CAP, Active packaging, Vacuum packaging.*

1. Introduction

Packaging is pervasive and essential to enhance and protect the product from processing and manufacturing, through handling and storage, finally to the consumer. It lies at the very heart of the modern food industry and successful food packaging technologies must bring to their professional duties a wide-ranging background drawn from a multitude of disciplines. Packaging has been defined as socio-scientific discipline which operates in society to ensure delivery of goods to the ultimate consumer of those goods in the best condition intended for their use. It includes a co-oridnated system of preparing goods for transport, distribution, storage, retailing and end-use, a means of ensuring safe delivery to the ultimate consumer in sound condition at optimum cost, and a techno-commercial function aimed at optimizing the costs of delivery while maximizing sales and profit. The desirable properties of a food product have to be maintained for its anticipated shelf life. Packaging may not improve the existing quality of a product but it should help in maintaining its keeping quality during storage, transport and deterioration.

2. Functions of Packaging

Food is packaged to preserve its quality and freshness, add appeal to consumers and to facilitate storage and distribution. It protects content from contamination and spoilage, makes it easier to transport and store goods and provides uniform measures of contents.

- *Containment*: The containment function of packaging makes a huge contribution to protect the environment from the myriad of products which are moved from one place to another on numerous occasions each day in any modern society.
- *Protection*: Packaging protects its content from outside environmental effects such as, water, moisture, vapour, gases, odour, micro-organisms, dust, shock and compressive forces. It also protects or conserves much of the energy expended during the production and processing of product.
- *Convenience*: Modernization and industrialization have participated tremendous changes in life styles and packaging industry has to respond to those changes. Package also functions by reducing the output from industrial production to a manageable, desirable "consumer size".
- *Communication*: Packaging material acts as silent sales man. The ability of consumers to instantly recognise products through distinctive branding and labelling enables supermarkets to function on a self –service basis. When international trade is involved and different languages are spoken, the use of unambiguous, really understood symbols on the package is imperative. Universal Product Codes are also used in warehouses where hand held barcode readers linked to a computer make stock taking quick and efficient.

The packaging has to perform its functions in three different environments:

- *Physical environment*: This causes physical damage to the product which includes shocks from drop, fall and bumps, damage from vibrations arising from transportation modes including road, rail, sea. It may also cause compression and crushing damage arising from stacking during transportation and storage.

- ✰ *Ambient environment*: This environment surrounds the package and may cause damage like absorption of gases, water vapour, light, exhaust fumes, dust and dirt from automobiles and infection from micro-organisms and macro-organisms.
- ✰ *Human environment*: This environment interacts with people and designing packages for this environment requires knowledge of the variability of consumers including vision, strength, weakness, dexterity and cognitive behaviour. It also includes knowledge of the results of human activity like liability, litigation, legislation and regulations.

The food contents can suffice with size of package according to requirement of an individual or a family in retail packaging. It is convenient to carry home, however, always increases cost of packaging. Most usually packaging materials are flexible packaging materials - plastic films, aluminum foil, paper *etc.* or their laminates with decorated exterior of package to attract the attention of consumers at the grocery shop or a food store. Bulk or wholesale packaging is done for safe transport of a product from the point of production to wholesale dealer and from there to the retailer. Hence packages should be easy to load or move from one place to the other at transit point either manually or by trolley. Bulk packaging is done in rigid packaging materials like wooden boxes, solid or corrugated cardboard boxes, plastic crates *etc.*

3. Types of Packaging Materials

3.1 Flexible

These packaging materials can be folded as per packaging needs and can be used for retail individual or family packs as wrappers, pouches, preformed bags *etc.* Presently, plastic films are most widely used flexible packaging materials. These are of several types and differ in their physical, chemical and functional properties. *e.g.* Plastic films, Paper, Aluminum foil (Nasa, 2014).

Polyethylene (PE) is the most commonly used plastic film of these days due to low cost, easy availability and unique properties. Low density polyethylene (LDPE) is prepared at a very high atmospheric pressure at about 150- 200°C, whereas high density polyethylene (HDPE) is prepared at comparatively low atmospheric pressure and temperature. Polyethylene is a tasteless, odourless and non-toxic film. It does not ruptures immediately; rather it will first stretch to some extent. It has the unique property of sealability of itself by the application of heat. LDPE film is transparent to translucent, highly flexible and has comparatively low permeability to water vapours, but it is fairly permeable to oxygen, carbon dioxide or odours. HDPE film is translucent to opaque and comparatively less permeable to water vapours and gases. It is fairly oil and grease resistant as compared to LDPE.

Polypropylene (PP) has a good gloss, high flex strength and resistance. It is also sometimes used for packaging those raw meat products which are subjected to heat treatment or cooked in the pack itself at a later stage. The film is readily heat sealable and has low water vapour permeability. Polyamide usually called Nylon film in the trade is inert, heat resistant and has excellent mechanical properties. Nylon-6 is a tasteless and odourless film and thus ideal for use in the packaging of fresh and processed foods. Polyester film is also inert and has excellent strength. PET (polyethylene terephthalate), polyester of importance is sold in United States of America by the name Mylar. Polyvinyl chloride is a plasticized film for packaging but has low folding endurance. It has good seal property and resistance to oil as well as grease.

Aluminium and steel foils are used for packaging food products. Thin gauge aluminium foils with pin holes are generally laminated to paper or plastic film with bonding agent to make suitable laminates. These laminates are used to package food products requiring protection against light, water vapour and gases especially dehydrated cooked meat. Paper Glassine or parchment papers have good grease resistance and high wet strength. A plasticizer may be added to make the paper still more soft and machinable. These opaque papers are sometimes used to wrap bacon and other fatty cuts of meat.

3.2 Semi-rigid

A semi-rigid container is intended to maintain a definite form or shape and is not influenced by the bulk of the contents *e.g.* Paperboard/cardboard containers, PET and PVC (polyvinyl chloride) containers, Aluminum containers, Molded containers. Paperboard sheets are cut, folded into desired form and glued. The material can be made as set up paper board boxes or folding carton or tray as per the demand. It provides convenience, strength and good product protection. PET and PVC plastic sheets can be moulded in shape, size and colour to suit specific product requirements. PET bottles and containers are extremely clear, virtually unbreakable and very light weight. They are ideal for the packaging of pickled meat products. Molded pulp containers are the cheapest packaging for the shell eggs. They allow wholesale trading of eggs along with the tray.

3.3 Rigid

These materials are solid and can not be molded into a shape easily *e.g.* glass containers, metal cans, fibre board containers, wooden boxes/crates/barrels. Glass containers are very old and versatile material for food packaging. Glass is chemically inert and is an excellent barrier to solids, liquids and gases. Glass bottles are used for packaging meat pickles *etc.* The main drawback of glass containers is the risk of breakage and comparatively heavy weight. Metal cans are primarily used for commercially sterilized food products. For canning of meat products, a sulphur-resistant lacquer is preferred to check black discolouration of the product and can bodies are soldered or welded. The product is hermetically (air tight) sealed in the can. Rigid thermoformed plastic containers are made by exposing the plastic sheet to heat and forming into various shapes either individual pieces or in combination. The plastic used in the thermoformed trays are high density polyethylene, polypropylene or polyvinyl chloride. In developed countries, thermoformed polystyrene foam trays are used for containing fresh meat pieces or chunks. Fiberboard containers, wooden boxes and plastic crates are used as wholesale or shipping containers.

4. Advances in Packaging Technology

In response to changing consumer lifestyles, large retail groups and food service industries have introduced highly competitive mix of marketing and trading strategies which depends on quality of packaging material and technology employed (Brody *et al.*, 2008). Despite the best of barriers, processing technologies and controls, foods are susceptible to biochemical and other forms of deterioration and thus there is need for appropriate packaging technology. Traditional packaging systems are refused since these systems do not provide any information about the quality of food products to the consumers and manufacturers at any stage of supply chain. The essence of a new technology to monitor the food spoilage from farm to fork has emerged to reduce hazards such as food borne diseases. Various methodologies of packaging technology have developed over the years for food quality monitoring during supply chain. The packaging requirements of a particular product are matched with the functional properties offered by a particular packaging material. New concepts of active packaging, intelligent packaging and nanotechnology offers innovative solutions which play an important role for improving or monitoring food quality and safety and extending shelf-life (Dobrucka and Cierpiszewski, 2014).

4.1 Active Packaging

Active packaging is an innovative packaging technology that incorporates certain additives into packaging film or within packaging containers by which package, product, and environment interact to prolong shelf life or enhance safety or sensory properties as well as maintain the quality of the food product (Priyanka and Anita, 2014). It also refers to the active materials, such as moisture absorbent, scavengers, antimicrobial and antioxidants releasing systems that are used in the food surrounding environment to enhance the performance of packaging system.

Table 12.1. Active Packaging Techniques

Absorbing System	*Releasing System*	*Other System*
Oxygen absorbers	Carbon dioxide emitters	Self-heating aluminium or steel cans and containers
Carbon dioxide absorbers	Ethanol emitters	Self-cooling aluminium or steel cans and containers
Ethylene absorbers	Antimicrobial releasers	
Humidity absorbers	Antioxidant releasers	
Off flavour Absorbers		
Lactose remover		
Cholesterol remover		

Source: Ahvenainen (2003).

Based on the European Union Guidance to the Commission Regulation No 450/2009 (EU, 2009), active packaging is a type of food packaging with an extra function, in addition to that of providing a protective barrier against external influence. Active packaging is intended to influence the packed food. The packaging absorbs food-related chemicals from the food or the environment within the packaging surrounding the food; or it releases substances into the food or the environment surrounding the food such as preservatives, antioxidants and flavourings. The "releasing active materials and articles" are those designed to deliberately incorporate components that would release substances into or onto the packaged food or the environment surrounding the food; and "released active substances" are those intended to be released from releasing active materials and articles into or onto the packaged food or the environment surrounding the food and fulfilling a purpose in the food (EU, 2009).

4.1.1 Antimicrobial Active Packaging

Antimicrobial active packaging is one of the most important concepts in active packaging because meat provides excellent nutrients for the growth of microorganism. Spoilage microorganisms including bacteria, yeast and molds, and pathogenic micrograms, specifically *Salmonella* spp., *S. aureus, L. monocytogenes, C. perfringens, C. botulinum,* and *E. coli* O157: H7 are the major concerns leading to quality deterioration and food safety issues in meat (Jayasena and Jo, 2013). The aims of using antimicrobial active packaging are to extend shelf-life and to ensure food safety of meat and meat products. There are three basic categories of antimicrobial packaging: 1) incorporation of an antimicrobial substance into a sachet/pad connected to the package from which the volatile bioactive substance is released during further storage; 2) direct incorporation of the antimicrobial agent into the packaging film; and 3) coating of packaging with a matrix that acts as a carrier for antimicrobial agents so that the agents can be released onto the surface of food through evaporation in the head-space (volatile substances) or migrate into the food (non-volatile additives) through diffusion.

4.1.2 Antioxidant Active Packaging (Oxygen-scavenging packaging)

High levels of oxygen present in meat packaging can facilitate microbial growth, lipid oxidation, development of off flavours and off odours, colour changes and nutritional losses. Lipid oxidation not only results in the development of off-flavours (rancidity), but also the potential formation of toxic aldehydes and the loss of nutritional quality because of polyunsaturated fatty acid (PUFA) degradation (Gomez-Estaca *et al.*, 2014). Therefore, control of oxygen levels in meat packaging is important to limit the rate of such deteriorative and spoilage reactions. Antioxidant active packaging can be used as a means of improving product quality and extending shelf life of meat and meat product by controlling the level of oxygen.

4.2 Intelligent Packaging

The essence of monitoring or checking the quality of food from farm to fork is increased due to the consumer satisfactions, reduction of food wastage and food poisoning. *Intelligent packaging* or "smart packaging" is proposed to carry out such tasks. Intelligent packaging is a packaging system that is capable of carrying out intelligent functions (such as detecting, sensing, recording, tracing, communicating, and applying scientific logic) to facilitate decision making to extend shelf life, enhance safety, improve quality, provide information, and warn about possible problems (Yam *et al.*, 2005). It is employed to provide information about the history of food handling and storage to enhance food products quality and meet consumer satisfactions. Many smart packaging technologies based on sensors have been proposed during last decades. The term sensor is defined as a device including control and processing electronics and software to measure physical or chemical quantity (Kuswandi *et al.*, 2012).

Smart packaging utilizes chemical sensor or biosensor to monitor the quality and safety of food from the producers to the costumers. This technology can lead to a variety of sensor designs, which are suitable for monitoring food quality and safety, such as freshness indication, pathogens, leakage, carbon dioxide, oxygen and pH detection as well as temperature fluctuations. Optoelectronic, colorimetric, enzymes, E-noses, dye based sensors, oxygen sensors *etc.* are widely used to evaluate the safety parameters of meat products available in market. The mentioned intelligent sensing methods are accurate in determining meat degradation (microbiological, sensorial and detection of metabolite concentrations) however, they are time-consuming, employ expensive instrumentation, require qualified panels and are not suitable to be used in meat packaging systems. Therefore, various indicators have been developed by meat technologist, which give some information about the presence or absence of a substance or the degree of interaction between two substances by changing in characteristics, like colour. The difference between sensors and indicators is that the latter do not have receptor and transducer components and instead communicate information through direct visual changes (Kuswandi *et al.*, 2011).

4.3 Modified and Controlled Atmosphere Packaging

Shelf life of meat products can be extended to several weeks by replacing the package air by a mixture of gases. The gases and their proportion to be used in modifying the atmosphere inside the package are carefully selected for a particular type of meat product in modified atmosphere packaging (MAP). MAP is defined as the enclosure of food products in a high gas barrier film in which the gaseous environment has been changed or modified to slow respiration rates, reduce microbial growth and retard enzymatic spoilage with the intent of extending shelf life. MAP is very important and practical preservation technique applied for extending shelf life of food, especially for fresh or minimally processed foods including meat that allow the retention of their "fresh" character (Sandhya, 2010).

In this technique, the air in the package is removed and exchanged with a gas of different composition that can inhibit growth of spoilage microorganisms and assist in maintaining high quality. Nitrogen, carbon dioxide and oxygen are the three most important gases. Nitrogen is an inert gas and does not react with the various constituents of a meat product. Oxygen is used when the development of a desirable colour becomes imperative. However, use of oxygen is avoided in meats that are highly susceptible to the development of oxidative rancidity. Carbon dioxide is used to create anaerobic environment inside the package. The initial headspace environment of MAP may change due to reactions of the meat and microorganisms during storage without any additional manipulation of the internal gas composition and humidity (McMillin, 2008). The most commonly used gases in MAP are oxygen (O_2), carbon dioxide (CO_2), and nitrogen (N_2). Carbon monoxide (CO) is also widely applied to meat packaging, especially red meat, although, relatively high costs, complicated standardisation of the gas mixture for each type of product and safety concerns have limited its application (Farber, 1991). Other gases such as argon (Ar), sulphur dioxide (SO_2) and nitrous and nitric oxides have also been investigated in meat packaging, but they have not been commercially applied mainly because of the potential safety issues and high costs (Sandhya, 2010).

In contrast with MAP, controlled atmosphere packaging (CAP) always maintains the same environmental conditions within the package by continuous monitoring and controlling gas atmosphere and humidity (McMillin, 2008). MAP at chill temperatures has been widely applied for red meat packaging, storage and transportation (Jeremiah, 2001), but CAP is more often used for packaging of fruits and vegetables (Prince, 1989). In controlled atmosphere packaging, the proportion of the gaseous mixture is checked every week and regulated as per the product requirements. In modified atmosphere packaging, we do not disturb the package environment at a later stage.

4.4 Aseptic Packaging

The packaging material is sterilized by passing through hydrogen peroxide with heat or ultraviolet radiation; whereas sterile environment in the packaging area is maintained by a positive pressure and outward flow of sterile air. There are various aseptic packaging systems which are successfully operating in different parts of the world. In India, 'Tetrapak' system is being followed for the aseptic packaging of milk and fruit juices at several places. Aseptic packages need not to be put in refrigerator. This packaging has been successfully used to prepare retail liquid packs. However, it requires a good amount of initial capital investment. So the technology has to be applied selectively after proper market survey for the packaging of intended liquid food product. It is likely to become still more popular in the coming years especially in countries with tropical climatic conditions. If we kill all the spoilage causing germs or microorganisms like bacteria, yeast, mould *etc.* present in the food product, then it is made sterile or germ free. If sterile food product is packaged in a germ free packaging material in germ free environment, then the product will not spoil for several months. Thus, aseptic packaging refers to the process of packaging pre-sterilized food product in the pre-sterilized packages under sterile environment. In this process, the product is sterilized by heating at a very high temperature (ultra high temperature) or at high temperature for a short time (HTST) or directly by steam injection.

4.5 Vacuum Packaging

Colour is the most important characteristic of fresh meat from the marketing point of view. It greatly influences the consumer selection. In cured meat also, cured meat colour is highly desired. Long term storage of the meat in permeable plastic films will alter its colour to undesirable dark brown. So the meat is stored for extended period in impermeable film laminates under vacuum. It will ensure retention of meat quality for a period of at least 8 weeks in fresh meat and 10 weeks in case of cured meat at a refrigerated storage of 0°C. Vacuum packaging machine is used to create vacuum in the filled-in package. Impermeable plastic film laminates - two layered or three layered or even five layered are used for pouch making. Thus, preventing the air from entering the packaged meat maintains it at a better level of quality. The aerobic bacteria are inactivated. The lipid oxidation and consequent rancidity development is checked. There is saving in the space during transport. The laminate pouches should have a good mechanical strength and allow perfect seals. The polyethylene (PE) layer is invariably used on the inner side. Some of the commonly used laminates are:

- Polyester/PE film laminate
- Polyamide/PE film laminate
- Aluminium foil/PE laminate
- PVDC/PE laminate

In the vacuum packaging, superficial spoilage of fresh meat caused mainly by *Pseudomonas* sp. is inhibited. However, lactic acid producing bacteria continue to grow at a slow pace for several weeks without immediately spoiling the meat.

4.6 Edible Films with Antimicrobials and Antioxidants

Edible films and coatings from renewable biopolymers can extend food shelf life by functioning as solute, gas, and vapour barriers. There is interest in the development of edible packaging due to food processors' needs for novel storage techniques, environmental concerns over disposal of non-renewable food packaging materials, and opportunities for creating new market outlets for film-forming ingredients derived from under-utilized agricultural commodities. The films can increase shelf-life and quality of oxygen sensitive food when used as packaging or as coatings. These are food grade suspensions which may be delivered by spraying, spreading, or dipping, which upon drying form a clear thin layer over the food surface. Coatings are a particular form of films directly applied to the surface of materials and are regarded as part of the final product. Biopolymers such as polysaccharides, proteins and lipids can be used alone or in combination to form coatings and films, the physical and chemical properties of the base materials, greatly influences the functionality of films and coatings produced. The choice of materials is generally based on their water solubility, hydrophilic and hydrophobic nature, easy formation into coatings and films, sensory properties, and targeted applications. Some minor components, usually plasticizers (such as glycerol and sorbitol) and acids (such as acetic or lactic acid), are added into the formulation for improving coating or film flexibility and regulating pH (Duan *et al.*, 2010).

4.7 Nano-composite Packaging

Nanotechnology involves the application of materials with at least one dimension of less than 100 nm. Particulates, platelets and fibers are the 3 main classes of nanomaterials (Thostenson *et al.*, 2005). Because of their nanoscale dimensions, these materials have proportionally larger surface area and consequently more surface atoms than their microscale counterpart. When added to compatible polymers, the nanomaterials can dramatically enhance the material properties of the nanocomposites, including improved mechanical strength, enhanced thermal stability and increased electrical conductivity (Uskokovic, 2007). Thus nanomaterials are increasingly being used in the food packaging industry due to the range of advanced functional properties they can bring to packaging materials.

5. Recent Uses of Various Packaging Systems in Meat and Meat Products

5.1 Fresh Meat

Fresh meat is a highly perishable commodity and supports the growth of several types of microorganisms due to higher nutrient content. Hence, fresh meat is stored in refrigerator at 4°C, it should be properly packaged to prevent moisture loss or dehydration. Colour reaction of meat pigments and limited exposure to air (oxygen) leads to the development of most desirable bright red colour; whereas prolonged exposure of meat to air results in the formation of undesirable brown stale colour. So, the packaging film should have some oxygen permeability while providing a good barrier for water vapours. Low density poly ethylene (LDPE) is the most preferred packaging material for fresh meat. The polyethylene should be food grade, transparent and fairly thick (150-200 gauge). Polyproplylene, polyvenylidine chloride or cellophane films can also be used for wrapping fresh meat cuts.

Shrink film is also used for wrapping large and uneven cuts of fresh meat and dressed poultry. The carcass cuts or dressed poultry are first wrapped in shrink film which is then immersed in hot water (90°C) for a few seconds. The film shrinks and makes a tight wrap over the meat. Shrink film packaging offers a neat appearance and is easy to handle. Vacuum packaging is preferred for long term storage of fresh meat under refrigeration. It is very effective for meat with an ultimate pH of less than 5.9, otherwise greening may occur in meat after a few weeks. The laminate for this packaging should have a good strength and proper sealing is necessary. Modified atmosphere packaging is used as an alternative to vacuum packaging. The gaseous mixture in MAP differs for various species. For mutton, a mixture of 70 per cent oxygen, 20 per cent carbon dioxide and 10 per cent nitrogen is generally used, whereas for pork, 70 per cent carbon dioxide, 20 per cent nitrogen and only 10 per cent oxygen is recommended. Again, for dressed poultry a mixture of 50 per cent carbon dioxide and 50 per cent nitrogen is considered ideal.

5.2 Frozen Meat

Freezing is the most preferred preservation method for long term storage of fresh or processed meat. Freezer chest of home refrigerator can be used for less than 5 kg, however commercial deep freezer must be used for sizeable quantity. The temperature of deep freezer should be ideally near -18°C. The meat to be stored in frozen condition must be properly packaged, otherwise it develops "freezer burn' due to continuous dehydration from the meat surface. The texture becomes tough and colour becomes very dark. For frozen storage of meat, the packaging material should have good strength even at freezer temperature. It should have very little permeability to water vapours. It should also have very good grease resistance. Low density polyethylene (150-200 gauge) is the least cost protective film which can withstand low temperature and maintain clarity. Polyester or nylon/PE laminate can also serve as ideal over-wrap. Heat shrinkable low density polyethylene also provides all the required functional properties for this purpose.

5.3 Cured Meat

Cured meat products prepared with table salt (sodium chloride) and sodium nitrite along with other additives and have a very desirable pink colour and cured flavour. These two specific characteristics need to be protected in the packaging of cured meat products. Cured meat pigment or pink colour is very susceptible to oxidation and is converted to dull brown in the presence of oxygen. Colour is also faded on exposure to bright light. So, such meat is stored in a dark room and a 15 watt bulb is lighted only when required. Lactic acid bacteria are frequently encountered in cured meat products. The packaging material should be a good oxygen and water vapour barrier. It should be flexible enough to make a close contact with meat. The packaging techniques for short term storage are over-wrapping in polyethylene or PVDC shrink packaging for irregular cuts like hams. Modified atmosphere packaging in gaseous mixture of 85-90 per cent nitrogen and 10-15 per cent carbon dioxide also keeps the cured meat well for about 12 weeks at 0-4°C.

5.4 Cooked Meat Products

Most meat products like meat patties, sausages, nuggets, meat balls *etc.* are cooked to an internal temperature of 72-82°C to kill most of the microorganisms. The meat becomes tender during cooking and its refrigerated shelf life is also enhanced. These meat products can be packaged in pouches of polyethylene, polypropylene, PVDC *etc.* for short term storage lasting 10-12 days in a refrigerator (0-4°C). Meat products like meat gravies, meat soups, chicken curry are cooked at elevated temperatures. If temperature of more than 100°C is required, then cooking has to be under pressure. A household pressure cooker (15 lb pressure) provides a temperature of 121°C. Such cooked products can be packaged as mentioned above. However, cooking in hermetically (air tight) sealed metal cans make the products commercially sterile. These canned products are shelf stable at ambient temperature for a period of 2 years. Thus, basically cooked products are packaged to maintain the delicate cooked flavour, hold the desired texture and preserve the typical appearance besides preventing weight loss and recontamination. These products have also undergone high temperature cooking under pressure. So, these are commercially sterile and keep well at ambient temperature for a few months even in tropical climatic conditions.

5.5 Dehydrated Meat

Since ancient period efforts have been made to preserve meat by drying. Many methods such as sun drying, oven drying *etc.* were tried, but freeze dried meat were accepted as the best because it could again absorb enough water on rehydration and stay tender. These days several crisp meat products like chips, curls *etc.* are being developed. Soup powder is also available in the market. All the dried meat products are susceptible to ingress of moisture and rancidity development. So, the packaging material should not allow any moisture or oxygen inside the product. Aluminium foil/polyethylene laminate is ideally suited for this purpose. If nitrogen is also filled in the package of a crisp product, it will protect the product against breakage by providing cushioning from all the sides.

6. Conclusion

The concept of 'package' as a simple instrument for the marketing of food is changing to match the needs of consumers and the food industry. Technological progress and revolutionary improvements with efficient economics in meat industry have led to innovation in the packaging line. New systems, materials, machinery, designs and environmental concerns are some innovations in packaging sector. Recently, a series of new packaging technologies and materials have been developed including active packaging, intelligent packaging, edible coatings/films, biodegradable packaging, and nano-material packaging. These technologies and materials have the potential to improve the quality and safety, prolong the shelf-life, reduce the environment impact, and increase the attractiveness of the packaged product to the retailers and consumers.

REFERENCES

Ahvenainen, R. (2003). Active and intelligent packaging: An introduction. In: Ahvenainen, R., Ed., Novel Food Packaging Techniques. CRC Press, Boca Raton, 5-21. http: //dx.doi.org/10.1533/9781855737020.1.5

Broody, A. L., Betty, B., Jung, H. H., Clarie, K. S., and Tara, H. M. (2008). Innovative food packaging solutions. Journal of Food Science, 73(8), R107- R116.

Dobrucka, R., and Cierpiszewki, R. (2014). Active and intelligent packaging food-research and development – a review. Polish Journal of Food and Nutrition Science, 64(1), 7-15.

Duan, J., Cherian, G., and Zhao, Y. (2010). Quality enhancement in fresh and frozen lingcod (Ophiodon elongates) fillets by employment of fish oil incorporated chitosan coatings. Food Chemistry, 119, 524-532.

Gomez-Estaca, J., Lopez-de-Dicastillo, C., Hernandez-Munoz, P., Catala, R., and Gavara, R. (2014). Advances in antioxidant active food packaging. Trends in Food Science and Technology, 35, 42-51.

Jayasena, D., and Jo, C. (2013). Essential oils as potential antimicrobial agents in meat and meat products: a review. Trends in Food Science and Technology, 34(2), 96-108.

Jeremiah, L. E. (2001). Packaging alternatives to deliver fresh meats using short-or long-term distribution. Food Research International, 34, 749–772.

Kuswandi, B., Wicaksono, Y., Jayus Abdullah, A., Heng, L., and Ahmad, M. (2011). Smart packaging: sensors for monitoring of food quality and safety. Sensing and Instrumentation for Food Quality and Safety, 5(3–4), 137–146.

Kuswandi, B., Jayus Larasati, T., Abdullah, A., and Heng, L. (2012). Real-time monitoring of shrimp spoilage using on-package sticker sensor based on natural dye of curcumin. Food Analytical Methods, 5(4), 881–889.

McMillin, K. W. (2008). Where is MAP going? A review and future potential of modified atmosphere packaging for meat. Meat Science, 80, 43–65.

Nasa, P. (2014). A review on pharmaceutical packaging material. Worlds Journal of Pharmaceutical Research, 3(5), 344-368.

Priyanka, P., and Anita, K. (2014). Active packaging in food industry: a review. IOSR Journal of Environmental Science, Toxicology and Food Technology, 8(5), 1-7.

Sandhya (2010). Modified atmosphere packaging of fresh produce: current status and future needs. Food Science and Technology, 43, 381-392

Thostenson, E. T., Li, C., and Chou, T.W. (2005). Nano-composites in context. Composites Science and Technology, 65(3), 491–516.

Uskokovic, V. (2007). Nanotechnologies: What we do not know. Technology in Society, 29, 43-61.

Yam, K.L., Takhistov, P.T., and Miltz, J. (2005). Intelligent packaging: concepts and applications. Journal of Food Science, 70, 1–10.

Chapter 13

Biosensors: A Tool for Real Time Monitoring of Quality and Safety in Meat Based Food

A.K. Verma*, V. Rajkumar and S.K. Jayant

Goat Products Technology Laboratory,
ICAR-Central Institute for Research on Goats, Makhdoom,
Farah, Mathura – 281 122, U.P.
****e-mail: arun.lpt2003@gmail.com***

ABSTRACT

Today's consumers are more inclined towards meat and meat products including fish which are fresh, nutritious and safe. In order to ensure such type of products, their quality monitoring, microbiological and chemical safety evaluation is warranted on regular basis. There are several methods and instruments being used conventionally to check quality and safety. However, these are complicated, time consuming and expensive. These limitations in the quality evaluation of meat based foods can be overcome through design and application of biosensors. Initially they were large in size but gradual developments and evolution have helped in their miniaturization even in the form of chip. Biosensors can be of various types depending upon application principles. The research work in the area of quality evaluation via application of biosensors has greatly enhanced making it possible to determine meat quality as well as microbial and chemical hazards associated with foods including meat and fish precisely in shortest possible time. The attitude of meat consumers for demand of various qualities will be a driving force for more research development in field of biosensor based real time monitoring of many more quality and safety parameters associated with meat foods in near future.

Keywords*: Meat, Biosensor, Quality, Pathogens, Chemical safety.*

1. Introduction

Quality of food products including meat attracts much attention in human nutrition. To improve and control products quality, knowing the quality of raw material and its assessment while processing is essential. One of the most used foods in the food diet is meat. Freshness of food is the most important criterion in a quality control approach to assure the consumer that the product meets this criterion on a regular basis. An intricate combination of physical, chemical and microbiological processes damages product freshness followed by spoilage (Alomirah *et al.*, 1998). Biochemical activities of microorganism as contaminants in the meat lead to loss of its freshness.

Meat is a nutrient dense food. Its production involves dynamic practices and processing steps starting from the animal farm to the slaughterhouses. These practices and steps expose the animal and finally meet with various elements which are regarded hazardous from meat quality as well as consumer's health point of view. Such hazards can be of physical, chemical and biological in nature. Additionally various other ante-mortem and post-mortem factors also influence the quality of meat foods. These quality alterations and other hazards should be monitored as early as possible for the supply of fresh, nutritious and healthy meat. Through evolution we have developed senses to differentiate good foods from bad. However, analytical approaches are more accurate and objective ways to assess the food quality and safety (Warriner *et al.*, 2014). We have several methods and analytical tools to evaluate quality and safety of meat and meat products precisely. Nevertheless such approaches are expensive, time consuming and need technical expertise. Thus a tool which screens the quality of meat and associated hazards quickly is needed in the meat industry to maintain safety and quality. As a result industries are greatly demanding biosensors for online quality evaluation of foods including meat and meat products. Biosensor-based approaches are very much advanced with wider utilities to assess the quality and safety of meat (Arora *et al.*, 2011).

Emergence of nanoscience and nanotechnology simulate the scientists to fabricate devices used in bioanalysis. Nanoparticles (NPs), in particular metal NPs, have shown excellent conductivity and catalytic properties that make them suitable to promote electrochemical reactions through the electron transfer between the redox centres in proteins and electrode surfaces. Among the various NPs, magnetic NPs such as Fe_3O_4 have been widely used in chemistry, medicine, and industry due to their unique magnetic and electronic properties. In electrochemical sensors, magnetic nanoparticles offer several advantages and have been used as a substitute to fix various biomolecules and enzymes. Biocompatibility, higher surface-to-volume ratio, greater catalyst loading ability, high dispersion and very high stability are some of qualities associated with nanoparticles (Lin and Leu, 2005; Liu *et al.*, 2013). Nanoscale approaches greatly change the optical behaviour as compared to bulk materials. In nanoparticles tightly packed electrons have higher energy gradient between the highest valence band and the low conduction band as a result higher energy being emitted upon arrival of electrons to the ground state. Such phenomenon is observed through shift in colour and semi-conductor properties. This chapter will deal with biosensors, their types and application in evaluating quality and safety of meat foods.

2. Biosensor

A biosensor has a sensing element for specific detection of target and a method to transduce the interaction as a measureable signal (Thakur and Ragavan, 2013). In the original definition of biosensor biological entity like antibody, enzyme, oligonucleotide and receptor were the recognition element. Among these recognition elements enzymes are preferred one due to selectivity and ability to amplify the interaction through generation of an electroactive product such as hydrogen peroxide or chromogenic substrate. The detection of target to antibodies, receptors or hybridization of nucleic acid is more difficult since no electroactive or chromogen is formed. Though, quartz crystal microbalance (QCM) or Surface Plasmon Resonance (SPR) could be helpful to assess such interactions. Earlier, radio-labels were used to detect the interaction, however now enzyme and fluorescent markers with new generation labels are available through nanoparticles. As regard to nanoparticles, nanotechnology has offered a substitute to the bio-recognition agents like molecular imprinted polymers or other artificial receptors (Reddy *et al.*, 2012; Reddy *et al.*, 2013). Direct catalytic nanoparticles which specifically interact with the target thus nullifying the requirement of biological component are in great demand. Thus newer definition of biosensors includes sensors detecting the biological systems. The metrics applied to biosensors can rapidly and selectively detect the targets of interest with least user input (Warriner *et al.*, 2014). The properties of nanoparticles have found application in identification and determination of analyte besides transducing the bioanalytical interactions. Over the years biosensors have witnessed great evolution. Initially developed biosensor was quite larger in size and had an oxygen electrode modified with glucose electrode coated with Teflon membrane. This sensor consisted of a pH probe size electrode linked to a potentiostat connected to an old pen recorder (Singh *et al.*, 2014).Gradually the size of biosensors started reducing and major achievement observed in late 1980's with the commercialization of glucose test strip consisting of a screen printed electrode along with enzyme layer and mediator fixed on the anode (Warriner *et al.*, 2014). Later sensor size was very much reduced and included needed user inputs to give the result. Though, such biosensors were not precise, the devices provided a way to monitor blood glucose for diabetic patients. Glucose sensors further scaled down with the developments in nanotechnology requiring only small sample volume for testing (Lim and Kim, 2013; Scognamiglio, 2013).

3. Techniques and Types of Biosensor

The transduction techniques used in biosensors can be electrochemical or optical based. The electrochemical transduction is simple in design, sensitive and amenable to miniaturization (Arugula and Simonian, 2014) while optical transduction testing require camera/detectors for measuring qualitatively and is similar or better in sensitivity than electrochemical method (Thakur and Ragavan, 2013).

There have been progresses in developing lab-on-chip devices or microfluidic sensors in addition to miniaturization of transduction techniques. Introduction of economical microfluidic devices are the key to develop nanosensor (Kumar *et al.*, 2013). Microfluidics offer handling of small sample volume to be precisely manipulated in terms of flow, mixing, heating and cooling, besides enabling fully integrated extraction and detection on a single chip. This technique is fast, needs lesser reagents and highly sensitive within an automated system. With the help of microfluidic system molecular reaction like polymerase chain reaction can be executed (Shu *et al.*, 2014). Though pathogens can be detected very fast in Real Time-PCR, the device is of bench scale and its compatibility with biosensors is lacking. A continuous flow PCR microfluidic chip coupled with hybridization enable detection of several pathogens in an automated device (Jiang *et al.*, 2014). In the future major advancement in pathogen monitoring can be seen with the developments in isothermal PCR techniques (Zhi *et al.*, 2014).

3.1 Electrochemical Sensors

These sensors are based on various principles and accordingly different terms are coined (Warriner *et al.*, 2014). Amperometry measures the current generated/ consumed from redox reactions at the working electrode surface. The arrangement can be composed of a two electrode arrangement whereby the counter electrode also functions as the reference. Three electrode systems have a separate reference electrode that enables the applied potential to be controlled. The potential difference between working and reference electrodes is measured by potentiometric sensors. It depends on the mobility besides activity of ions. The examples of potentiometric sensors are ion selective electrodes and gas sensors although other types of sensors have also been reported (Sliwinska *et al.*, 2014).The differences in electrical resistances between two electrodes are determined by conductometric sensors. Here AC impedance can be used to improve sensitivity and selectivity. Mass and viscoelastic changes induced differences in frequency of a quartz crystal resonator at the surface of quartz crystal microbalance (QCM) are detected using piezoelectric sensors. Here also AC impedance is used to monitor resonance of quartz crystal. The QCM is highly sensitive and can measure mass changes in the order of 1 mg/ cm^2 although can be susceptible to interference from non-specific binding effects.

3.2 Optical Sensors

Surface plasmon resonance (SPR) is frequently referred to as a mass sensor and an optical based technique. SPR immunosensor monitors biomolecular interactions based on the change in refractive index (RI) on the sensor. In this technique antibodies are immobilized by a coupling matrix on the surface of a thin film of metal like gold deposited on the reflecting surface of an optically transparent wave guide. When visible or near-infrared radiation is made to totally internally reflect at the interface of the metal and the reflecting surface by a prism, SPR occurs. The angle of incidence at which this occurs is called the resonance angle (Rand *et al.*, 2002). Refractive index of medium adjacent to sensor changes due to antigen antibody interaction resulting in changes in the angle of resonance. This change in resonance angle is proportional to the change in the concentration of antigens bound on the surface.

3.3 Magnetoresistance Sensors

Magnetoresistive sensors measure the changes in metal resistivity under the influence of a magnetic field. This is a quite newer technique but it is expected that this will prove as a powerful tool in the field of nanosensors. These sensors utilize the changes in magnetism of paramagnetic beads for better selectivity and sensitivity than traditional fluorescent labels (Zhi *et al.*, 2014).

4. Biosensing of Meat Quality

4.1 Measurement of Tissue Moisture

A low muscle pH coupled with high temperature results in an unusual condition in meat called pale soft and exudative (PSE) caused by meat protein denaturation and subsequent reduction in water holding capacity (WHC) (Castro-Giraldez *et al.*, 2010). Consequently meat loses its water-soluble proteins, flavour and appearance; which make it unsuitable for both processors and consumers. Several methods have been advocated to detect this meat quality loss and dielectric technique is one of the most used approaches. A significant research work has been carried-out to study meat and its products by using different dielectric techniques; mainly with coaxial probe, transmission-line and electrode-impedance method (Damez and Clerjon, 2013). Apart from these approaches, a ring-resonator technique is being recently used to evaluate meat quality (Jilani *et al.*, 2013). This relatively newer method offers an edge over the conventional methods, such as minimal sample preparation, rapid measurements, easy sample loading and lower fabrication cost. Jilani *et al.* (2014) used highly sensitive microstrip ring-resonator based biosensor for the evaluation of meat moisture. Incorporation of high impedance ring in this resonator exhibits significant change in resonance frequency with respect to normal resonator and gives better performance.

4.2 Meat Differentiation

Different techniques for classification of meat species have been used, for example immunochemical and electrophoretic analysis of meat proteins and more modern techniques that allow identification of species-specific DNA sequences or alternative DNA detection systems based on polymerase chain reaction (PCR). DNA-techniques, even though very reliable, are quite expensive and time consuming.

The well-known biotin–streptavidin chemistry has been utilized to produce lectin panels on microcontact printed gold surfaces. Biotin (Vitamin H) binds streptavidin with extremely high affinity. Biotin is a small molecule and can be conjugated with lectin proteins without altering their functions. A tetrameric protein streptavidin can bind one biotin molecule. According to Dubois *et al.* (1987) patterning and functionalization of surfaces for example microcontact printing, self-assembled monolayers (SAM) of alkanethiols on gold are significant for the stability and easy preparation leading to high quality monolayers. Surface properties can easily be changed with the use of various chain-terminating moieties. In a study

Carlsson *et al.* (2005) used a microcontact printed gold-wafers to make a lectin panel to identify and differentiate various meat juices from fresh beef, chicken, pork, cod, turkey and lamb. These workers used streptavidin-biotin approach to attach lectins to gold surface. Lectins recognize and bind specifically to carbohydrate structures present on different proteins. Null ellipsometry and scanning ellipsometry were applied to evaluate bio-recognition and visualization of the binding pattern of the lectins and the meat juice proteins, respectively. It was concluded that with this method the studied meat juices could be discriminated from each other.

4.3 Determination of Drip Loss

Glycogen replenishes the ATP during post-mortem muscle metabolism by converting into lactic acid. The level of muscle glycogen is determined by genetic as well as ante-mortem and post-mortem factors (Scheffler and Gerrard, 2007). It is a decisive factor for the ultimate pH of meat and other quality attributes which are affected by muscle pH like drip loss, colour, processing yield and organoleptic properties (Hamilton *et al.*, 2003; Enfält and Hullberg, 2005; Costa *et al.*, 2009). The evaluation of glycogen content in live muscle means to operate a muscle biopsy ante-mortem. Monin and Sellier (1985) proposed a formula to determine glycogen content in muscle as glycolytic potential (GP). It is sum of all compounds which are generated during post-mortem glycolysis and expressed in μmol of lactate. Lundström and Enfält (1997) found application of glucose estimation in meat juice to screen pig with RN-gene which are stress susceptible.

According to Hamilton *et al.* (2003) free glucose concentration in the longissimus muscle exudate was significantly linked with pork freshness than glycolytic potential. This method was quicker and less expensive to measure. Bowker *et al.* (2014) and Zelechowska *et al.* (2012) stated that muscle exudates may be a good source of protein markers which can be helpful in developing fast, non-invasive approaches to assess pork quality and beef tenderness respectively. By evaluating glucose and lactate via biosensor in drip Przybylski *et al.* (2016) established their association with muscle glycolytic potential. They recorded the highest correlation between glucose and glycolytic potential in the muscle. The lactate in muscle and drip as measured by enzymatic method and biosensor, respectively were found related. There was significant association between glycogen, glucose, lactate and glycolytic potential and meat quality traits. The result showed possibility of drip loss prediction through evaluation of muscle glycolytic potential.

4.4 Assessment of Meat Freshness

Xanthine evaluation could be a useful tool for monitoring of actual condition and quality of preserved food including meat. It is derived from guanine and hypoxanthine with the help of enzyme guanine deaminase and xanthine oxidase, respectively. Xanthine has been able to establish itself as a marker to monitor post-mortem fish and meat freshness. Its concentration increases with the storage of meat due to gradual degradation of adenosine triphosphate (Shan *et al.*, 2009). It is mostly analysed by following methods: enzymatic colorimetric (Berti *et al.*, 1988), enzymatic fluorometric, fluorometric mass spectrometric and fragmentography (Olojoa *et al.*, 2005), high-pressure liquid chromatography (Kock *et al.*, 1993) and capillary column gas chromatography (Renata *et al.*, 2002). The biosensing methods can deliver many advantages over these routine techniques in terms of simplicity, sensitivity, speed and cost.

Nakatani *et al.* (2005) used amperometry to evaluate hypoxanthine in fish meat. The sensor consisted of xanthine oxidase immobilized with glutaraldehyde and bovine serum albumin on nafion-coated platinum (Pt) disc electrode. The sensor received hypoxanthine responses in 0.1 mol/l phosphate buffer (pH 7.0), at potential of 600mV vs. Ag/AgCl. There was linear response from biosensor in 30 second for hypoxanthine in the concentration range 2 $x10^6$ - $1.85x10^4$ mol/l. The workers advocated use of this sensor to measure hypoxanthine in fish meat without any matrix interference. Devi *et al.* (2011) also devised a xanthine biosensor by electropolymerizing mixture of zinc oxide nanoparticles (ZnO-NPs) and pyrrole on Pt electrode to form a ZnO-NPs–polypyrrole (PPy) composite film and immobilization of xanthine oxidase. The biosensor responded within 5 second at pH 7.0 and temperature 35 °C and showed linearity from 0.8M to 40M for xanthine. The biosensor lost 40 per cent of its activity after its 200 uses over 100 days at 4 °C.

Again Devi *et al.* (2013a) constructed an amperometric xanthine biosensor by covalently immobilizing xanthine oxidase (XOD) onto citrate coated silver nanoparticles deposited on gold electrode surface through cysteine self-assembled monolayers (SAM) to measure xanthine in fish, chicken, pork and beef. Biosensor needed less than 5 second to respond at pH 7.0 and 35 °C upon polarization at 0.5 V vs. Ag/AgCl. The linearity for xanthine ranged from 2 to 16 M with 0.15 M detection limit and 0.17 $mA/M/cm^2$ sensitivity. Further amperometric xanthine biosensor was designed by Sadeghi *et al.* (2014) by covalently immobilizing crude XOD extracted from bovine milk onto hybrid nanocomposite film. The biosensor needed 8 second to respond for xanthine and linearity for xanthine ranged from 0.2 to 36.0 µM with 0.1 µM detection limit and 13.58 $\mu A/\mu M/cm^2$sensitivity. The biosensor was successful in determining fish and chicken meat freshness.

Dervisevic *et al.* (2015) prepared an enzyme electrode by immobilizing XOD via nanocomposite host matrix incorporated with multi-walled carbon nanotube (MWCNT) in poly (GMA-co-VFc) copolymer film. The optimum conditions for the electrode were pH 7.0 and 45 °C +0.35 V and response time and sensitivity were ~4second 16 mA/M, respectively. The limit of detection for xanthine biosensor was 0.12 µM and it measured xanthine concentration in fish meat reliably. It was observed that addition of MWCNT in the polymeric mediator film was linked with good response and better performance of biosensor. An electrochemical biosensor with nano-interface for determination of xanthine and fish freshness was developed by Kavitha *et al.* (2013). Linearity for the sensor ranged from 0.4 to 2.4 Nm and response time for sensor was 2 second. There was no interference from ascorbic acid, urea and sucrose. The limit of detection (LOD) was found to be 2.5 pM and limit of quantification as 8.3 pM.

Devi *et al.* (2013b) described fabrication of xanthine biosensor by covalently immobilizing xanthine oxidase onto chitosan bound gold coated iron nanoparticles and electrodepositing on the surface of pencil graphite electrode (PGE). Upon polarization at 0.5V vs Ag/AgCl biosensor needed 3 second response time with 0.1

to 300 μM linearity range at pH 7.4 and 35 °C temperature. The limit of detection for biosensor was 0.1 μM and activity of biosensor was reduced by 25 per cent after 100 uses over a period of 100 days when stored at 4 °C. In order to evaluate meat freshness most recently Albelda *et al.* (2017) reported the fabrication of a graphene/titanium dioxide nanocomposite (TiO2-G) and used it as an electrode in hypoxanthine (Hx) sensor. The Hx sensor showed good catalytic activity with linearity range 20 μM to 512 μM, 9.5 μM detection limit and 4.1 nA/μM sensitivity. Moreover the biosensor was found resistant to the interference by uric acid, ascorbic acid and glucose. This biosensor was used to detect Hx in pork tenderloin stored at room temperature for a week. It was also stated that graphene/titanium dioxide nanocomposite could be an effective electrode material for electrochemical biosensors to assess meat freshness.

4.5 Assessment of Meat Tenderness

The most important quality attributes for meat, that consumers desire and what the meat industry is trying to supply are taste, tenderness and leanness (Steenkamp, 1990). Out of these qualities, tenderness is one of most significant parameters to attract consumer so meat industry should step up to determine and control meat tenderness (Huffman *et al.*, 1996). Though molecular testing to detect polymorphism in the calpastatin and/or calpain genes are present to help in the breeding programmes (Van Eenennaam *et al.*, 2007), the fast, accurate and objective approach to evaluate meat tenderness is lacking. Traditional approaches to determine meat tenderness are sensory evaluation and Warner Bratzler Shear Force (WBSF) method. These are time consuming, costly, subjective and indirect ways to assess tenderness (Boccard *et al.*, 1981). According to Wheeler *et al.* (2002) WBSF method is more accurate in tenderness identification than other approaches based on digital imaging or colorimetric detection (Wheeler *et al.*, 2002).

It is now very clear that increase in meat tenderness due to conditioning is mostly due to degradation of cytoskeletal proteins by calpain proteolytic system (Koohmaraie, 1996; Geesink *et al.*, 2000). Of all the biochemical variables that contribute ultimate meat tenderness and make prediction of carcass quality difficult, a number of groups have identified firstly calpastatin and secondly μ-calpain activity levels at slaughter as the best predictors of beef tenderness after 10–14 days of conditioning (Whipple *et al.*, 1990; Dransfield, 1993). Thus early information about these predictors of ultimate carcass quality within 24 h post-slaughter should potentially enable carcass grading and adjust the conditioning phase. The determination of calpastatin quantity is used as a specific, reliable and fast tool to assess meat tenderness and can replace the tedious activity measurements. Sensors based on immunological recognition are specific, easy to handle, economical and have the possibility for multiplexing.

Capillary biosensors serve as a sampling device and optical waveguide (Weigl and Wolfbeis, 1994). The capillary tubes support fluid flow through the interior and light propagation in the walls of the capillary tube. This allows for both sampling and analysis in one location. According to Thomas (2006) capillary tubes can easily be adapted to automated devices so capillary biosensors will be small and portable having sampling and analysis facility and need least manual work. Capillary tubes reduce the sample volume, incubation time, sampling cost and are less expensive. These are cheaper and produced in mass so are disposable item of the biosensor system (Misiakos and Kakabakos, 1998).

Bratcher *et al.* (2008) came with a capillary based biosensor to detect calpastatin a meat tenderness indicator. The muscle longissimus was analysed for calpastatin using traditional method, biosensor, WBSF and crude protein and results were compared. The responses of capillary biosensor were also evaluated against earlier developed optical biosensor. Capillary tube biosensor was found moderately accurate for prediction of calpastatin activity when 0 and 48 h sampling time were combined. When the 0 and 48 h sampling periods were combined, the capillary tube biosensor was moderately accurate in predicting calpastatin activity. The developed biosensor was found to have better precision to measure calpastatin activity than optical fibre biosensor. Zor *et al.* (2011) described immunological bio-recognition based analytical method and evaluation system for amperometric quantification of calpastatin activity. The method was claimed to be accurate, easier to use and economical. Through the integration of this assay with portable electrochemical device called Tenercheck analysis of meat tenderness was done.

5. Assessment of Meat Safety

5.1 Microbiological Quality

Conventional microbiological approaches and molecular techniques have limitations as longer time is needed for microbial enrichment and DNA polymerase contaminants can influence the PCR results by generating false negative results. Additionally failure of heat shocked microbes to grow in culture media makes plate count method unsuitable for testing. The application of molecular techniques has greatly shortened the time to get results. Molecular biology has greatly improved the techniques by reducing the time required to obtain results. PCR tests are normally run in a laboratory context thus evaluation food quality on spot can improve the safety of food distribution. In this connection the recent advances in biosensor technology promise sensitive and specific point-of-care tests with rapid results. Application of biosensor techniques could be increasingly significant in detection of pathogens due their potential of being rapid, sensitive, easy to use, portable, and economical (Liébana *et al.*, 2009; Afonso *et al.*, 2013). Additionally combination of biosensor technology nanotechnology can be advantageous as compared to traditional methods in terms of time of analysis, sensitivity and simplicity (Chen *et al.*, 2009; Mao *et al.*, 2011).

The combination of antibodies, cells and DNA have been used as probes while developing biosensors (Lei *et al.*, 2006). In DNA-biosensors also known as genosensors DNA act as a biological element (Cecchini *et al.*, 2012), where it hybridize with target DNAs and allows rapid monitoring. Such oligonucleotide sequences probe based sensors are specific and sensitive. Bai *et al.* (2010) detected 11 food-borne pathogens in buffer and pork using microarray approach with specific DNA probes fixed on a sensor surface. In order to detect meat spoilage, Boka *et al.* (2012) reported development of enzyme based amperometric biosensors for biogenic amines. These workers used monoxidase A, diamine oxidase and putrescine oxidase enzymes immobilized on graphite electrodes to determine tyramine, tryptamine, phenylethylamine, putrescine contents. Pork and fish samples stored at different conditions were also extracted to measure biogenic amines by HPLC to correlate the results.

Very recently an innovative approach was made through exploiting endonuclease-controlled aggregation of plasmonic gold nanoparticles for label-free and ultrasensitive detection of bacterial DNA (McVey *et al.*, 2017). Here RNA-functionalized gold nanoparticles were used to form DNA-RNA heteroduplex structures with target DNA. This heterduplex was vulnerable to cleavage by RNAse H enzyme so that released target DNA could hybridize with another RNA probe. This process goes on continuously till cleavage of total RNA probes. The leftover unprotected nanoparticles aggregated upon exposure to high electrolytic medium. This approach is very sensitive and femtomolar concentration of target DNA can be detected within 3 h.

An emerging and promising method to record functionality or biological activity of an analyte is cell-based detection systems (Banerjee and Bhunia, 2010). This system senses disturbances in the physiological activity of mammalian cells following exposure. Such biosensors are fully capable to detect pathogens in different matrices including foods (Rider *et al.*, 2003; Tencza and Sipe, 2004; Liu *et al.*, 2009). Most of the cell based biosensors(CBBs) measure optical properties of cellular metabolites or intracellular enzymes (Gray *et al.*, 2005; Curtis *et al.*, 2008). In order to monitor food safety Banerjee and Bhunia (2010) came up with a mammalian CBB using B lymphocyte Ped-2E9 cell line encapsulated in collagen matrix in 3 D frame as sensing element. This biosensor was unique in that it detects analyte and mammalian cell interactions, differentiate pathogens from non-pathogens as well as active and inactive toxins to determine the risk accurately. The sensor was found applicable for broad ranges of bacteria and toxins and results can be obtained within 2 h for toxins and 4-6 h for bacterial pathogens. In ready-to-eat meat the sensor can detect 10^2-10^4 CFU/g of *L. monocytogenes* and *B. cereus* and in milk even nanogram of active toxins can be detected.

Daszczuk *et al.* (2014) reported use of *Bacillus subtilis* as a chassis in cell-based biosensor to monitor meat freshness. The workers identified promotors activated upon release of volatile compounds from spoiled meat by analysing transcriptome. The promotor PsboA was the most activated one indicating the gene expression required for bacteriocin subtilosin. A novel BioBrick compatible integration plasmid for *B. subtilis* was created and cloned PsboA as a BioBrick in front of the gene encoding the chromoprotein amilGFP within this vector. The identified promotor initiated fluorescent protein production in *B. subtilis* in response to meat spoilage proving its utility as a biosensor.

The outbreaks of *E. coli* O157: H7 food poisoning due to contaminated ground beef such as hamburgers have raised growing interest in rapid pathogen identification on ground beef (USDA-FSIS, 1996a; Mohle-Boetani *et al.*, 2001) and beef carcasses (USDA-FSIS, 1996b). Number of detection techniques which are capable of providing reliable identification of *E. coli* O157: H7 at low cell concentrations in ground beef and beef products have been reported. These methods are pulsed-field gel electrophoresis (Lee *et al.*, 1996), immuno-magnetic beads (Irwin *et al.*, 2002; Tu *et al.*, 2003), epifluorescence microscopy, time-resolved fluoroimmunoassay and light addressable potentiometric sensor (Tu *et al.*, 2002).

Rapid and accurate monitoring of antibody-antigen interaction can be accomplished using optical-based SPR biosensors (Woodbury *et al.*, 1998). SPR phenomenon can be obtained using target specific antibody to measure sensitivity and specificity of bindings (Kretschmann and Raether, 1968). Waswa *et al.* (2007) used the SpreetaTM, an SPR based biosensor, to detect *Escherichia coli* O157: H7 spiked food items like milk, apple juice and ground beef extract using specific antibodies. Here the sensor surface immobilized with pathogen specific antibody was injected with pathogen spiked food samples. This biosensor was found specific to *E. coli.* O157: H7 and had 10^2-10^3 CFU/ml sensitivity.

Campbell *et al.* (2007) reported fabrication of piezoelectric-excited millimeter-sized cantilever (PEMC) sensors consisting of a piezoelectric and a borosilicate glass layer. They immobilized anti-E.coli O157: H7 antibody sensor and exposed it with samples like broth, ground beef containing *E.coli* O157: H7.The pathogen concentration in samples was determined by both plating and by pathogen modeling program and it was found that the PEMC sensor detects *E. coli* O157: H7 reliably at 50–100 cells/ml with a 3 ml sample. The detection of same pathogen in ground beef sample using flow-based biosensor was reported by Chen *et al.* (2008). Here gene specific oligonucleotide was immobilized onto a piezoelectric transducer. This was followed by gene amplification after extraction from the samples using synthetic primers and denaturation of the amplicons before hybridization with the immobilized oligonucleotide. The researchers also inducted gold-nanoparticle-labelled secondary oligonucleotide to get amplified signals and improve detection limit.

Liang *et al.* (2014) introduced a smartphone-based biosensor as a preliminary screening tool to monitor microbial spoilage on ground beef. The meat sample spiked with *Escherichia coli* K12 solutions (10^1–10^8 CFU/ml) was irradiated with an 880 nm near infrared LED perpendicular to the surface. The scatter signals at various angles were evaluated utilizing the gyro sensor and the digital camera of a smartphone. The 15°, 30° and 45° scatter angles represented 10^8 CFU/ml, 10^4 CFU/ml and 10 CFU/ml *E.coli* concentrations, respectively.

A rapid detection as well as enumeration of *E.coli* O157: H7 in beef sample using immunosensing assay has been reported (Hassan *et al.*, 2015). The approach did not need broth enrichment and was based on the electrocatalytic properties of gold nanoparticles (AuNPs) towards hydrogen evolution reaction and superparamagnetic microbeads (MBs) as pre-concentration/purification platforms. Pathogen inoculated samples were tested using anti-*E. coli* O157–magnetic beads conjugate (MBs–pECAb) as a capture platform. The microbes were detected using chronoamperometric measurement with screen-printed carbon electrode (SPCEs) after sandwiching with modified gold nanoparticles. The reported limit of detection of pathogen in minced beef was 457 CFU/ml.

Campylobacter, which can cause diarrhoea and fever, is mostly found in raw poultry meat. In the developing countries about 40-60 per cent children under the age of 5 year will encounter at least one symptomatic infection of *Campylobacter jejuni* (Coker *et al.*, 2002). Nowadays campylobacteriosis is considered the most common foodborne illness. Isolation of *Campylobacter* spp. is difficult because the bacteria are usually present in very few numbers.

DNA electrochemical biosensors are faster, simpler and cheaper than traditional diagnostic methods. Morant-Miñana and Elizalde (2015) reported detection of *Campylobacter* spp. in food matrices using electrochemical genosensor. The sensor was based on thin-film gold electrodes deposited onto Cyclo Olefin Polymer (COP). The genosensor had 1-25 nM linearity range for amplicon and 90 pM detection limit. The results of poultry meat samples and PCR product samples were found comparable. It was stated that the development of this kind of sensor is a step towards fabrication of Lab on Chip (LOC) which believed to perform an automated and complete assay, including sample preparation, PCR amplification, and electrochemical detection of *Campylobacter* spp. in raw poultry meat samples. One of the most common methods to quantify a specific biological probe on biochip is fluorescence detection using an optical source. An organic light-emitting diode (OLED) based DNA biochip was designed to detect *Campylobacter* spp. (Manzono *et al.*, 2015). The analysis of the real samples by classical plate method and molecular method were also performed to validate the results from DNA-biochip. The biochip was found highly sensitive and fast.

*Listeria monocytogenes*is one of the causative agents of gastroenteritis (Vazquez-Boland *et al.*, 2001). Its detection is commonly achieved by culture methods, which typically requires few days to confirm the results. During infection, this pathogen invades host cell by lysis of phagocytic vacuoles and proliferate in the cytosol. This invasion is helped by listeriolysin O (LLO) and phospholipase C enzymes. The enzyme phospholipase C is released as an inactive propeptide and is activated by a metalloprotease (MMP) at low pH (Marquis and Hager, 2000). This MMP is used as a biomarker and its higher level indicates *L. monocytogenes* infection. This virulence protease used as a biomarker for the detection of food contamination. In a minced meat, Loessner *et al.* (1996) used A511: : luxAB recombinant phage for the detection of Listeria. Through this approach 10 bacteria per gram could be detected in 24 h. Banerjee *et al.* (2008) reported detection of Listeria, LLO and *Bacillus* spp. enterotoxin using a multi-well plate-based biosensor containing B-cell hybridoma, Ped-2E9, encapsulated in type I collagen matrix.

A colorimetry based assay using magnetic nanoparticles for the detection Listeria was established (Alhogail *et al.*, 2016). Here the protease from the Listeria was detected using D-amino acid substrate which is associated with carboxylic acid on the nanoparticles. In order of form self-assembled monolayer on gold sensor surface, cysteine residue at the C-terminal of the substrate was used turning magnetic nanobeads black colour. Listeria protease cleaves specific peptide sequence of the monolayer and changes black colour to golden. It was observed that the appearance of the gold surface area is proportional to the bacterial concentrations in CFU/ml.

The developed sensor had $2.17x10^2$ CFU/ml detection limit. Artificially inoculated meat and milk samples were also tested for Listeria using this sensor.

Application of quartz crystal microbalance (QCM) immunosensor to detect *S. typhimurium* in chicken meat was investigated (Su and Li, 2005). The resonance frequency and motional resistance were recorded proportional to the microbial concentration in the range of 10^5–10^8 and 10^6–10^8 cells/ml, respectively. The use of anti-Salmonella-magnetic beads improved the detection limit up to 10 cells/ml. The identification of the same pathogen in beef sample using Cyranose-320™electronic nose system has been observed (Balasubramanian *et al.*, 2005). An array of 32 conducting polymer sensors was used to detect the odour patterns of the headspace of the meat samples. The volatile organic compounds emanated from vacuum packaged beef strip were also analysed. The electronic nose could be able to detect *S. typhimurium* contaminated meat samples having population level ≥ 0.7 log10 CFU/g.

5.2 Chemical Quality

Indiscriminate and abundant use of antibiotics for the management of livestock and their diseases have resulted in development of resistance among microbial populations against these drugs as well as their accumulation in the muscles. Thus a fast, sensitive and specific tests need to be devised to assess the antimicrobial and other chemical residues in the animal products. The detection of antibiotic residue like streptomycin and dihydrostreptomycin in whole milk, honey, pork meat and kidney following an immunosensor inhibition assay has been reported (Ferguson *et al.*, 2002). Later the same workers conducted a study to precisely detect chloramphenicol and chloramphenicol-glucuronide residues in meat, milk, honey and prawn with the help of an SPR-biosensor assay (Ferguson *et al.*, 2005). A membraned-based immunosensor was used to detect chloramphenicol residue in pork, beef, chicken and shrimp following chemiluminescence phenomenon (Park and Kim 2006). An optical-SPR biosensor assay for quinolones in egg, fish, and poultry meat was reported by Huet *et al.* (2008). A linearity range of 0.1–100 ppb in fish and 0.1–10 ppb in egg or poultry meat for norfloxacin was achieved.

The detection of eight sulphonamides through rapid biosensor immunoassay in chicken serum was reported (Haasnoot *et al.*, 2003). Four years later a fast, specific biosensor immunoassay to detect flumequine in broiler muscle and serum was performed by same workers (Haasnoot *et al.*, 2007). It was stated that the assay offered simplified sample preparation and appropriate measurement ranges. Another SPR-based assay was described which can detect at least 19 sulfonamides in porcine muscle (McGrath *et al.*, 2005). Here it was claimed that this approach reduces the risk of false positive results due to non-recognition of acetylated derivatives of sulphonamides.

For the marine products cases of poisoning are very common and their timely monitoring can help avoiding consumer suffering. An optical biosensor was developed for the detection of paralytic shellfish poisoning toxins in mussels, clams, cockles, scallops and oysters with limit of detections between 2 and 50 ppb (Fonfria *et al.*, 2007). A six-channel portable optical biosensor with integrated fluidics was used to detect domoic acid in clams (Stevens *et al.*, 2007). Here detection limit for the analyte was reported to be around 3 ppb.For diarrhoeic shellfish poisoning toxins Llamas *et al.* (2007) developed an optical antibody-based competitive method for okadaic acid determination in mussel and limit of detection was found 126 ppb.

6. Conclusion

The complex value chain for the processing of meat foods expose them with numerous physical, chemical and biological contaminants. These contaminants can adversely affect the quality as well as safety of meat products. Modern educated and conscious consumers are very much concerned about the quality of meat food products they consume to protect their health and also demand for the food which is nutritious, fresh and healthy. This has created enormous pressure on meat industry as well as research to come with tools and techniques for monitoring of meat quality. There are several methods in practice to check the quality and safety of meat and fish. Nevertheless, such methods are time consuming, expensive, complex and confined to the laboratory. Thus there is need to develop methods and devices like biosensors which are simple, easy to use and offer real time monitoring of meat quality. So far various attempts have been made to evaluate meat and fish quality using different types of biosensors. These biosensors offer convenience, simplicity in operation and are fast. We expect many more developments and application of biosensors in near future for online monitoring of quality and safety of meat foods at ground level.

REFERENCES

Afonso, A. S., Pérez-López, B., Faria, R. C., Mattoso, L. H., Hernández-Herrero, M., Roig-Sagués, A. X., Maltez-da Costa, M., and Merkoçi, A. (2013). Electrochemical detection of Salmonella using gold nanoparticles. Biosensors and Bioelectronics, 40(1), 121-126.

Albelda, J. A., Uzunoglu, A., Santos, G. N. C., and Stanciu, L. A. (2017). Graphene-titanium dioxide nanocomposite based hypoxanthine sensor for assessment of meat freshness. Biosensors and Bioelectronics, 89, 518-524.

Alhogail, S., Suaifan, G. A., and Zourob, M. (2016). Rapid colorimetric sensing platform for the detection of Listeria monocytogenes foodborne pathogen. Biosensors and Bioelectronics, 86, 1061-1066.

Alomirah, H., Alli, I., Gibbs, B., and Konishi, Y. (1998). Identification of proteolytic products as indicators of quality in ground and whole meat. Journal of Food Quality, 21(4), 299-316.

Arora, P., Sindhu, A., Dilbaghi, N., and Chaudhury, A. (2011). Biosensors as innovative tools for the detection of food borne pathogens. Biosensors and Bioelectronics, 28(1), 1-12.

Arugula, M. A., and Simonian, A. (2014). Novel trends in affinity biosensors: current challenges and perspectives. Measurement Science and Technology, 25(3). 1-22.

Bai, S., Zhao, J., Zhang, Y., Huang, W., Xu, S., Chen, H., Fan, L.M., Chen, Y., and Deng, X. W. (2010). Rapid and reliable detection of 11 food-borne pathogens using thin-film biosensor chips. Applied Microbiology and Biotechnology, 86(3), 983-990.

Balasubramanian, S., Panigrahi, S., Logue, C., Marchello, M., and Sherwood, J. (2005). Identification of Salmonella inoculated beef using a portable electronic nose system. Journal of Rapid Methods and Automation in Microbiology, 13(2), 71-95.

Banerjee, P., and Bhunia, A. K. (2010). Cell-based biosensor for rapid screening of pathogens and toxins. Biosensors and Bioelectronics, 26(1), 99-106.

Banerjee, P., Lenz, D., Robinson, J. P., Rickus, J. L., and Bhunia, A. K. (2008). A novel and simple cell-based detection system with a collagen-encapsulated B-lymphocyte cell line as a biosensor for rapid detection of pathogens and toxins. Laboratory Investigation, 88(2), 196-206.

Berti, G., Fossati, P., Tarenghi, G., Musitelli, C., and Melzi d'Eril, G. (1988). Enzymatic colorimetric method for the determination of inorganic phosphorus in serum and urine. Clinical Chemistry and Laboratory Medicine, 26(6), 399-404.

Boccard, R., Buchter, L., Casteels, E., Cosentino, E., Dransfield, E., Hood, D., Joseph, R., MacDougall, D., Rhodes, D., and Schön, I. (1981). Procedures for measuring meat quality characteristics in beef production experiments. Report of a working group in the commission of the European communities'(CEC) beef production research programme. Livestock Production Science, 8(5), 385-397.

Bóka, B., Adányi, N., Virág, D., Sebela, M., and Kiss, A. (2012). Spoilage detection with biogenic amine biosensors, comparison of different enzyme electrodes. Electroanalysis, 24(1), 181-186.

Bowker, B., Eastridge, J., and Solomon, M. (2014). Measurement of muscle exudate protein composition as an indicator of beef tenderness. Journal of Food Science, 79(7), C1292-C1297.

Bratcher, C., Grant, S., Vassalli, J., and Lorenzen, C. (2008). Enhanced efficiency of a capillary-based biosensor over an optical fiber biosensor for detecting calpastatin. Biosensors and Bioelectronics, 23(11), 1674-1679.

Campbell, G. A., Uknalis, J., Tu, S.I., and Mutharasan, R. (2007). Detect of Escherichia coli O157: H7 in ground beef samples using piezoelectric excited millimeter-sized cantilever (PEMC) sensors. Biosensors and Bioelectronics, 22(7), 1296-1302.

Carlsson, J., Winquist, F., Danielsson, B., and Lundström, I. (2005). Biosensor discrimination of meat juice from various animals using a lectin panel and ellipsometry. Analytica Chimica Acta, 547(2), 229-236.

Castro-Giráldez, M., Aristoy, M.C., Toldrá, F., and Fito, P. (2010). Microwave dielectric spectroscopy for the determination of pork meat quality. Food Research International, 43(10), 2369-2377.

Cecchini, F., Manzano, M., Mandabi, Y., Perelman, E., and Marks, R. S. (2012). Chemiluminescent DNA optical fibre sensor for *Brettanomyces bruxellensis* detection. Journal of Biotechnology, 157(1), 25-30.

Chen, S.H., Wu, V. C., Chuang, Y.C., and Lin, C.S. (2008). Using oligonucleotide-functionalized Au nanoparticles to rapidly detect foodborne pathogens on a piezoelectric biosensor. Journal of Microbiological Methods, 73(1), 7-17.

Chen, X., Zhang, J. J., Xuan, J., and Zhu, J. J. (2009). Myoglobin/gold nanoparticles/carbon spheres 3-D architecture for the fabrication of a novel biosensor. Nano Research, 2(3), 210-219.

Coker, A. O., Isokpehi, R. D., Thomas, B. N., Amisu, K. O., and Obi, C. L. (2002). Human campylobacteriosis in developing countries1. Emerging Infectious Diseases, 8(3), 237-243.

Costa, L. N., Tassone, F., Davoli, R., Fontanesi, L., Dall'Olio, S., Colombo, M., Buttazzoni, L., and Russo, V. (2009). Glycolytic potential in semimembranosus muscle of Italian Large White pigs. Journal of Muscle Foods, 20(4), 392-400.

Curtis, T., Naal, R. M. Z., Batt, C., Tabb, J., and Holowka, D. (2008). Development of a mast cell-based biosensor. Biosensors and Bioelectronics, 23(7), 1024-1031.

Damez, J.L., and Clerjon, S. (2013). Quantifying and predicting meat and meat products quality attributes using electromagnetic waves: An overview. Meat Science, 95(4), 879-896.

Daszczuk, A., Dessalegne, Y., Drenth, I., Hendriks, E., Jo, E., van Lente, T., Oldebesten, A., Parrish, J., Poljakova, W., and Purwanto, A. (2014). Bacillus subtilis biosensor engineered to assess meat spoilage. ACS Synthetic Biology, 3(12), 999-1002.

Dervisevic, M., Custiuc, E., Çevik, E., and [a]enel, M. (2015). Construction of novel xanthine biosensor by using polymeric mediator/MWCNT nanocomposite layer for fish freshness detection. Food Chemistry, 181, 277-283.

Devi, R., Batra, B., Lata, S., Yadav, S., and Pundir, C. (2013a). A method for determination of xanthine in meat by amperometric biosensor based on silver nanoparticles/cysteine modified Au electrode. Process Biochemistry, 48(2), 242-249.

Devi, R., Thakur, M., and Pundir, C. (2011). Construction and application of an amperometric xanthine biosensor based on zinc oxide nanoparticles–polypyrrole composite film. Biosensors and Bioelectronics, 26(8), 3420-3426.

Devi, R., Yadav, S., Nehra, R., Yadav, S., and Pundir, C. (2013b). Electrochemical biosensor based on gold coated iron nanoparticles/chitosan composite bound xanthine oxidase for detection of xanthine in fish meat. Journal of Food Engineering, 115(2), 207-214.

Dransfield, E. (1993). Modelling post-mortem tenderisation–IV: Role of calpains and calpastatin in conditioning. Meat science, 34(2), 217-234.

Dubois, L. H., Zegarski, B. R., and Nuzzo, R. G. (1987). Fundamental studies of the interactions of adsorbates on organic surfaces. Proceedings of the National Academy of Sciences, 84(14), 4739-4742.

Enfält, A. C., and Hullberg, A. (2005). Glycogen, glucose and glucose6phosphate content in fresh and cooked meat and meat exudate from carriers and noncarriers of the RN–Allele. Journal of Muscle Foods, 16(4), 330-341.

Ferguson, J., Baxter, A., Young, P., Kennedy, G., Elliott, C., Weigel, S., Gatermann, R., Ashwin, H., Stead, S., and Sharman, M. (2005). Detection of chloramphenicol and chloramphenicol glucuronide residues in poultry muscle, honey, prawn and milk using a surface plasmon resonance biosensor and Qflex® kit chloramphenicol. Analytica Chimica Acta, 529(1), 109-113.

Ferguson, J., Baxter, G., McEvoy, J., Stead, S., Rawlings, E., and Sharman, M. (2002). Detection of streptomycin and dihydrostreptomycin residues in milk, honey and meat samples using an optical biosensor. Analyst, 127(7), 951-956.

Fonfría, E. S., Vilariño, N., Campbell, K., Elliott, C., Haughey, S. A., Ben-Gigirey, B., Vieites, J. M., Kawatsu, K., and Botana, L. M. (2007). Paralytic shellfish poisoning detection by surface plasmon resonance-based biosensors in shellfish matrixes. Analytical Chemistry, 79(16), 6303-6311.

Geesink, G., Ilian, M., Morton, J., and Bickerstaffe, R. (2000). Involvement of calpains in post-mortem tenderisation: A review of recent research. In Proceedings New Zealand Society of Animal Production (Vol. 60, pp. 99-102). New Zealand.

Gray, K. M., Banada, P. P., O'Neal, E., and Bhunia, A. K. (2005). Rapid Ped-2E9 cell-based cytotoxicity analysis and genotyping of Bacillus species. Journal of clinical Microbiology, 43(12), 5865-5872.

Haasnoot, W., Bienenmann-Ploum, M., and Kohen, F. (2003). Biosensor immunoassay for the detection of eight sulfonamides in chicken serum. Analytica Chimica Acta, 483(1), 171-180.

Haasnoot, W., Gerçek, H., Cazemier, G., and Nielen, M. W. (2007). Biosensor immunoassay for flumequine in broiler serum and muscle. Analytica Chimica Acta, 586(1), 312-318.

Hamilton, D., Miller, K., Ellis, M., McKeith, F., and Wilson, E. (2003). Relationships between longissimus glycolytic potential and swine growth performance, carcass traits, and pork quality. Journal of Animal Science, 81(9), 2206-2212.

Hassan, A.R.H.A.A., de la Escosura-Muñiz, A., and Merkoçi, A. (2015). Highly sensitive and rapid determination of *Escherichia coli* O157: H7 in minced beef and water using electrocatalytic gold nanoparticle tags. Biosensors and Bioelectronics, 67, 511-515.

Huet, A.C., Charlier, C., Singh, G., Godefroy, S. B., Leivo, J., Vehniäinen, M., Nielen, M., Weigel, S., and Delahaut, P. (2008). Development of an optical surface plasmon resonance biosensor assay for (fluoro) quinolones in egg, fish, and poultry meat. Analytica Chimica Acta, 623(2), 195-203.

Huffman, K., Miller, M., Hoover, L., Wu, C., Brittin, H., and Ramsey, C. (1996). Effect of beef tenderness on consumer satisfaction with steaks consumed in the home and restaurant. Journal of Animal Science, 74(1), 91-97.

Irwin, P., Damert, W., Brewster, J., Gehring, A., and Tu, S. I. (2002). Immunomagnetic bead mass transport and capture efficiency at low target cell densities in phosphate buffered saline. Journal of Rapid Methods and Automation in Microbiology, 10(2), 129-147.

Jiang, X., Shao, N., Jing, W., Tao, S., Liu, S., and Sui, G. (2014). Microfluidic chip integrating high throughput continuous-flow PCR and DNA hybridization for bacteria analysis. Talanta, 122, 246-250.

Jilani, M. T., Wen, W. P., Zakariya, M. A., and Cheong, L. Y. (2013). Microstrip ring resonator based sensing technique for meat quality. In Wireless technology and applications (ISWTA), IEEE Symposium (pp. 220-224): IEEE. Pullman Kuching Malaysia.

Jilani, M. T., Wen, W. P., Zakariya, M. A., Cheong, L. Y., and Rehman, M. Z. U. (2014). An improved design of microwave biosensor for measurement of tissues moisture. In RF and wireless technologies for biomedical and healthcare applications (IMWS-Bio). IEEE MTT-S International Microwave Workshop Series on (pp. 1-3): IEEE. London, United Kingdom.

Kavitha, T., Sakthivel, G., Swaminathan, S., John, B., and Uma, M. (2013). Development of electrochemical biosensor with nano-interface for xanthine sensing ea novel approach for fish freshness estimation. Food Chemistry, 139, 963-969.

Kock, R., Delvoux, B., and Greiling, H. (1993). A high-performance liquid chromatographic method for the determination of hypoxanthine, xanthine, uric acid and allantoin in serum. Clinical Chemistry and Laboratory Medicine, 31(5), 303-310.

Koohmaraie, M. (1996). Biochemical factors regulating the toughening and tenderization processes of meat. Meat Science, 43, 193-201.

Kretschmann, E., and Raether, H. (1968). Radiative decay of non-radiative surface plasmons excited by light. Zeitschrift für Naturforschung A, 23(12), 2135-2136.

Kumar, S., Kumar, S., Ali, M., Anand, P., Agrawal, V. V., John, R., Maji, S., and Malhotra, B. D. (2013). Microfluidic integrated biosensors: Prospects for pointofcare diagnostics. Biotechnology Journal, 8(11), 1267-1279.

Lee, M.S., Kaspar, C. W., Brosch, R., Shere, J., and Luchansky, J. B. (1996). Genomic analysis using pulsed-field gel electrophoresis of *Escherichia coli* 0157: H7 isolated from dairy calves during the United States national dairy heifer evaluation project (1991–1992). Veterinary microbiology, 48(3-4), 223-230.

Lei, Y., Chen, W., and Mulchandani, A. (2006). Microbial biosensors. Analytica Chimica Acta, 568(1), 200-210.

Liang, P.S., San Park, T., and Yoon, J.Y. (2014). Rapid and reagentless detection of microbial contamination within meat utilizing a smartphone-based biosensor. Scientific Reports, 4, 5953.

Lieìbana, S., Lermo, A., Campoy, S., Barbeì, J., Alegret, S., and Pividori, M. A. I. (2009). Magneto immunoseparation of pathogenic bacteria and electrochemical magneto genosensing of the double-tagged amplicon. Analytical Chemistry, 81(14), 5812-5820.

Lim, B., and Kim, Y.P. (2013). Enzymatic glucose biosensors based on nanomaterials. In M.B. Gu and H.S. Kim (Eds.), Biosensors Based on Aptamers and Enzymes (pp. 203-219). Springer.

Lin, M. S., and Leu, H. J. (2005). A Fe_3O_4based chemical sensor for cathodic determination of hydrogen peroxide. Electroanalysis, 17(22), 2068-2073.

Liu, S., Tran, K. K., Pan, S., and Shen, H. (2009). Detecting and differentiating microbes by dendritic cells for the development of cell-based biosensors. Biosensors and Bioelectronics, 24(8), 2598-2603.

Liu, Y., Zhou, J., Gong, J., Wu, W.P., Bao, N., Pan, Z.Q., and Gu, H.Y. (2013). The investigation of electrochemical properties for Fe_3O_4@ Pt nanocomposites and an enhancement sensing for nitrite. Electrochimica Acta, 111, 876-887.

Llamas, N. M., Stewart, L., Fodey, T., Higgins, H. C., Velasco, M. L. R., Botana, L. M., and Elliott, C. T. (2007). Development of a novel immunobiosensor method for the rapid detection of okadaic acid contamination in shellfish extracts. Analytical and Bioanalytical Chemistry, 389(2), 581-587.

Loessner, M. J., Rees, C., Stewart, G., and Scherer, S. (1996). Construction of luciferase reporter bacteriophage A511: : luxAB for rapid and sensitive detection of viable Listeria cells. Applied and Environmental Microbiology, 62(4), 1133-1140.

Lundström, K., and Enfält, A.C. (1997). Rapid prediction of RN phenotype in pigs by means of meat juice. Meat Science, 45(1), 127-131.

Manzano, M., Cecchini, F., Fontanot, M., Iacumin, L., Comi, G., and Melpignano, P. (2015). OLED-based DNA biochip for *Campylobacter* spp. detection in poultry meat samples. Biosensors and Bioelectronics, 66, 271-276.

Mao, S., Yu, K., Lu, G., and Chen, J. (2011). Highly sensitive protein sensor based on thermally-reduced graphene oxide field-effect transistor. Nano Research, 4(10), 921-930.

Marquis, H., and Hager, E. J. (2000). pHregulated activation and release of a bacteria associated phospholipase C during intracellular infection by *Listeria monocytogenes*. Molecular Microbiology, 35(2), 289-298.

McGrath, T., Baxter, A., Ferguson, J., Haughey, S., and Bjurling, P. (2005). Multi sulfonamide screening in porcine muscle using a surface plasmon resonance biosensor. Analytica Chimica Acta, 529(1), 123-127.

McVey, C., Huang, F., Elliott, C., and Cao, C. (2017). Endonuclease controlled aggregation of gold nanoparticles for the ultrasensitive detection of pathogenic bacterial DNA. Biosensors and Bioelectronics, 92, 502-508.

Misiakos, K., and Kakabakos, S. (1998). A multi-band capillary immunosensor. Biosensors and Bioelectronics, 13(7), 825-830.

Mohle-Boetani, J. C., Farrar, J. A., Werner, S. B., Minassian, D., Bryant, R., Abbott, S., Slutsker, L., and Vugia, D. J. (2001). *Escherichia coli* O157 and Salmonella infections associated with sprouts in California, 1996–1998. Annals of Internal Medicine, 135(4), 239-247.

Monin, G., and Sellier, P. (1985). Pork of low technological quality with a normal rate of muscle pH fall in the immediate post-mortem period: The case of the Hampshire breed. Meat Science, 13(1), 49-63.

Morant-Miñana, M. C., and Elizalde, J. (2015). Microscale electrodes integrated on COP for real sample *Campylobacter* spp. detection. Biosensors and Bioelectronics, 70, 491-497.

Nakatani, H. S., dos Santos, L. V., Pelegrine, C. P., Terezinha, S., Gomes, M., Matsushita, M., de Souza, N. E., and Visentainer, J. V. (2005). Biosensor based on xanthine oxidase for monitoring hypoxanthine in fish meat. American Journal of Biochemistry and Biotechnology, 1(2), 85-89.

Olojo, R., Xia, R., and Abramson, J. (2005). Spectrophotometric and fluorometric assay of superoxide ion using 4-chloro-7-nitrobenzo-2-oxa-1, 3-diazole. Analytical Biochemistry, 339(2), 338-344.

Park, I.S., and Kim, N. (2006). Development of a chemiluminescent immunosensor for chloramphenicol. Analytica Chimica Acta, 578(1), 19-24.

Przybylski, W., Sionek, B., Jaworska, D., and Santé-Lhoutellier, V. (2016). The application of biosensors for drip loss analysis and glycolytic potential evaluation. Meat Science, 117, 7-11.

Rand, A. G., Jianming, Y., Brown, C. W., and Letcher, S. V. (2002). Optical biosensors for food pathogen detection. Food Technology, 56(3), 32-39.

Reddy, S. M., Hawkins, D. M., Phan, Q. T., Stevenson, D., and Warriner, K. (2013). Protein detection using hydrogel-based molecularly imprinted polymers integrated with dual polarisation interferometry. Sensors and Actuators B: Chemical, 176, 190-197.

Reddy, S. M., Phan, Q. T., El-Sharif, H., Govada, L., Stevenson, D., and Chayen, N. E. (2012). Protein crystallization and biosensor applications of hydrogel-based molecularly imprinted polymers. Biomacromolecules, 13(12), 3959-3965.

Renata, S., Pagliarussi, L., Luis, A.P., Freitas, L., and Bastos, J. K. (2002). A quantitative method for the analysis of xanthine alkaloids in *Paullinia cupana* (guarana) by capillary column gas chromatography. Journal of Separation Science, 25(5 6), 371-374.

Rider, T. H., Petrovick, M. S., Nargi, F. E., Harper, J. D., Schwoebel, E. D., Mathews, R. H., Blanchard, D. J., Bortolin, L. T., Young, A. M., and Chen, J. (2003). AB cell-based sensor for rapid identification of pathogens. Science, 301(5630), 213-215.

Sadeghi, S., Fooladi, E., and Malekaneh, M. (2014). A nanocomposite/crude extract enzyme-based xanthine biosensor. Analytical Biochemistry, 464, 51-59.

Scheffler, T., and Gerrard, D. (2007). Mechanisms controlling pork quality development: The biochemistry controlling postmortem energy metabolism. Meat Science, 77(1), 7-16.

Scognamiglio, V. (2013). Nanotechnology in glucose monitoring: advances and challenges in the last 10 years. Biosensors and Bioelectronics, 47, 12-25.

Shan, D., Wang, Y., Xue, H., and Cosnier, S. (2009). Sensitive and selective xanthine amperometric sensors based on calcium carbonate nanoparticles. Sensors and Actuators B: Chemical, 136(2), 510-515.

Shu, B., Zhang, C., and Xing, D. (2014). Segmented continuous-flow multiplex polymerase chain reaction microfluidics for high-throughput and rapid foodborne pathogen detection. Analytica Chimica Acta, 826, 51-60.

Singh, R., Mukherjee, M. D., Sumana, G., Gupta, R. K., Sood, S., and Malhotra, B. (2014). Biosensors for pathogen detection: A smart approach towards clinical diagnosis. Sensors and Actuators B: Chemical, 197, 385-404.

Sliwinska, M., Wisniewska, P., Dymerski, T., Namiesnik, J., and Wardencki, W. (2014). Food analysis using artificial senses. Journal of Agricultural and Food Chemistry, 62(7), 1423-1448.

Steenkamp, J.B. E. (1990). Conceptual model of the quality perception process. Journal of Business Research, 21(4), 309-333.

Stevens, R. C., Soelberg, S. D., Eberhart, B.T. L., Spencer, S., Wekell, J. C., Chinowsky, T., Trainer, V. L., and Furlong, C. E. (2007). Detection of the toxin domoic acid from clam extracts using a portable surface plasmon resonance biosensor. Harmful Algae, 6(2), 166-174.

Su, X.L., and Li, Y. (2005). A QCM immunosensor for Salmonella detection with simultaneous measurements of resonant frequency and motional resistance. Biosensors and Bioelectronics, 21(6), 840-848.

Tencza, S. B., and Sipe, M. A. (2004). Detection and classification of threat agents via high content assays of mammalian cells. Journal of Applied Toxicology, 24(5), 371-377.

Thakur, M., and Ragavan, K. (2013). Biosensors in food processing. Journal of Food Science and Technology, 50(4), 625-641.

Thomas, T. S. (2006). Development of a capillary based helicobacter hepaticus biosensor. University of Missouri—Columbia.

Tu, S. I., Uknalis, J., Gore, M., and Irwin, P. (2002). The capture of *Escherichia coli* O157: H7 for light addressable potentiometric sensor (LAPS) using two different types of magnetic beads. Journal of Rapid Methods and Automation in Microbiology, 10(3), 185-195.

Tu, S. I., Uknalis, J., Gore, M., Irwin, P., and Feder, I. (2003). Factors affecting the bacterial capture efficiency of immuno beads: a comparison between beads with different size and density. Journal of Rapid Methods and Automation in Microbiology, 11(1), 35-46.

USDA-FSIS. (1996a) National beef microbiological baseline data collection program: cows and bulls, December 1993–November 1994.

USDA-FSIS. (1996b). National federal plant raw ground beef microbiological survey, August 1993–March 1994.

Van Eenennaam, A., Li, J., Thallman, R., Quaas, R., Dikeman, M., Gill, C., Franke, D., and Thomas, M. (2007). Validation of commercial DNA tests for quantitative beef quality traits. Journal of Animal Science, 85(4), 891-900.

Vázquez-Boland, J. A., Kuhn, M., Berche, P., Chakraborty, T., Domýìnguez-Bernal, G., Goebel, W., González-Zorn, B., Wehland, J., and Kreft, J. (2001). Listeria pathogenesis and molecular virulence determinants. Clinical Microbiology Reviews, 14(3), 584-640.

Warriner, K., Reddy, S. M., Namvar, A., and Neethirajan, S. (2014). Developments in nanoparticles for use in biosensors to assess food safety and quality. Trends in Food Science and Technology, 40(2), 183-199.

Waswa, J., Irudayaraj, J., and DebRoy, C. (2007). Direct detection of *E. coli* O157: H7 in selected food systems by a surface plasmon resonance biosensor. LWT-Food Science and Technology, 40(2), 187-192.

Weigl, B. H., and Wolfbeis, O. S. (1994). Capillary optical sensors. Analytical Chemistry, 66(20), 3323-3327.

Wheeler, T., Vote, D., Leheska, J., Shackelford, S., Belk, K., Wulf, D., Gwartney, B., and Koohmaraie, M. (2002). The efficacy of three objective systems for identifying beef cuts that can be guaranteed tender. Journal of Animal Science, 80(12), 3315-3327.

Whipple, G., Koohmaraie, M., Dikeman, M., and Crouse, J. (1990). Predicting beef-longissimus tenderness from various biochemical and histological muscle traits. Journal of Animal Science, 68(12), 4193-4199.

Woodbury, R. G., Wendin, C., Clendenning, J., Melendez, J., Elkind, J., Bartholomew, D., Brown, S., and Furlong, C. E. (1998). Construction of biosensors using a gold-binding polypeptide and a miniature integrated surface plasmon resonance sensor. Biosensors and Bioelectronics, 13(10), 1117-1126.

Zelechowska, E., Przybylski, W., Jaworska, D., and Santé-Lhoutellier, V. (2012). Technological and sensory pork quality in relation to muscle and drip loss protein profiles. European Food Research and Technology, 234(5), 883-894.

Zhi, X., Deng, M., Yang, H., Gao, G., Wang, K., Fu, H., Zhang, Y., Chen, D., and Cui, D. (2014). A novel HBV genotypes detecting system combined with microfluidic chip, loop-mediated isothermal amplification and GMR sensors. Biosensors and Bioelectronics, 54, 372-377.

Zór, K., Castellarnau, M., Pascual, D., Pich, S., Plasencia, C., Bardsley, R., and Nistor, M. (2011). Design, development and application of a bioelectrochemical detection system for meat tenderness prediction. Biosensors and Bioelectronics, 26(11), 4283-4288.

Nanotechnology: A Future of Safety and Quality for Meat and Meat Products

Rajkumari Sanjukta

Division of Animal Health, ICAR Research Complex for Northeastern Hill Region, Umiam, Ribhoi, Meghalaya

**e-mail: sanjukta.rajkumari@gmail.com*

ABSTRACT

Nanotechnology has a substantial impact on the processing of meat and meat products so as to ensuring safety and quality of meat and biosecurity. Development of novel nanomaterials and generation of advance processing techniques and products enable its wide and successful applications in meat and its products. The main advantages of employing nanomaterials in meat and its products; it is able to increase antimicrobial efficiency, improve organoleptic properties, enhance bioavailability, and site-specific delivery of active components. Nanotechnology applications in meat and its products are wide and varied not withstanding only to the products by direct application rather by indirect application for facilities and personnel involved in the production units. In meat industry it has been acclaimed for encapsulation and targeted delivery of active compounds, enhance flavours, incorporation of antibacterial agents, increase and improving storing capability, as nanosensors for detecting contamination, as nanotracers for source tracking and good presentation of the products for overall safety and quality assurance. In this chapter nanotechnology application in meat and meat products is broadly classified into direct and indirect applications for ease of understanding. Direct application includes nanopackaging, nanotracer (traceability and authencity of meat and its products), nano-enabled sensors (quick detection of pathogen/contaminant) and nanomaterials as nutraceuticals, functional foods, additives, supplements and as other active components. Indirect applications are applications of nanotechnology as disinfectants for equipment and production rooms, surface biocides, filters (air/water) and protective clothing used in the meat industry. However, nanotechnology can be best applied in meat products completely by addressing the challenges in reception by the public, economics, legislation and regulations to ensure biosafety of human and environment.

Keywords: *Nanotechnology, Meat, Meat products.*

1. Introduction

The United Nation, Food and Agricultural Organization (FAO) foresee that the world has to upsurge the food production by 70 per cent with the anticipated population explosion up-to 9 billion by 2050. Further, King *et al.* (2018) opined that with this tremendous population growth along with the economic improvement in the developing and underdeveloped populations would lead to increase demand of protein from animal sources. Thus this will force an increased food production from the existing agro-ecosystems, further pressurizing the environment.

Further, in an era of climate change and the shrinking of the environment in its totality inclusive of its renewable and non-renewable resources, results in the challenge for food security worldwide. Therefore, there is an urgent need for an alternative and source of food or change in the food production, distribution system for sustainable food security. The overall need for food is even more increasing than ever particularly the animal source protein. The increasing demand of food in a deteriorating ecology is well emphasized with time but it needs to be prioritized from sustainable resources. The growing consumer awareness and demand for affordable and safe food, producers will be compelled for the exploration of a sustainable technology fulfilling production need, safety and economic viability. Innovative and scientific technologies will be inevitable for achieving these challenges.

Nanotechnological innovations and advances are promising technology with potential to bring about change in the food industry in various aspects like increasing food productions, nutritional enhancement, food safety and security worldwide (King *et al.*, 2018). Various applications of nano technological advances either singly or in combinations with other technologies are noticeable in the specific food delivery system, encapsulation, flavour, nutrient enhancement (supplements, additives, nutrients) food packaging, antibacterial enhancement, food spoilage and contamination sensors, increasing shelf enhancement, tracing and tracking food, healthier and sustainable food system and consumer satisfaction and acceptability and environmental safety.

2. Meat and Meat Products: Safety and Quality

"Meat is normally regarded as the edible parts of the food animals, which consume mainly grass and other arable crops, namely, sheep, goats, pigs, horses, deer and others" (David and Robert, 2014). Meat is relished worldwide as an important dietary component having both macronutrient and micronutrients (all essential nutrients; proteins, fats, vitamins, minerals) for good health and optimal nutrition. The future for meat and its meat products will depend mainly on consumer demand and the prices at which they can be profitably produced. Meat is consumed fresh or after preservations. Preservations may be for shorter or longer one like drying, curing, cold, heat, chemicals, irradiation and high pressure to also prevent from shrinkage, sweating, loss of bloom and drying. Chemical preservation can be achieved by smoking, salting and pickling.

Meat consumption is increasing worldwide and this brings along the challenges of meat hygiene and safety for better health and nutrition. One of the greatest challenges of meat hygiene and safety is the control of foodborne diseases particularly meat borne diseases. These diseases may be caused by bacteria, virus or parasites. Major and most common pathogens are bacterial in nature like *Salmonella, E.coli* 0157: H7, *Camplylobacter, Listeria, Staphylococcus, Clostridium, Bacillus etc.* Other hindrances to food safety also includes toxins, pesticides, antibiotic residues, drugs, heavy metals which are posing a major concerns to public health with severe implications to health and well being. Besides the common pathogens causing food borne diseases, food spoilage is also a major concern.

Meat plants required a strict HACCP protocols to ensure production of quality and safe meat. There are various procedure to be undertaken starting from farm to consumers. Assessment on the cleanliness and hygiene are routine procedures and maintaining microbiological, chemical analysis reports of the plant is crucial for tracking or tracing the source of contamination or pathogens. Application of nanotechnology can be incorporated in various stages of meat plant right from the primary production to consumers which can lessen the burden of some of these challenges.

3. Basics of Nanotechnology

"Nanotechnology" is a very novel area having potential to turn basic research into successful innovations. The term "nanotechnology" was coined by Norio Taniguchi in 1974. Nano is a Greek word meaning "dwarf". Nanotechnology is technology in nanosize ranging from 1nm to 100 nm and a nanometer is 10^{-9}m *i.e.* one billionth of a meter. The concept of "nanotechnology" was introduced by Richard P. Feyman (Nobel Laureate in Physics, 1965).

"Nanotechnology is the understanding and control of matter at dimensions of roughly 1 to 100 nanometers, where unique phenomena enable novel applications, encompassing nanoscale science, engineering and technology, nanotechnology involves imaging, measuring, modeling, and manipulating matter at this length scale" (National Nanotechnology Initiative). Nanomaterials can be of various forms like nanoparticles, nanopores, nanoshells, nanotubes, quantam dots, fullerenes, nanofibres, nanosheets, nanowhiskers, cantilevers, liposomes, dendimers, nanosensors *etc.* There are two ways of synthesizing nanomaterials *viz.* bottom up and top down approaches. Applications of nanotechnology are varied including almost all fields of science and engineering. In biological system they can be applied for diagnostics, therapeutics, vaccines, drug discovery, as nutraceuticals, food industry, delivery systems, microfluidics and various others. Nanotechnology has a great potential to solve many problems and enhance human and animal health and welfare.

4. Potential Applications of Nanotechnology in Enhancing Meat Quality and Safety

"Nanotechnology offers different ways of application in meat industry *viz.* as preservatives, possibility of reducing preservatives and other undesirable or potentially harmful substances in food products, affects development of new or improved tastes, textures and bioavailability of nutrients and supplements, packaging which will in turn extend shelf–life and keep products safe from microbial pathogens"(Chaudhry and Castle, 2011).

Application of nanotechnology to ensure safety and quality of meat can be broadly classified into direct and indirect applications for ease of understanding. Direct application will include all those applications which are directly applied on meat and meat product to ensure safety and retain or enhance its quality like the applications stated by King *et al.* (2018), like "nano-enabled packaging, biosensors and rapid detection methods for contaminants and technologies that assure the authenticity and traceability of products". Indirect applications are those which are not directly applied on meat or meat products but rather on the environment, facilities, equipments, used by personnel so as to ensure safety of the meat and meat product like "disinfectants, surface biocides, protective clothing, air and water filters" (King *et al.*, 2018). Recent advances in nanotechnology application in meat industry are listed in Table 14.1.

Table 14.1. Recent Advances in Nanotechnology Application in Meat Industry

Sl. No.	*Nanomaterial*	*Applications*	*Highlights*	*References*
1	Gold nanoparticles	Pork meat (meat speciation)	Colourimetric detection and distinguished single base of difference in DNA.	He and Yang, 2018
2	Nanocomposites (chitosan and carboxymethyl cellulose films)	Camel Meat (antibacterial effect and delay spoilage)	Increased the shelf life and inhibit bacterial growth retarded spoilage by chemical changes in the biomolecules and enhanced the sensory attributes	Khezrian *et al.*, 2018
3	Nanosensors	Meat (spoilage detection)	"An amine sensor was developed based on 2D assembly for meat spoilage monitoring".	Han *et al.*, 2018
4	Nanocarriers	Food- against (antibacterial effect)	"The loaded systems showed antimicrobial efficacy against pathogens in real foods however depended on food matrices' composition".	Chatzidaki *et al.*, 2018
5	Nanosensors	Pork meat	"Porcine meat detection by higly sensitive genosensor detecting upto 1 per cent pork meat in mixtures within 2.5hr including DNA extraction".	Torelli *et al.*, 2017
6	Nanoencapsulated system	Pork meat	"Enhanced diffusion of encapsulated fatty acids in pork meat, improved nutritional profile of meat, lipid quality indices were significantly influenced by cooking".	Ojha *et al.*, 2017

Sl. No.	*Nanomaterial*	*Applications*	*Highlights*	*References*
7	Nano-composite	Meat against spoilage	"Inhibited the growth of spoilage and pathogenic bacteria, improved organoleptic properties and increased shelf life of meat samples".	Sani *et al.*, 2017
8	Nano-composite sensor	Meat spoilage detection	"Favorable microenvironment for direct electrochemistry of xanthine oxidase. It exhibited excellent electro catalytic activity towards hypoxanthine".	Albelda *et al.*, 2017
9	Nanosenor	Fish, Chicken, Beef	"Highly sensitive for the determination of xanthine".	Dervisevic *et al.*, 2017
10	Nano additives	Meat (food additives)	"BNC resulted in increased water-binding properties, hardness, cohesiveness, and chewiness and suitable fat mimetic, good stability".	Marchetti *et al.*, 2017
11	Quantam dots	Food (detection of pathogens)	"Detect bacterial growth (*E.coli*) on food products".	Mohamadi *et al.*, 2017
12	Nanosensors	Food (detection of pathogens)	"Ultra-rapid colorimetric biosensor developed for the detection of *E. coli*".	Ghadeer *et al.*, 2017
13	Nanocarrier	Food (bacteriosatic effect)	"Bacteriostatic effect against *Listeria monocytogenes*".	de Almeida *et al.*, 2018
14	Nanoparticles	Meat (shelf life)	"Superior film when compared with other, significant action against the microbes extends the shelf life".	Rahman *et al.*, 2017
15	Nanosensors	Poultry meat (detection of pathogens)	"Able to detect pathogens using cotton-tips. Lower sensitivity of each bacterial cell was determined".	Alamer *et al.*, 2017
16	Nanosensors	Pork (adulteration detection)	"Excellent specificity against albumins from other species and on-site pork adulteration detection in raw meat".	Lim and Ahmed, 2016
17	Nano-scale antibiotic	Fish (antibiotic and preservation)	"Edible coatings from Nano-N present a high antimicrobial activity and extended the shelf life".	Wu *et al.*, 2016
18	Nanocomposites	Pork (packaging)	"Advantages over current high barrier commercial films enhanced by nanoclays".	Lloret *et al.*, 2016
19	Nanosensors	Food (antibiotic detection)	"Analysis of foodstuffs for the penicillins' detection".	Karaseva *et al.*, 2016
20	Nanocomposites	Meat (antibiotic determination)	Antibiotic determination in meat and environmentally friendly procedure.	Castillo-García *et al.*, 2015
21	Nanoparticles (Chitosan)	Pork (Preservatives)	Improved the antioxidant and antimicrobial property and prolonged the shelf-life.	Hu *et al.*, 2015
22	Nanoparticle (Silver)	Packaging (antibacterial)	Nanoscale silver affected most bacteria *in vitro* although not *in situ*.	Kuuliala *et al.*, 2015

Sl. No.	Nanomaterial	Applications	Highlights	References
23	Nanocomposites (ZnO and Silver)	Poultry (antimicrobial effect and spoilage delay)	Antimicrobial effect and delayed lipid oxidation.	Panea *et al.*, 2014
24	Nanoparticles (ZnO)	Poultry meat (antibacterial)	Inhibited the growth of pathogens	Akbar *et al.*, 2014
25	Nanoparticle (paprika oleoresin)	Poultry meat	Enhanced marinating performance of poultry meat.	Yusop *et al.*, 2012

5. Direct Application

5.1 Nanopackaging

Meat and meat products are highly perishable and preservation is one of the solutions to increase the shelf life and safe meat. Nano-packaging systems are unquestionably the most promising use of nanotechnology in meat and meat products. There are primarily three ways of nano-technological applications to food packaging *viz.* "direct incorporation into food products, incorporation in food packaging material, and application in food processing" (Sharma *et al.*, 2017).

Nanopackaging includes various packaging methods like biodegradable packaging, active packaging, edible coatings/films, smart/intelligent packaging *etc.* Nanomaterial packaging has the potential to safeguard food quality and safety, decreases environmental influences, prevents microbial and lipid oxidation spoilage, contamination, enhance organoleptic properties like tenderness, colour, aroma, increases shelf life and also increases the overall presentation and more appealing to the consumers (Panea *et al.*, 2014; Majid *et al.*, 2016). Nanomaterials provide intelligent/smart and active packaging by enhancing the mechanical and barrier parameters of food packages. "The use of polymer nanotechnology in packaging of food aims to improve the principal features of traditional packaging systems *i.e.*, containment (ease of transportation and handling), convenience (being consumer friendly), protection and preservation (avoids leakage or break-up and protects against microbial contaminants, offering longer shelf life), marketing and communication (real-time information about the quality of enclosed food stuffs, besides the nutritional constituents and preparatory guidelines" (Vanderroost *et al.*, 2014). It may be categorized as "improved, intelligent/smart and active packaging' (Sharma *et al.*, 2017).

5.2 Improved Packaging through Nanocomplexes

Polymer nanocomplexes are improved nanopackaging materials prepared by mixing polymers with inorganic or organic fillers having specific geometrical shapes like fibers, flakes, spheres (Prateek *et al.*, 2016). Packaging fillers are influenced by the ratio of the maximum to minimum dimension of the fillers; fillers with higher ratio have more defined surface area with accompanying strengthening features (Rafieian and Simonsen, 2014). Several nanomaterials like silica, clay, organo-clay, graphene, polysaccharide nanocrystals, carbon nanotubes, chitosan, cellulose-based, and other metal nanoparticles, such as, ZnO_2, colloidal Cu, or Ti are being increasingly employed as fillers (Sharma *et al.*, 2017).

Nano-packaging system was also developed using blueberry extract to interspace between silicate interlayer spaces of clay (Gutierrez *et al.*, 2017). All-cellulose nanocomposite was prepared using sugarcane bagasse nanofiber and N, N-dimethylacetamide/lithium chloride (Ghaderi *et al.*, 2014). Carbon nanotubes along with allyl-isothiocynante could inhibit *Salmonella choleraesuis* in storage upto 40 days (Dias *et al.*, 2013). Nanocomposite films made using chitosan in combinations with either metal (silver or gold) or plants component like cinnamaldehyde have been proven to exhibit antimicrobial property against various microbes like activity against *Aspergillus niger, Candida albicans, E. coli,P. aeruginosa* and *S. aureus* (Rieger *et al.*, 2015).

5.3 Active Packaging

Active packaging is novel packaging system that incorporates active compounds like antimicrobial agents, additives, supplements, preservatives, ethylene remover or absorbs water or oxygen from the packaged food items so as to ensure quality, safety of the product and ensure increased shelf life. Several metals (silver, gold, zinc) and metal oxide (titanium dioxide, zinc oxide, silicon oxide and magnesium oxide) based nanomaterials have been exploited in various active packaging development (Sharma *et al.*, 2017). The active compounds used in active packaging can either act directly on contact or they can interact with the components of food materials. The antimicrobial activity they exhibit can be attributed to direct contact with microbes (interrupting transmembrane electron transfer, disrupting/penetrating the cell envelope); oxidizing cell components; and production of secondary products (*e.g.*, reactive oxygen species - ROS or heavy metal ions), leading to cell damage (Sharma *et al.*, 2017). Orsuwan *et al.* (2016) prepared binary blendfilms comprising of agar and banana powder (A/B) compositefilms reinforced with silver nanoparticles (A/B/AgNPs). Thesefilms exhibit improved antibacterial property against *E. coli* and *L. monocytogenes.*

Titanium dioxide nanoparticles based food packagings are highly efficient, as they possess UV blocking potential and photo-catalytic activity (Farhoodi, 2016). Iron-containing kaolinite was used to develop oxygen scavenging packaging films by altering high-density polyethylene (HDPE) films (Busolo and Lagaron 2012). Holding of enzymes at nanoscale has been exploited to increase the surface area, increase efficiency, increase stability, pH, temperature, resist degradation by enzymes (proteases) and sustained release of enzymes in the food (Brandelli *et al.*, 2017).

5.4 Intelligent/Smart Packaging System

Intelligent/smart packaging systems boost the communication aspect of a package and such type of advanced packaging recognize any characteristics of the packaged food and apply different mechanisms to register and convey information regarding the existing quality or condition of the food with regard to its safety and digestibility. Intelligent/smart packaging systems use diverse methods to communicate *viz.*, nano-sensors, time temperature indicators, oxygen sensors, CO_2 sensors, humidity indicator, spoilage indicator, pathogens indicator, freshness indicators *etc.* (Sharma *et al.*, 2017). Nanosensors usually incorporate an intelligent function by means of labels or coatings inorder to ensure leakage proof (like vacumn packaging), indicates time-temperature (freeze-thaw-refreeze) or safety from microbes (spoilage/food borne microbes contaminations) (Fuertes *et al.*, 2016). Ultra-violet based activated oxygen indicator membrane was developed by electrospinning method by encapsulating the core component of TiO_2 nanoparticles, glycerol and methylene blue (Mihindukulasuriya and Lim, 2014). Chitosan based CO_2 indicator was developed for screening storage shelf life of packaged food (Jung *et al.*, 2012). Fluorescent nanoparticle was developed for selective detection of *E.coli* by conjugating antibodies against *E. coli* with citrate modified oleic acid-capped NaYF4: Yb, Er (Ong *et al.* (2014).

A polyaniline film was developed for real-time detection of fish spoilage at different temperatures by exploiting the chemicals released during spoilage of fish (basic volatile amines) (Kuswandi *et al.*, 2012). Monitoring of temperature of food during storage, handling, and distribution can be done by time temperature indicators like times trip, gold nanoparticles based strips for chilled foods giving colour indication at temperature above freezing indicating spoilage (Robinson and Morrison, 2010). Humidity indicator has also been explored using iridescent technology like the use of nanocrystalline cellulose iridescent film by casting method and adjusting colour when interacting with electromagnetic fields (Zhou, 2013).

5.5 Nanomaterials as Food Components/Additives or Functional Foods

Nanomaterials as food components/additives or functional foods are placed directly into food, or as a part of food packaging (Coles and Frewer, 2013; Rhim *et al.*, 2013). There are various advantages of using nano-encapsulated food additives and supplements like increasing bioavailability of fat soluble ingredients in food, improves taste, reduce the use of other food constituents like sugar, salt, fats and preservatives (Chaudhry and Castle, 2011).

Ramachandraiah *et al.* (2018) had reviewed comprehensively various strategies like nutritional strategy, meat products reformulation and meat storage/consumption strategy, which directly or indirectly correlate with the enhancement of safety and meat quality and also its products. Under nutritional strategy, used of nano-chelating technology for micronutrients delivery like Zn-nano-max, Zn-nano-methionine, and magnesium oxide nanoparticle improve growth performance and enhance the activities of digestive enzymes (Srinivasan *et al.*, 2017). Incorporation of nanomaterials (bioactive nanoparticels) in feed functioned as an antibiotic replacer, reducing the burden of antimicrobial resistance. Nanocomposites, nanoemulsions as delivery of vitamins or minerals, nanoencapsulation of linoliec acid in conjugated form reduces fat and cholesterol and improve of fatty acids in meat (Wenjuan, *et al.*, 2010; Heo *et al.*, 2016). Meat products reformulation strategy included for (i) Incorporation of probiotics (*Bifidobacterium longum* and *Lactobacillus reuteri*) and prebiotics by microencapsulations in meat products like sausages (Jahne *et al.*, 2013); nanoencapsulated antimicrobial in chicken broth (Mate *et al.*, 2016), nanoparticles loaded with thyme essential oil in chitosan of beef burgers (Ghaderi-Ghahfarokhi *et al.*, 2016) which reduces lipid oxidation, improves antimicrobial effect and sensory properties; (ii) Size dependent incorporation of calcium in fish product induces transglutaminase activity thereby enhancing textural properties (Yin *et al.*, 2017) (iii) Sodium reduction by addition of salt chitin nanofibers to reduce sharpness and improve saltiness (Jiang *et al.*, 2017) (iv) Phosphate and fat reduction/replacement like *e.g.* phosphate encapsulation reduces the quantity and retain its beneficial effect (Sickler *et al.*, 2013).; starch nanoparticles mimic fat (Kim *et al.*, 2015) nanoparticulated whey proteins can act as fat replacer (Liu *et al.*, 2017) (v) nitrite replacement like use of nanoencapsulated cinnamon oil, nanomicelle paprika, nanoformulation using ginger extract in spent hens and others like nanoencapsulation of antioxidants, antimicrobials, flavours *etc.* (Ghaderi-Ghahfarokhi *et al.*, 2016).

5.6 Nanosensors for Rapid Detection of Pathogen/Contaminant

There are many hindrances to food safety; mentioned may be pathogens, pesticides, antibiotics, adulterants, toxins, heavy metals, organic compounds and others (Zeng *et al.*, 2016). Food safety authority should take utmost care in rendering a food product safe, as food borne illness is a major threat to human health. Thus, ruling out the presence of any of these contaminants need to be taken in priority before it reaches the market and ultimately to the consumers table. Nanotechnology has culminated various technological advances for detections and quantification of such contaminants with high sensitivity, specificity, selectivity, simplicity, fast, low cost, portable, ability to multiplex and many other advantages (Ali *et al.*, 2014).

Nanosensor is a device that consists of bioreceptor and a transducer that can detect biological interaction (analyte with bioreceptor) and is measured by transducer as electric signal, which be interpreted as the final results. The commonly used biorceptor in such nanosensors are based on various cellular interactions ranging from antigen-antibody, nucleic acid, enzyme-substrate, microbes, synthetic substances and others. The most common metal oxide and metal nanoparticles used for sensors and biosensors are gold, silver, magnetic, silica nanoparticles, carbon nanotubes, gold nanorods and quantum dots. These are mainly employed in nanosensors in order to detect, quantify analytes most often used in meat industry like "gasses, vapors, ions, small biomolecules and various microbes involve in foodborne diseases (Banerjee *et al.*, 2017; Wang and Duncan, 2017).

Foodborne illnesses pose a significant health-risk worldwide, affecting upwards of 600 million people every year (WHO, 2016). Nanosensors has been targeted for detection and of food borne pathogens relevant and most often reported in meat industry like *Salmonella* Typhimurium, *Salmonella* Enteritidis (Joo *et al.*, 2012), *Staphylococcus aureus, Listeria monocytogenes,* E coli O157: H7 (Mao *et al.*, 2006;Maurer *et al.*, 2012; Suaifan *et al.*, 2017) and *Campylobacter jejuni*. Immuno based nanosensor with lower sensitivity was developed to detect *Salmonella* Typhimurium, *Salmonella* Enteritidis, *Staphylococcus aureus* and *Campylobacter jejuni* on chicken surfaces using cotton tips (Saleh and Zourob, 2017). Amine-functionalized mangnetic nanoparticles based rapid and efficient assays was developed for simultaneous detection of *B. cereus, B. subtilis, E. coli, P. vulgaris, P. aeruginosa, Sarcinalutea, S.*

aureus and *Salmonella* (Huang *et al.*, 2010). The simultaneously detection of multiple toxins (cholera toxin, ricin, shiga like toxin, and Staphylococcal enterotoxins B) by multiplexed fluoroimmunoassays were developed by conjugating highly luminescent semiconductor nanocrystals and antibodies (Goldman *et al.*, 2004). Nanosensors have been employed for antibiotics detection in chicken tissue (Ahn and Lim, 2015), detection of biogenic amines to monitor meat spoilage (Liu *et al.*, 2015) and detection of melamine (Duncan, 2011), pesticide residues (Xiang *et al.*, 2011) and detection of mycotoxins has been broadly reviewed (Rai *et al.*, 2015).

5.7 Nanotracer

Tracing the authenticity of meat and its products is a primary factor in ensuring the safety and quality related decision by consumers. The major concerns arising to ascertain the source of meat or its products and various associated information are basically the malpractices of adulteration of meat or meat products by lower quality or cheaper meat or ingredients mimicking meat, speciation of meat, the rise of demand from consumers for healthy and quality meat entailing its origin, organic status, food safety, adhering to global standards and biosecurity issues. Furthermore, this will aid in the existing system of product investigation, handling crisis like disease outbreak, trade embargo or any product recall or management of any predicament. Many researchers have also noted the importance of tracing application of nanotechnology either during the development of technology or reviewed its usage in meat industry (King *et al.*, 2018).

Nanobarcode technology is one such technology which can address the various issues arising due to unknown origin and uncertainty of the authenticity of the products. Nanobarcode enables to monitor the supply chain, through the unique product information provided citing various data from farm to fork, leaving the customers more aware and enable them to take an informed decision with regards to the quality and safety of food they consumed. Such products are already in market in some countries like United Kingdom, Oxonica, with the provision of unique reading strips consisting of silver, gold and platinum with different reflectivity (King *et al.*, 2018).

6. Indirect Applications

6.1 Disinfectants for Facilities and Infrastructures

Food processing plants particularly meat processing plant adhere to various microbiological and chemical standards as per the introduction of Hazard Analysis Critical Control Points (HACCP) for assessing critical control points in production so as to ensure safety of the products from microbial contaminations, chemical hazards of the product, purity of the air, cleanliness and hygiene of production area, where these hazards can be controlled, monitored, and eliminated or minimalized (Konopka *et al.*, 2009).

Many chemical disinfectants used in the meat industry contain ammonia, aldehydes, acrolein, acetaldehyde (formaldehyde), hydrogen sulfide, mercaptans and others. The use of strong chemicals as disinfectant in food industry is restricted due to its adverse effect on health like emerging microbial resistance, carcinogenic effect, toxicity, irritation on skin, mucous membrane, palpitation *etc.* besides its difficulties with solubility and the possibility of direct application.

This necessitates the requirement for a novel solution in this field. Many researchers have investigated numerous nanomaterials with broad antimicrobial activity for the development of an efficient disinfectant. King *et al.* (2018), has thoroughly reviewed this aspect, detailing numbers of commercially available nanosilver based antimicrobial products. Further, many metal nanoparticles poses a threat due to its haphazard disposal in the environment, the extent and effect of which is currently under discourse (EPA, 2012). One plausible answer to this problem could be the use of Engineered Water Nanostructures (EWNS), produced by electrospraying water vapor, for applications in air and on surfaces. The major advantages of EWNS are that it is safe, ecofriendly, and in contrast to the existing chemical disinfectant its mode of action is through reactive oxygen species (ROS). Therefore has strong external charges, high mobility, airborne and interactive to inactivate microbes on the surfaces, further, no chemical residues are deposited as it vaporizes without any apparent ill effect on health if inhaled (Pyrgiotakis *et al.*, 2015).

6.2 Antimicrobial Agents/Biocides

In meat processing plant, the usage of nanoengineering based surface biocides on equipment or machines will reduce the need for harmful cleaning and disinfecting agents, stop the development of persistent biofilms, prevent cross contamination from surfaces, prevent clogging of machines or other equipments, reduce cost of production and above all ensure food safety (Griffith *et al.*, 2015). Eleftheriadou *et al.* (2017) reported the usage of nanometals like silver, titanium and zinc for creation of antimicrobial surfaces with anti-fouling properties. Cited may be the commercially available Nano-silver refrigerators, cutting boards or self-sanitizing coatings (PEN, 2013). Titanium oxide has been used in combination with UV light for potential ecofriendly uses in several meat processing plants and facilities to prevent surface and air contaminations (Green Earth Nano Science Inc., 2016). Visible-light-active photocatalysis was also developed by chemical modification of TiO_2 (Banerjee *et al.*, 2017). Other combinations like doping of TiO_2 with copper under visible light has also ben reported to possess antibacterial property (Yadav *et al.*, 2014).

6.3 Filtrations of Air and Water Used in Meat Industry

Air and water are the major source of spread and transmission of pathogens leading to contamination in meat processing unit. Nanomaterials are currently used extensively for water treatment due to its efficiency, quickness and cost effectiveness when compared to the traditional technologies. Some of the nanomaterials used in water treatment currently available are nanoadsorbents, nano-enabled membranes, nanophotocatalysts or nano-enabled disinfection systems (Rodrigues *et al.*, 2017). Nanotechnology has been also used for removal of phosphorus from wastewater before releasing in the environment, a main problem often encountered in poultry meat processing plant. Silver nanoparticles are widely used in air filters by various companies in the USA, Korea, Taiwan, China (PEN, 2013).

6.4 Personnel Protective Equipment (PPE) Used in Meat Industry

In order to address the issue of contaminations of meat or its products from the clothing's of handlers as well as for their own safety a variety of nano-based personal protective wears and accessories are available ranging from trousers, socks, coats, facemask *etc.* manufactured and marketed by several companies in the developed countries like USA, UK, Japan, Taiwan, China, Canada, Germany, Czech Republic (PEN, 2013). However, determining the antibacterial performance and lifespan of each nano-enabled clothing item will require further assessments and research works.

7. Safety Concerns and Regulation

Safety concerns and regulations are one of the most important aspects when it comes to the application of nanotechnology or its products in food technology. Assuring safety concerns and regulations will enable to ascertain that the nanomaterial used in foods does not persist, do not accumulate and not lead to any toxicity.

Risk management as well as legislation in the field of nanotechnology needs to be executed for improving the current risk assessing techniques, good governance, implementing framework for incorporation of nanomaterials in foods. Properties of the nanomaterials will determine the nature of effect it will have on the human body. Some of the plausible adverse effects of nanomaterial are toxicity, genotoxicity and carcinogenicity, which is also an emerging concern. The main safety concern of nanomaterials raised by many researchers is its accumulation from the packaging material to the food, which may impact the health of the consumer (Jain *et al.*, 2016). Most of the substances used in the synthesis of nanomaterials are regarded as GRAS (Generally Regarded as Safe). However this consideration is on its original state rather than the nano-state, which definitely might have undergone changes in its physico-chemical characteristics, therefore there is a requirement for more detailed studies on the risk of the nanomaterials. A number of regulatory gap studies have shown that development in nanotechnologies are not taking place in a regulatory vacuum, as the potential risks will be controlled under the existing frameworks (Amenta *et al.*, 2015).

"There are efforts worldwide to address and regulate the production and safe handling/use of nanomaterials (NMs) and nanotechnology either by legislation or by (non-binding) recommendations and guidances" (Van der Meulen *et al.*, 2014). Currently, there is no legislation laid down exclusively for regulation of nanomaterials (Arts *et al.*, 2014). In the EU, they do not have legislation exclusively for nanomaterials regulation, however some regulations pertaining to "Provision of food information to consumers, plastic food contact materials and articles, active and intelligent materials and articles, biocidal products and cosmetic products" exist. In the USA, the Food and Drug Administration (FDA) is accountable to ensure safety of food additives/food contact materials/feed additives. Considerations with regard to nanosize range in food were laid down by FDA in the "Draft Guidance for Industry on Use of Nanomaterials in Food for Animals" (US-FDA, 2014). Japan and Korea are actively participating to working party of nanomaterials manufacturing.

In India, the only regulation for food safety is the Food Safety and Standards Act (2006) and we lack any legislation targeted for nanotechnology risk management and we don't have the resources or expertise for such risk (Barpujari, 2011). In China food safety is regulated under the Food Safety Law, which does not include any NM specifications. Guidance and standards on *e.g.* appropriate (test) methods for risk assessment can be harmonized at international level and periodically adapted to technical progress. FAO and WHO have jointly created the Codex Alimentarius Commission, an intergovernmental agency that aims at creating international food standards, guidelines, codes of practice and advisory texts, which could also cover nanotechnology-based products.

8. Conclusion

Nanotechnology's application in meat industry ranges from farm-to fork for various purposes; quality control, functional food, novel food supplements, additives, packaging, sensors, increase shelf life and others. Nano-form has unique physico-chemical properties, which enhances in various ways with respect to the native form in terms of biological, antimicrobial, nutritional, safety limit of the meat products. As the research and development in the field of nanotechnology is emerging, it will also bring similar opportunities for application towards meat industry; however, more and more research is required for its proper application. Despite the numerous advantages of nanotechnology application in meat industry for ensuring meat safety and quality, associated risk or safety regulations of nanomaterial should not be ignored. Therefore, all issues related to regulations and legal aspects need to be strictly adhered upon so as to make aware of the impact of nanoparticles on human health and environment, for an informed public perceptions and decision-making.

REFERENCES

Ahn, J., and Lim, H. B. (2015). Drop-type chemiluminescence (DCL) system and sample treatment platform using magnetic nanoparticles to determine enrofloxacin and its metabolite in a chicken meat. Food Analytical Methods, 8(1), 79–85. http: //dx.doi.org/10.1007/s12161-014-9871-1.

Akbar, A. and Anal, A.K. (2014). Zinc oxide nanoparticles loaded active packaging, a challenge study against *Salmonella* Typhimurium and *Staphylococcus aureus* in ready-to-eat poultry meat. Food Control, 38, 88-95.

Alamer, S., Chinnappan, R., and Zourob, M. (2017). Development of Rapid Immuno-based Nanosensors for the Detection of Pathogenic Bacteria in Poultry Processing Plants, Procedia Technology,27, 23-26.

Albelda, J.A.V., Uzunoglu,A., Santos, G.N.C., and Stanciu, L.A. (2017). Graphene-titanium dioxide nanocomposite based hypoxanthine sensor for assessment of meat freshness. Biosensors and Bioelectronics, 89, 518-524.

Ali, M.E., Hashim, U., Mustafa, S., Che-Man, Y.B., Adam, T., and Humayun, Q. (2014). Nanobiosensor for the detection andquantification of pork adulteration in meatball formulation. Journal of Experimental of Nanoscience. 9,152-160.

Amenta, V., Aschberger, K., Arena, M., Bouwmeester, H., Moniz, H., Brandhoff, P., Gottardo S., Marvin, H.J.P., Mech, A., Pesudo, L.Q., Rauscher, H., Schoonjans, R., Vettori, M.V., Weigel, S., and Peters, R.J. (2015). Regulatory aspects of nanotechnology in the agri/feed/food sector in EU and non-EU countries. Regulatory Toxicology and Pharmacology, 73, 463-476. http: //dx.doi.org/10.1016/j.yrtph.2015.06.016

Arts, J.H.E., Hadi, M., Keene, A.M., Kreiling, R., Lyon, D., Maier, M., Michel, K., Petry, T., Sauer, U.G., Warheit, D., Wiench, K., and Landsiedel, R. (2014). A critical appraisal of existing concepts for the grouping of nanomaterials. Regulatory Toxicology Pharmacology, 70 (2), 492-506.

Banerjee, T., Shelby, T., and Santra, S. (2017). How can nanosensors detect bacterial contamination before it ever reaches the dinner table? Future Microbiology, 12(2),97–100.

Barpujari, I. (2011). Attenuating risks through regulation: issues for nanotechnology in India. Journal of Biomedicine and Nanotechnology, 7 (1), 85e86.

Brandelli, A., Brum, L. F. W., and dos Santos, J. H. Z. (2017). Nanostructured bioactive compounds for ecological food packaging. Environment Chemistry Letter, 15, 193–204. doi: 10.1007/s10311-017-0621-7.

Busolo, M. A., and Lagaron, J. M. (2012). Oxygen scavenging polyolefin nanocomposite films containing an iron modified kaolinite of interest in active food packaging applications. Innovative. Food Science Emerging. Technology. 16, 211–217. doi: 10.1016/j.ifset.2012.06.008

Castillo-García, M.L., Aguilar-Caballos, M.P., and Gómez-Hens, A. (2015). A europium- and terbium-coated magnetic nanocomposite as sorbent in dispersive solid phase extraction coupled with ultra-high performance liquid chromatography for antibiotic determination in meat samples. Journal of Chromatography A, 1425, 73-80

Chatzidaki, M.D., Papadimitriou, K., Alexandraki, V., Balkiza, F., Georgalaki, M., Papadimitriou, V., Tsakalidou E, and Xenakis, A. (2018). Reverse micelles as nanocarriers of nisin against foodborne pathogens. Food Chemistry,255, 97-103

Chaudhry, Q., and Castle, L. (2011). Food applications of nanotechnologies: an overview of opportunities and challenges for developing countries. Trends Food Science Technology, 22, 595–603. doi: 10.1016/j.tifs.2011.01.001

Coles, D., and Frewer, L. J. (2013). Nanotechnology applied to European food production – a review of ethical and regulatory issues. Trends in Food Science and Technology, 34(1), 32–43. http: //dx.doi.org/10.1016/j.tifs.2013.08.006

David, S. C., and Robert J. H. (2014) Gracey's Meat Hygiene, ISBN: 978-1-118-65002-8, Wiley-Blackwell, 11th Edition. doi10.1002/9781118649985.ch13.

De Almeida Roger, J., Magro, M., Spagnolo, S., Bonaiuto, E., Baratella, D., Fasolato, L., and Vianello, F. (2018). Antimicrobial and magnetically removable tannic acid nanocarrier: A processing aid for *Listeria monocytogenes* treatment for food industry applications Food Chemistry, 267, 430-436.

Dervisevic, M., Dervisevic, E., Cevik, E and Senel, M. (2017). Novel electrochemical xanthine biosensor based on chitosan–polypyrrole–gold nanoparticles hybrid bio-nanocomposite platform. Journal of Food and Drug Analysis, 25, 510-519.

Dias, M. V., Nilda, F. S., Borges, S. V., de Sousa, M. M., Nunes, C. A., de Oliveira,I. R. N., and Medeiros, E.A.A. (2013). Use of allylisothiocyanate and carbon nanotubes in an antimicrobial film to package shredded, cooked chicken meat. Food Chemistry, 141,3160–3166. doi: 10.1016/j.foodchem.2013.05.148.

Duncan, V. T. (2011). Applications of nanotechnology in food packaging and food safety: Barrier materials, antimicrobials and sensors, Journal of Colloid and Interface Science, 363, 1–24.

Eleftheriadou, M., Pyrgiotakis, G., and Demokritou, P. (2017). Nanotechnology to the rescue: Using nano-enabled approaches in microbiological food safety and quality. Current Opinion in Biotechnology, 44, 87–93. http: //dx.doi. org/10.1016/j.copbio. 2016.11.012.

Farhoodi, M. (2016). Nanocomposite materials for food packaging applications: characterization and safety evaluation. Food Engineering Review, 8, 35–51. doi: 10.1007/s12393-015-9114-2.

Fuertes, G., Soto, I., Carrasco, R., Vargas, M., Sabattin, J., and Lagos, C. (2016). Intelligent packaging systems: sensors and nanosensors to monitor food quality and safety. Journal of Sensors, 2016,8. doi: 10.1155/2016/4046061.

Ghadeer, A.R.Y., Alhogail, S.S., and Zourob, M. (2017). Paper-based magnetic nanoparticle-peptide probe for rapid and quantitative colorimetric detection of Escherichia coli O157: H7, Biosensors and Bioelectronics, 92, 702-708

Ghaderi, M., Mousavi, M., and Labbafi, M. (2014). All-cellulose nanocompositefilm made from bagasse cellulose nanofibers for food packaging application. Carbohydrate Polymer, 104, 59–65. doi: 10.1016/j.carbpol.2014.01.013

Ghaderi-Ghahfarokhi, M., Barzegar, M., Sahari, M. A., and Azizi, M. H. (2016). Nanoencapsulation approach to improve antimicrobial and antioxidant activity of thyme essential oil in beef burgers during refrigerated storage. Food and Bioprocess Technology, 9, 1187–1201.

Goldman, E.R., Clapp, A.R., Anderson, G.P., Uyeda, H.T., Mauro, J.M.,Medintz, I.L., and Mattoussi, H. (2004). Multiplexed toxin analysis using four colours of quantum dot fluororeagents. Analytical Chemistry, 76,684e8.

Green Earth NanoScience Inc (2016). Home. Retrieved from http: //www. greenearthnanoscience.com/.

Griffith, A., Neethirajan, S., and Warriner, K. (2015). Development and evaluation of silver zeolite antifouling coatings on stainless steel for food contact surfaces. Journal of Food Safety, 35(3), 345–354.

Gutierrez, T. J., Ponce, A. G., and Alvarez, A. V. (2017). Nano-clays from naturaland modified montmorillonite with and without added blueberry extract foractive and intelligent food nanopackaging materials. Material Chemistry and Physics, 194,283–292. doi: 10.1016/j.matchemphys.2017.03.052.

Han, J., Li, Y., Han, J., Li, Y., Yuan, J., Li, Z., Zhao, R., Han, T. and Han, T. (2018). To direct the self-assembly of AIEgens by three-gear switch: Morphology study, amine sensing and assessment of meat spoilage Sensors and Actuators B: Chemical, 258, 373-380

He, Z., and Yang, H. (2018). Colourimetric detection of swine-specific DNA for halal authentication using gold nanoparticles, Food Control, 88, doi 10.1016/j.foodcont.2018.01.001

Heo, W., Kim, E. T., Cho, S. D., Kim, J. H., Kwon, S. M., Jeong, H. Y., Ki, K.S., Yoon, H.B., Ahn Y.D., Lee, S.S., and Kim Y.J. (2016). The in vitro effects of nano-encapsulated conjugated linoleic acid on stability of conjugated linoleic acid and fermentation profiles in the rumen, Asian-Australasian Journal of Animal Science, 29, 365–371.

Hu J, Wang X, Xiao Z, Bi W. (2015). Effect of chitosan nanoparticles loaded with cinnamon essential oil on the quality of chilled pork LWT - Food Science and Technology, 63,519-526

Huang, Y.F., Wang, Y.F., Yan, X.P. (2010). Amine-functionalized magnetic nanoparticles for rapid capture and removal of bacterial pathogens. Environment Science and Technology,44,7908-7913.

Jahne, J., Bonaparte, C., Kuhne, M., and Klein, G. (2013). Viability of microencapsulated probiotic lactobacilli during storage at different temperatures. Berliner und Munchener Tierarztliche Wochenschrift, 126, 10–15.

Jain, A., Shivendu, R., Nandita, D., and Chidambaram, R. (2016). Nanomaterials in food and agriculture: an overview on their safety concerns and regulatory issues. Critical Review in Food Science Nutrition. doi: 10.1080/10408398.2016.1160363.

Jiang, W. J., Tsai, M. L., and Liu, T. (2017). Chitin nanofiber as a promising candidate for improved salty taste. LWT - Food Science and Technology, 75, 65–71.

Joo, J., Yim, C., Kwon, D., Lee, J., Shin, H.H., Cha, H.J., and Jeon, S. (2012).A facile and sensitive detection of pathogenic bacteria using magnetic nanoparticles and optical nanocrystal probes. Analyst, 137,3609-3612.

Jung, J., Puligundla, P., and Ko, S. (2012). Proof-of-concept study of chitosan based carbon dioxide indicator for food packaging applications. Food Chemistry, 135, 2170–2174. doi: 10.1016/j.foodchem.2012.07.090.

Karaseva,N., Ermolaeva, T. and Mizaikoff, B. (2016). Piezoelectric sensors using molecularly imprinted nanospheres for the detection of antibiotics Sensors and Actuators B: Chemical, 225, 199-208

Khezrian, A., and Shahbazi, Y. (2018). Application of nanocomposite chitosan and carboxymethyl cellulose films containing natural preservative compounds in minced camel'smeat. International Journal of Biological Macromolecules, 106, 1146–1158

Kim, H. Y., Park, S. S., and Lim, S. T. (2015). Preparation, characterization and utilization of starch nanoparticles. Colloids and surfaces B. Biointerfaces, 126, 607–620.

King, T., McLeodb, O. M. J., and Duffyc, L.L. (2018). Nanotechnology in the food sector and potential applications for the poultry Industry. Trends in Food Science and Technology, 72, 62-73. https: //doi.org/10.1016/j.tifs.2017.11.015

Konopka, M., Kowalski, Z. and Wzorek, Z. (2009). Disinfection of meat industry equipment and production rooms with the use of liquids containing silver nano-particles. Archives of Environmental Protection, vol. 35, no. 1, pp. 107–115.

Kuswandi, B., Jayus, Restanty, A., Abdullah, A., Heng, L. Y., and Ahmad, M. (2012). A novel colorimetric food package label for fish spoilage based on polyaniline film. Food Control. 25, 184–189. doi: 10.1016/j.foodcont.2011. 10.008

Kuuliala, L., Pippuri, T., Hultman, J., Auvinen, S. M., Kolppo, K., Nieminen, T., and Jääskeläinen, E. (2015). Preparation and antimicrobial characterization of silver-containing packaging materials for meat. Food Packaging and Shelf Life, 6, 53-60

Lim, S. A and Ahmed,M. (2016). A label free electrochemical immunosensor for sensitive detection of porcine serum albumin as a marker for pork adulteration in raw meat Food Chemistry, 206, 197-203

Liu, G., Jæger, T. C., Nielsen, S. B., Ray, C. A., and Ipsen, R. (2017). Interactions in heated milk model systems with different ratios of nanoparticulated whey protein at varying pH. International Dairy Journal, 74, 57–62.

Liu, S. F., Petty, A. R., Sazama, G. T., and Swager, T. M. (2015). Single-Walled carbonnanotube/metalloporphyrin composites for the chemiresistive detection of aminesand meat spoilage. AngewandteChemie International Edition, 54(22), 6554–6557.http: //dx.doi.org/10.1002/anie.201501434.

Lloret,E., Trbojevich, A.F., Arnau, J. and Picouet, P.A. (2016). Relevance of nanocomposite packaging on the stability of vacuum-packed dry cured ham Meat Science, 118, 8-14.

Majid, I., Nayik, G. A., Dar, M. S. and Nanda, V. (2016). Novel food packaging technologies: innovations and future prospective. Journal of Saudi Society of Agricultural. Science. doi: 10.1016/j.jssas.2016.11.003

Mao, X., Yang, L., Su, X. and Yi, Y. (2006). A nanoparticle amplification based quartz crystal microbalance DNA sensor for detection of *Escherichia coli* O157: H7. Biosensor and Bioelectronics. 21: 1178e85.

Marchetti, L., Muzzio, B., Cerrutti, P., Andrés, S. C. and Califano, A. N. (2017) Bacterial nanocellulose as novel additive in low-lipid low-sodium meat sausages. Food Structure,14: 52-59.

Mate, J., Periago, P. M., and Palop, A. (2016). Combined effect of a nanoemulsion of dlimonene and nisin on *Listeria monocytogenes* growth and viability in culture media and foods. Food Science and Technology International, 22, 146–152.

Maurer, E.I., Comfort, K.K., Hussain, S.M., Schlager,J.J. and Mukhopadhyay, S.M. (2012). Novel platform development using an assembly of carbon nanotube, nanogold and immobilized RNA capture element towards rapid, selective sensing of bacteria. Sensors (Basel);12: 8135e44.

Mihindukulasuriya, S., and Lim L. (2014). Nanotechnology development in food packaging: A review, Trends in Food Science and Technology, 40(2),doi 10.1016/j.tifs.2014.09.009

Mohamadi, E., Moghaddasi, M., Farahbakhsh, A. and Kazemi, A. (2017). A quantum-dot-based fluoroassay for detection of food-borne pathogens. Journal of Photochemistry and Photobiology B: Biology, 174, 291-29

Ojha, K. S., Perussello, C.A., García, C.A., Kerry, J.P., Pando, D and Tiwari, B.K. (2017). Ultrasonic-assisted incorporation of nano-encapsulated omega-3 fatty acids to enhance the fatty acid profile of pork meat Meat Science. 132, 2017, 99-106.

Ong, L. C., Ang, L. Y., Alonso, S., and Zhang, Y. (2014). Bacterial imaging with photostable up conversion fluorescent nanoparticles. Biomaterials 35, 2987–2998. doi: 10.1016/j.biomaterials.2013.12.060

Orsuwan, A., Shankar, S., Wang, L. F., Sothornvit, R., and Rhim, J. W. (2016). Preparation of antimicrobial agar/banana powder blend films reinforced with silver nanoparticles. Food Hydrocoll. 60, 476–485. doi: 10.1016/j.foodhyd.2016.04.017

Panea, B., Ripoll, G., González, J, Fernández-Cuello, Á and Albertí, P. (2014). Effect of nanocomposite packaging containing different proportions of ZnO and Ag on chicken breast meat quality. Journal of Food Engineering, 123, 104-112

PEN (2013). Consumer products inventory. http: //www.nanotechproject.org/cpi.

Prateek, Thakur, V. K., and Gupta, R. K. (2016). Recent progress on ferroelectric polymer-based nanocomposites for high energy density capacitors: synthesis,dielectric properties, and future aspects. Chemistry Review. 116, 4260–4317.doi: 10.1021/acs.chemrev.5b00495.

Pyrgiotakis, G., Vasanthakumar, A., Gao, Y., Eleftheriadou, M., Toledo, E., DeAraujo, A., McDevitt, J., Han, T., Mainelis, G., Mitchell, R., and Demokritou, P. (2015). Inactivation of foodborne microorganisms using engineered water nanostructures (EWNS). Environmental Science and Technology, 49(6), 3737–3745. http: //x.doi.org/10.1021/es505868ahttp: //www.nature.com/articles/srep21073#supplementary-information.

Rafieian, F., and Simonsen, J. (2014). Fabrication and characterization ofcarboxylated cellulose nanocrystals reinforced gluten in nanocomposite.Cellulose 21, 4167–4180. doi: 10.1007/s10570-014-0305-4.

Rahman, P. M., Mujeeb, V. M., and Muraleedharan, K. (2017). Flexible chitosan-nanoZnO antimicrobial pouches as a new material for extending the shelf life of raw meat. International Journal of Biological Macromolecules, 97: 382-391

Rai, M., Jogee, P. S., and Ingle, A. P. (2015). Emerging nanotechnology for detection ofmycotoxins in food and feed. International Journal of Food Sciences and Nutrition,66(4), 363–370. http: //dx.doi.org/10.3109/09637486.2015.1034251.

Ramachandraiah K., Choi, M. J., and Hong, G.P. (2018). Micro- and nano-scaled materials for strategy-based applications in innovative livestock products: A review. Trends in Food Science and Technology 71 (2018) 25–35. http: //dx.doi.org/10.1016/j.tifs.2017.10.017

Rhim, J. W., Park, H. M., and Ha, C. S. (2013). Bio-nanocomposites for food packaging applications. Progress in Polymer Science. 38, 1629–1652. doi: 10.1016/j.progpolymsci.2013.05.008

Rieger, K. A., Eagan, N. M., and Schiffman, J. D. (2015). Encapsulation ofcinnamaldehyde into nanostructured chitosan films. Journal of Applied Polymer Science.132: 41739. doi: 10.1002/APP.41739.

Robinson, D. K. R., and Morrison, M. J. (2010). Nanotechnologies for Food Packaging: Reporting the Science and Technology Research Trends. Observatory NANO. Availableonlineat: http: //www.observatorynano.eu/project/filesystem/files/Food per cent 20Packaging per cent 20 Report per cent 202010 per cent 20DKR per cent 20Robinson.pdf.

Rodrigues, S. M., Demokritou, P., Dokoozlian, N., Hendren, C., Karn, B., Mauter, M. S., Sadik O.A., Safarpour M., Unrine J.M., and Viers J. (2017). Nanotechnology for sustainable food Production: Promising opportunities and scientific challenges. Environmental Science: Nano, 4(4), 767–781. http: //dx.doi.org/10.1039/C6EN00573J.

Saleh,A., R., and Zourob, C.M. (2017). Development of rapid immuno-based nanosensors for the detection of pathogenic bacteria in poultry processing plants. Procedia Technology 27, 23 – 26.

Sani,M., Ehsani, A., and Hashemi, M. (2017). Whey protein isolate/cellulose nanofibre/TiO2 nanoparticle/rosemary essential oil nanocomposite film: Its effect on microbial and sensory quality of lamb meat and growth of common foodborne pathogenic bacteria during refrigeration. International Journal of Food Microbiology. 251, 8-14.

Sharma, C., Dhiman, R., Rokana, N., and Panwar, H. (2017). Nanotechnology: An Untapped Resource for Food Packaging. Frontier Microbiology. 8: 1735. doi: 10.3389/fmicb.2017.01735

Sickler, M. L., Claus, J. R., Marriott, N. G., Eigel, W. N., and Wang, H. (2013). Antioxidative effects of encapsulated sodium tripolyphosphate and encapsulated sodium acid pyrophosphate in ground beef patties cooked immediately after antioxidant incorporation and stored. Meat Science, 94, 285–288.

Srinivasan, V., Bhavan, P. S., Rajkumar, G., Satgurunathan, T., and Muralisankar, T. (2017). Dietary supplementation of magnesium oxide (MgO) nanoparticles for better survival and growth of the freshwater prawn *Macrobrachium rosenbergii* post-larvae. Biological Trace Element Research, 177, 196–208.

Suaifan, G., Alhogail, S. and Zourob, M. (2017). Paper-based magnetic nanoparticle-peptide probe for rapid and quantitative colorimetric detection of *Escherichia coli* O157: H7. Biosensors and Bioelectronics, 92, 15, 702-708.

Torelli, E., Manzano, M., and Marks R (2017). Chemiluminescent optical fibregenosensor for porcine meat detection. Sensors and Actuators B: Chemical, 247, 868-874."Chemiluminescent optical fibregenosensor for porcine meat detection." Sensors and Actuators B: Chemical, 247, 868-874.

Van der Meulen, B., Bremmers, H., Purnhagen, K., Gupta, N., Bouwmeester, H., and Geyer, L.L. (2014). Governing Nano Foods: Principles-based Responsive Regulation: EFFoST Critical Reviews# 3. Academic Press. Jayanthi, A.P., Beumer, K., Bhattacharya, S., 2012. Nanotechnology: 'Risk Governance' in India. Econ. 34 Political Wkly. XLVII (4).

Vanderroost, M., Ragaert, P., Devlieghere, F., and Meulenaer, B. D. (2014). Intelligent food packaging: The next generation. Trends in Food Science and Technolgy, 39,47–62. doi: 10.1016/j.tifs.2014.06.009.

Wang, Y., and Duncan, T. V. (2017). Nanoscale sensors for assuring the safety of food products. Current Opinion in Biotechnology, 44, 74–86. http: //dx.doi. org/10.1016/j. copbio.2016.10.005.

Wenjuan, Z., Wuqing, O., and Shuai, H. (2010). Effects of composite vitamin nanoemulsion on growth_performance and immunity in broilers. Chinese Academy of Agricultural Sciences(CAAS)http: //agris.fao.org/agrissearch/ search.do;jsessionid

Wu, C., Shiguo,Y., Chen, J., Liu, D., and Ye, X. (2016). Formation mechanism of nano-scale antibiotic and its preservation performance for silvery pomfret. Food Control, 69: 331-338

Xiang, L., Zhao, C., and Wang, J. (2011). Nanomaterials-based electrochemical sensors and biosensors for pesticide detection. Sensor Letters, 9(3), 1184–1189.

Yadav, H. M., Otari, S. V., Koli, V. B., Mali, S. S., Hong, C. K., Pawar, S. H., and Delekar S.D. (2014). Preparation and characterization of copper-doped anatase TiO2 nanoparticles with visible light photocatalytic antibacterial activity. Journal of Photochemistry and Photobiology A: Chemistry, 280, 32–38. http: //dx.doi.org/10.1016/j.jphotochem. 2014.02.006.

Yin, T., Park, J. W., and Xiong, S. (2017). Effects of micron fish bone with different particle size on the properties of silver carp (*Hypophthalmichthys molitrix*) surimi gels. Journal of Food Quality, 2017 Article ID 8078062.

Yusop, S.M., O'Sullivan, M.G., Preu, M., Weber, H., Kerry, J.F and Kerry, J.P. (2012) Assessment of nanoparticle paprika oleoresin on marinating performance and sensory acceptance of poultry meat LWT - Food Science and Technology, 46,349-355.

Zeng, Y., Zhu, Z., Du, D., and Lin, Y. (2016). Nanomaterial-based electrochemical biosensors for food safety. Journal of Electroanalytical Chemistry, 781, 147–154. http: //dx.doi. org/10.1016/j.jelechem.2016.10.030.

Zhou, C. (2013). Theoretical analysis of double-microfluidic-channels photonic crystal fiber sensor based on silver nanowires. Optics. Communications. 288, 42–46. doi: 10.1016/j.optcom.2012.09.060.

Chapter 15

Biogenic Amines: An Issue in Meat Quality and Safety

Gauri Jairath*, Y.P. Gadekar and A.K. Shinde

Department of Livestock Products Technology,
ICAR-Central Sheep and Wool Research Institute,
Avikanagar – 304 501, Rajasthan
**e-mail: gaurilpt@gmail.com*

ABSTRACT

Many public health associated issues have been raised due to low weighted nitrogenous bases termed as biogenic amines, when present at certain levels. The presence of substrate and the decarboxylase enzymes are prerequisite for the production of these amines. As meat and meat products are rich repository of proteins and are very perishable (readily available free amino acids), they act as suitable substrates. As they show consistent presence with microbial spoilage, they are also utilized as spoilage/freshness indicator (quality) of meat and meat products. They can be detected in our food by employing various new rapid analytical techniques. The key to control biogenic amines is the good manufacturing practices. Regulatory bodies have also prescribed the threshold limits of these amines in various foods for the safety of public health. Further, by practicing various controlling methods, the levels can be reduced to permissible/safe limits.

Keywords: *Public health significance, Free amino acid, Good manufacturing practices, Biogenic amines.*

1. Introduction

Biogenic amines (BAs) are non-volatile organic (basic) nitrogenous compounds of low molecular weight. On the basis of source, these have been grouped into endogenous and exogenous BAs. Endogenic amines also called as natural polyamines are formed during *de novo* polyamine biosynthesis and are produced naturally by animal, plant, and microorganism metabolism. This group mainly consists spermidine and spermine along with the putrescine-diamine, and in the case of plants and microorganisms, cadaverine and agmatine also become the member of the group. These polyamines play an important role in nucleic acid regulation, protein synthesis and possibly in the stabilization of membranes (Smith, 1980; Bardocz, 1995). In nutshell, polyamines are neurotransmitters which are produced by different tissues and get transmitted either locally or via the blood system. These are further divided into three classes *i.e.*, catecholamines (dopamine, norepinephrine, epinephrine), indolamines (serotonin, 5-hydroxytryptamine, melatonin) and histamines. They have many physiological functions like growth regulation, neural transmission, mediators of inflammation *etc.* (Onal, 2007). Exogenic amines are natural anti-nutritional factors detected in both raw and processed foods (Santos, 1996). They are formed during final processes of protein breakdown by the bacterial decaboxylases mediated removal of the alpha-carboxyl group from amino acids and they are usually named after corresponding precursor amino acid *e.g.*, histidine is decarboxylated to produce histamine, tryptophan to tryptamine, tyrosine to tyramine and lysine to cadaverine. However, putrescine can be produced by decarboxylation of three amino acids which are glutamine, arginine and agmatine (Stadnik and Dolatowski, 2010). Arginine gets easily converted to agmatine or as a result of bacterial activity can also be degraded to ornithine from which putrescine is formed by decarboxylation (Figure 15.1).

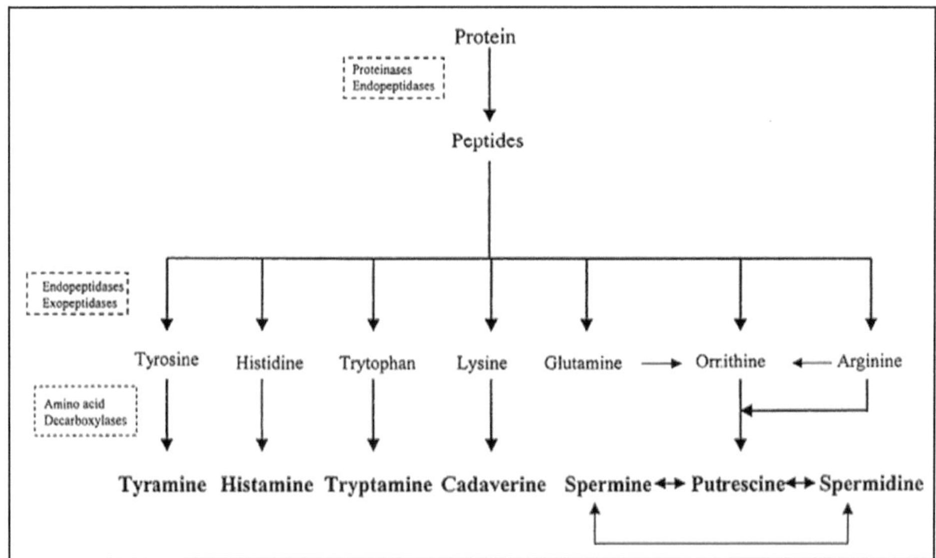

Figure 15.1. Conversion of Various Free Amino Acids into BAs (Ruiz-Capillas and Jiménez-Colmenero, 2004).

According to their chemical structure (Figure 15.2) BAs can either be aliphatic (putrescine, cadaverine, spermine, spermidine), aromatic (tyramine, β-phenylethylamine) or heterocyclic (histamine, tryptamine) (Karovicova and

Kohajdova, 2005; Onal 2007). They have also been classified as monoamines, diamines and polyamines.

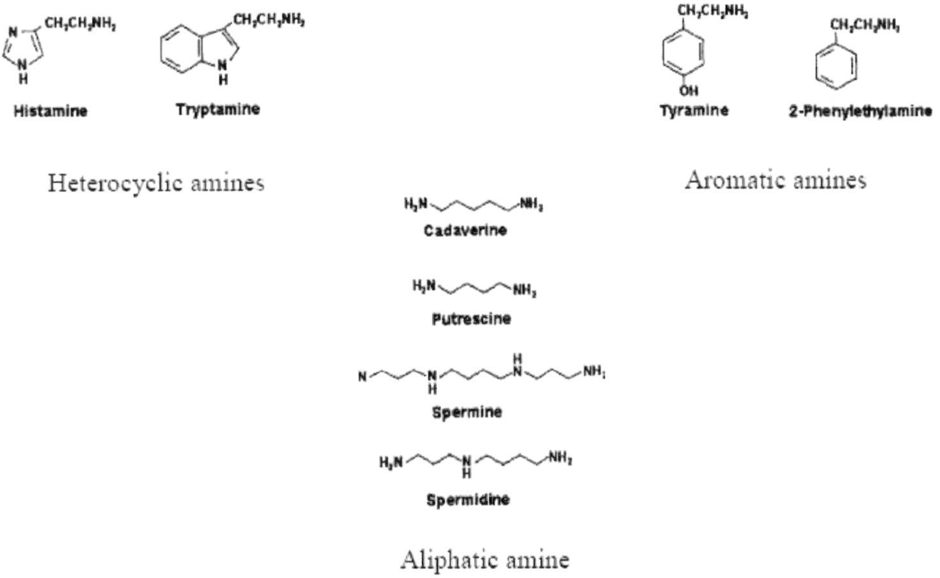

Histamine Tryptamine

Heterocyclic amines

Tyramine 2-Phenylethylamine

Aromatic amines

Cadaverine

Putrescine

Spermine

Spermidine

Aliphatic amine

Figure 15.2. Structure of Putrescine, Tyramine, Cadaverine, Histamine and 2-phenylethylamine.

These nitrogenous bases are found in various concentrations in diverse foods like meat, fermented foods, fish and fish product and cheese and are important for two reasons in food industry. Firstly, biogenic amines poisoning; the consumption of such foods which contain high concentrations of BAs, can lead to a health hazard due to direct toxic effect of these compounds and their interaction with some medicaments (Bardocz, 1995; Shalaby, 1996). They may cause poisoning with vasoactive or psychoactive effects, if too large quantity of biogenic amine is consumed orally. According to Fathi *et al.* (2013), biogenic amine may be exclusively present in various foods that are rich in protein like fish and meat. The major and well known example of food borne illness from BAs poisoning is 'scombroid poisoning' from fish of Scombridae family consumption associated with histamine (El-Habib, 2011, Stadnik and Dolatowski, 2010). The other type of BA's poisoning is 'cheese reaction' first reported in Netherland in 1967 (Gouda cheese) and is associated with tyramine (Stratton *et al.*, 1991). The presence of these BAs has been reported in widely varying concentrations in different variety of muscle foods, both raw and processed (Ntzimani *et al.*, 2008). Secondly, quality/freshness indicators; they may have a role as indicators of quality and/or acceptability in some foods (Hernandez-Jover *et al.*, 1997). Thus, the safety of muscle foods present a major concern for industry and consumers. The quality and freshness of meat depends significantly on the concentration of BAs. For example, presence of histamine in raw fish or fish products is the most important quality indicator as chemical indicators for spoilage of fish and is considered as a threat to public health bodies. The levels of BAs tend to be highest in fermented meat products, but considerable concentrations have also been detected in fresh and cooked meat. Meat being rich in protein content is more important substrate with respect to amine decarboxylation and thus, consequently, a good BAs producer.

Further, global meat production is projected to be 16 per cent higher in 2025 than in the base period (2013-15). This compares with an increase of almost 20 per cent in the previous decade. Developing countries are projected to account for the vast majority of the total increase including India. Global annual per capita meat consumption is expected to reach 35.3 kg retail weight equivalent (r.w.e.) by 2025, an increase of 1.3 kg r.w.e. compared to the base period (OECD/FAO, 2016). The consistent increase in meat consumption further increases the risk of biogenic amines related health issues.

Therefore, this chapter will detail the BAs content in meat and meat products, BAs genesis, associated risk and health hazards, detection of these amines, reduction or control measures, recommended limits and their use as a quality/freshness indicator.

2. Biogenic Amines in Meat and Meat Products

BAs are naturally present as natural metabolic products/intermediates in variety of foodstuffs such as fruits and vegetables in low concentrations. Meat and meat products are important components among daily diet and are the natural source of the substrate (free amino acids) for decarboxylases to produce biogenic amines. They contain different types of biogenic amines in varying contents. Tyramine, cadaverine, putrescine and histamine are the most commonly found BAs in muscle foods (Stadnik and Dolatowaski, 2010). Spermidine and spermine are the only amines which are present at significant levels in fresh meat (Hernández-Jover *et al.*, 1997). The spermine content is usually between 20 and 60 $mg{\cdot}kg^{-1}$ in meat and meat products of warm-blooded animals and spermidine level in meat rarely exceeds 10 $mg.kg^{-1}$. The concentrations of different amines tend to vary depending on the different types of products (Table 15.1) as well as species. For example, putrescine, tyramine and spermine content in beef and pork are 4.7, 24.7, 28.4 $mg{\cdot}kg^{-1}$and 2.3, 1.3, 31.3 $mg{\cdot}kg^{-1}$, respectively (Min *et al.*, 2007b). Histamine, tyramine, putrescine, cadaverine, phenylethylamine and tryptamine in meat products were found as 4.8, 6.7, 123, 25, 1-1.1 and 1.1 $mg{\cdot}kg^{-1}$ (European Food Safety Authority Scientific Panel, 2011).

Though BAs are found in almost all muscle foods, but fermented meat products constitute considerably higher amount of biogenic amines as they contain higher amount of non-protein nitrogen fraction (free amino acids), the main precursors of biogenic amines. The concentration of free amino acids (FAA) increases during fermentation because of microbial activity and also due to the activity of endogenous meat enzymes. These meat enzymes are in turn favoured by the denaturation of proteins as a consequence of acidity increase, dehydration and action of sodium chloride (Suzzi and Gardini, 2003). This shows the complex interaction of factors responsible for the concentration of biogenic amines and this may also be the reason

Table 15.1. Levels of Biogenic Amines in Meat and Meat Products

	Biogenic Amines (mg/kg)								
Meat and Meat Product	*Histamine*	*Tyramine*	*Cadaverine*	*Putrescine*	*Tryptamine*	*Phenyl-alanine*	*Spermidine*	*Spermine*	*Reference*
Minced beef and pork	ND-8	ND-39	ND-96	ND-96	-	-	ND-5	14.39	Wortberg and Woller, 1982
Leg lamb storage at 5°C for 5 days	-	-	1.3	3.3	-	-	-	-	Edwards *et al.*, 1983
Raw ground beef at 4°C for 12 days	31.8	12.4	ND	74.1	-	-	113.3	331.3	Sayem-El-Daher *et al.*, 1984
Pork storage in CO_2/air at 2±1°C for 21 days	-	-	0.3	1	-	-	-	-	Edwards *et al.*, 1985
Vacuum packed beef at 1°C for 7 weeks	3	6	54	18	-	-	3	25	Edwards *et al.*, 1987
Pork storage in CO_2 at 2±1°C for 21 days	-	0.7	39.6	6.6	ND	-	3.2	26.5	Ordonez *et al.*, 1991
Fresh vacuum packed beef at 1°C for 120 days	ND	286	-	-	49	ND	-	-	Smith *et al.*, 1993
Pork raw	4.7	-	13.3	7.8	-	-	7.0	67.1	Halasz *et al.*, 1994
Beef raw	ND-1.1	ND	ND	ND-1.75	-	-	1.9-4.2	28.7-44.6	Hemandez-Jover *et al.*, 1996a
Mortadella	ND-4.8	ND-66.0	0.6-7.0	ND-3.9	ND-1.0	ND-1.4	1.9-8.9	7.8-32.2	Hernandez-Jover *et al.*, 1996b
Cooked ham	ND	ND-11.9	ND-0.9	ND-3.9	ND	ND	1.7-3.0	18.1-25.4	Hernandez -Jover *et al.*, 1996b
Dry sausages	<1-200	3-320	<1-790	<1.850	<10-91	>1-48	<1-14	19-48	Eerola *et al.*, 1997
Spanish ripened sausage Mini-salami	4-16	3-12	ND	42-139	-	-	5.10	25	Trevino *et al.*, 1997
Spanish ripened sausage Fuet	15.2	156.9	367	64.7	10.0	10.1	10.3	30.6	Bover-Cid and Holzapfel, 1999

	Biogenic Amines (mg/kg)								
Meat and Meat Product	*Histamine*	*Tyramine*	*Cadaverine*	*Putrescine*	*Tryptamine*	*Phenyl-alanine*	*Spermidine*	*Spermine*	*Reference*
Fresh pork (CO_2) at 1-1.5°C for 13 weeks	16	60	68	20	-	40	9	600	Nadon *et al.*, 2001
Hamburger	5.9-16.1	1.5-35.5	1.3-10.2	0.3-12.3	2.3-13.5	ND-2.5	1.6-5.1	2.1-6.1	Durlu-Ozkaya *et al.*, 2001
Adult bovine meat samples	-	10.71	18.54	2.08	20.42	-	2.20	27.15	Vinci and Antonelli, 2002
Salami	<LD	-	-	-	17	-	-	-	Lange *et al.*, 2002
Ham	-	-	-	-	7.5	-	-	-	Lange *et al.*, 2002
Pepperoni sausage (fermented)	-	0.9	-	2.6	-	-	-	9.6	Kim *et al.*, 2005
Beef	-	24.7	-	4.7	-	-	-	28.4	Min *et al.*, 2007b
Pork	-	1.3	-	2.3	-	-	-	31.3	Min *et al.*, 2007b
Smoked turkey fillets (aerobically packed)	32.9	25.0	-	-	4.1	-	-	-	Ntzimani *et al.*, 2008
Chicken raw thigh	1.09	-	1.27	0.45	-	-	-	-	Ibrahim *et al.*, 2017
Chicken raw breast	0.45	-	0.99	0.99	-	-	-	-	Ibrahim *et al.*, 2017

ND: Not detected, <LD: Below the limit of detection.

of different concentration of BAs in fermented products with comparable microbial flora. For example, the tryptamine level of 17 mg·kg^{-1} in salami was detected by Lange *et al.* (2002) and putrescine, tyramine and spermine content in fermented pepperoni sausages as 2.6, 0.9 and 9.6 mg·kg^{-1}, respectively by Kim *et al.* (2005). Papavergou *et al.* (2012) detected tyramine, putrescine, histamine and cadaverine at high concentrations in Greek retail market fermented meat products in range of 0 to 510, 0 to 505, 0 to 515 and 0 to 690 mg·kg^{-1}, respectively.

3. Genesis of BAs

The availability of FAA, the presence of microorganisms with amino acid decarboxylases, and favourable conditions for their growth are prerequisite for the genesis of BAs. In the release process of FAA from tissue proteins, proteolysis either autolytic or bacterial, which offer a substrate for decarboxylases reactions, plays an important role (Shalaby, 1996). Further, availability of substrate (FAA) is associated with the raw material (meat composition, pH, handling conditions, *etc.*) as the substrate source and reaction medium. These FAA are taken up by microorganisms through membrane antiporter protein (Figure 15.3) into the cell. Then FAA are acted upon by amino acid decarboxylase (aaDC) present inside the cell cytoplasm and free the CO_2 from the FAA. The membrane antiporter protein then, removes (excretes) the decarboxylated product from the cytoplasm and completes the biogenic amines genesis process.

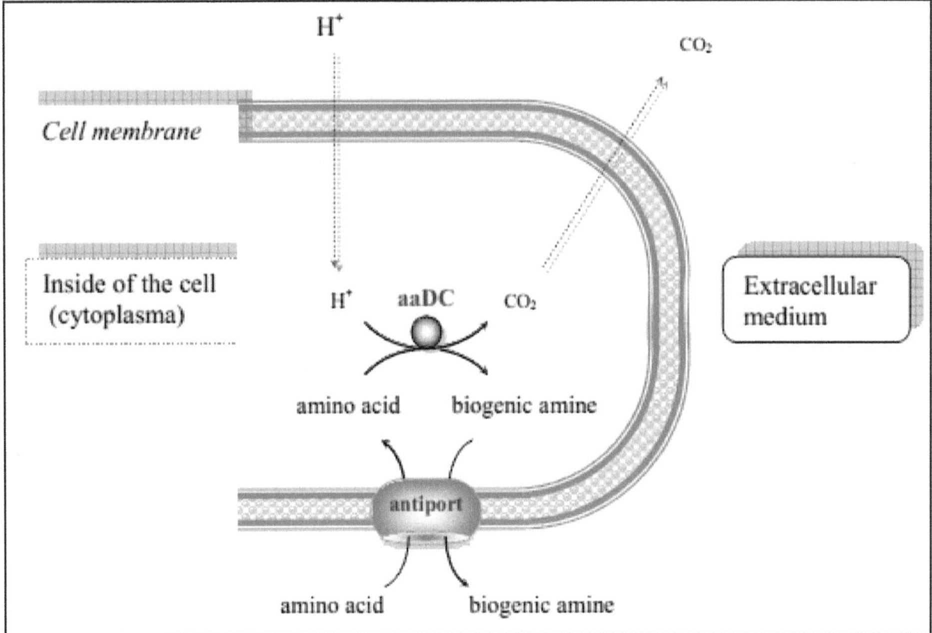

Figure 15.3. Biogenic Amine Biosynthesis Pathways in Bacteria (Adopted from Bover-Cid *et al.*, 2000).

4. Factors Associated with BAs Production in Meat and Meat Products

The above mentioned process clears the role of substrate (FAA) and enzyme (decarboxylase) interaction and thus reflects that; all the factors such as availability of the substrate (FAAs), the enzyme, and also their level of activity, affect the type and amount of biogenic amines present in each case (Ruiz-Capillas and Jimenez-Colmenero, 2004).

The first factor *i.e.* availability of FAA in the food including meat products, depends upon the quality of raw materials and type of raw material. These factors are closely related to each other and are further influenced by the technological processes associated with types of meat derivative (steak, roast, ham, ground, restructured, comminuted, fresh, cooked, smoked, fermented, *etc.*). The second factor *i.e.* presence of decarboxylases is closely tied to microbiological aspects which thereby influence the final level of biogenic amines. These microbiological aspects are further related to the bacterial species, strain (decarboxylase +ve) and their growth. The growth is further related to storage conditions (time/temperature, packaging, temperature abuses, *etc.*). Therefore, the activity of substrate and enzyme is also related with the combined action of these factors which ultimately determines the final concentrations of biogenic amines (Figure 15.4). These factors have been briefed below:

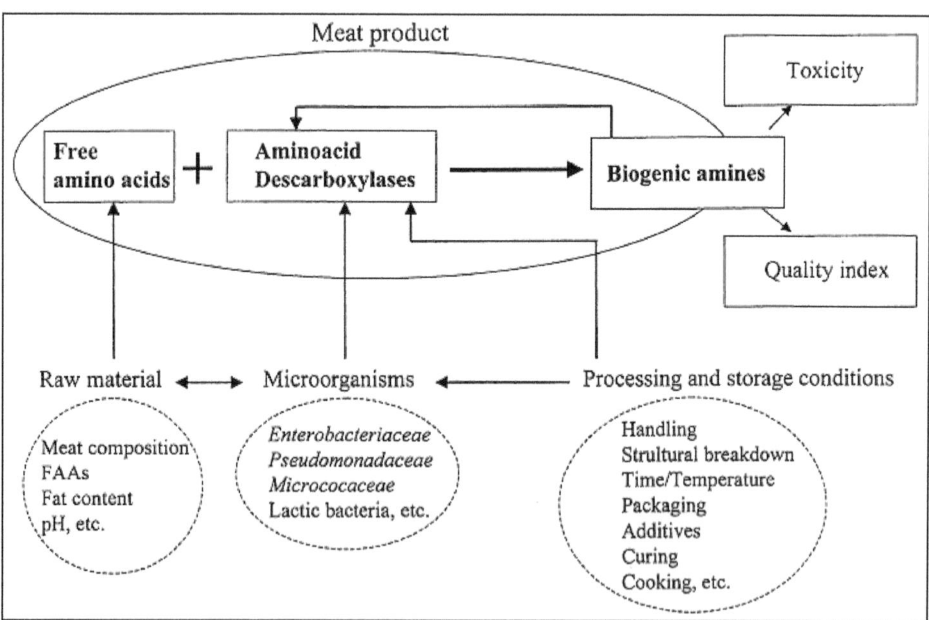

Figure 15.4. Factors Influencing the Formation of Biogenic Amines (Adopted from Ruiz-Capillas and Jimenez-Colmenero, 2004).

4.1 Raw Materials

The composition (presence of FAAs, their concentrations, fat content in meat), pH, and source of raw material determines the availability and level of substrate (FAA) availability which is directly linked up with the genesis and level of BAs. To provide substrate for BAs formation, proteolysis is a crucial factor, because it is directly related to availability of FAA. During storage and treatment, the concentrations of FAA increase as a result of proteolytic events, due essentially to many of the microorganisms present (Fernandez *et al.*, 2007; Komprda *et al.*, 2008). It has also been documented that fat content negatively influences the formation of biogenic amines (Kebary *et al.*, 1999). This phenomenon has been most intensively studied in cheese, where it has been observed that the concentration of BAs decreases along with increase in fat content. This is probably due to fact that fat and moisture are inversely related and higher fat content lowers the water activity (a_W) rather than the actual amine content; a_w causes some inhibition of growth of proteolytic bacteria and, hence, falls in the level of FAA in the medium (Banwart, 1981).

Further, the level of biogenic amines is greatly influenced by net pH balance as pH affects the production by two mechanisms (European Food Safety Authority Scientific Panel, 2011). One is affecting the growth by acidity which inhibits the growth of microorganisms. The other affects the production and activity of the enzyme because in low pH environment, bacteria are more stimulated to produce decarboxylase as a part of their defence mechanisms against the acidity. As acidity increases, decarboxylase activity of enzymes increases and thereby increases the production of BAs(Fernandez *et al.*, 2007). For example, activity of histidin decarboxylase enzyme increases in acid media, with an optimum pH range of 4 to 5.5 (Halasz *et al.*, 1994). The final pH of the meat, which itself depends upon multiple factors, could, therefore, have a considerable influence on the genesis and level of BA. For example, high pH of dark, firm and dry meat (DFD) favors microbial proliferation, which promotes FAAs availabilty, but limits decarboxylase activity. The opposite conditions exist for PSE (pale, soft, exudative) meat. Therefore, the meat type also (PSE and DFD) affects factors directly involved in the activity of amino decarboxylases enzymes and the final presence of biogenic amines in these meat (Ruiz-Capillas and Jimenez- Colmenero, 2004).

Besides above, type of meat source can influence the formation of biogenic amines. For example, histamine and tyramine formation tends to be found less in packed meats made only with pork (*e.g.*, cooked ham and cured ham) than in other meat derivatives containing mixtures of beef and pork (salami, salchichon, chorizo or Bologna sausages), where there was a greater tendency to form tyramine (Vidal-Carou *et al.*, 1990).

4.2 Microbial Contaminant

Microorganisms with decarboxylase activity play crucial role in the genesis of BAs (Figure 15.3). However, this decarboxylase activity is associated with certain groups of microorganisms. For example, enterobacteria is responsible for putrescine and cadaverine production and enterococci is for tyramine. However, within the microbial groups, strain specificity plays an important role to produce

BAs. In general, the ability to produce BA is more widely distributed among certain genera or species which suggest that horizontal gene transfer may account for their dissemination between strains (Coton and Coton, 2009). The synthesis of BA in bacteria may be associated with the supply of energy and to help protect from acid stress (Foster, 2004). Additional roles in DNA regulation and as free radical scavengers have also been suggested (Wortham *et al.*, 2007). In addition, the strains which harbours BA related plasmids or who's enzymes constitute the pathways encoded by unstable plasmids are able to produce BA (Lucas *et al.*, 2005). Some strains are even able to simultaneously produce more than one amine simultaneously, either due to the presence of different decarboxylases or to the action of a single enzyme which decarboxylates different amino acids (Bover-Cid and Holzapfel, 1999).

Though species of many genera such as *Bacillus, Citrobacter, Clostridium, Klebsiella, Escherichia, Proteus, Pseudomonas, Salmonella, Shigella, Photobacterium* and the lactic bacteria *Lactobacillus, Pediococcus* and *Streptococcus* can decarboxylate one or more amino acid (Huis in't Veld *et al.*, 1990) but, *Enterobacteriaceae, Pseudomonadaceae, Micrococcaceae* and lactic bacteria are the only microorganisms which chiefly attribute to decarboxylase activity in muscle foods.

In meat and meat products certain types of BA are produced by certain strains of bacteria. For example, in fresh meat, *Enterobacteriaceae* have been identified as the main producers of cadaverine, while, putrescine has been associated with high total aerobic viable counts (Bauer *et al.*, 1996; Bover-Cid *et al.*, 2001a). In fermented products, lactic acid bacteria contribute the most in the production of biogenic amines. Edwards *et al.* (1987) documented that tyramine formation is restricted to Lactobacilli particularly, *L. divergens* and *L. carnis*. The histamine-producing strains, especially *Hafnia alvei*, increased in the minced meat during incubation at 20-22°C (Maijala *et al.*, 1993).

4.3 Starter Culture

Starter cultures with the inability of the culture to form BAs, but also having ability to grow well at the temperature intended for processing of the product and competitiveness in suppressing the growth of such microflora which promote wild amine production should be considered during the selection of starter cultures (Suzzi and Gardini, 2003). While selecting, the culture with aminooxidase activity is fundamental in preventing the formation of high levels of BAs in fermented meat products (Karovicova and Kohajdova, 2005). A rapid increase in acidity caused by amine negative starter cultures can largely prevent BAs accumulation in sausages (Bover-Cid *et al.*, 2001a).

In meat industry, lactic acid bacteria are the most widely used microorganisms as starter culture along with micococci and/or coagulase-negative staphylococci. This popularity is because of their contribution in colour formation and aroma development owing to their acidification property, proteolytic and lipolytic activities (Suzzi and Gardini, 2003; Latorre-Moratalla *et al.*, 2010). Though, starter LAB is able to compete with nonstarter bacteria during the later phase of ripening and throughout the storage period and thus, can further avoid excessive biogenic amines production. But, the selection of lactic acid bacteria with application in meat fermentation has to take in consideration the various, specific requirements of the fermentation process (Roig-Sagues and Eerola, 1997).

Starter cultures are not always able to control the decarboxylase-positive strains as factors like raw meat microbiological quality and the characteristics of natural microflora, in particular amine positive nonstarter LAB (Roig-Sagues and Eerola, 1997) are responsible of the discrepancies. The amine positive nonstarter LAB is often responsible for BA formation in fermented sausages (Paulsen and Bauer, 1997). In dry sausages, *Carnobacterium, Lactobacillus curvatus* and *L. plantarum* strains are responsible of tyramine (Masson *et al.*, 1996) which in turn recommends the use of amino-negative starters *viz. L. sakei* or *Pediococus pentosaceus* to prevent the formation of BA in dry sausage (Bover-Cid *et al.*, 2001b), although any reduction would always depend on other factors influencing formation, especially the raw material. The use of mixed starter can also be effective to reduce the BA production. For example, in Turkish sausages (Soudjoucks), the addition of mixed starter cultures (*L. sakei, Pediococcus pentosaceus, Staphylococcus xylosus* and *S. carnosus*) avoided the formation of putrescine in comparison to the natural microflora producing high levels of tyramine and putrescine (Ayhan *et al.*, 1999). Mixed starter cultures like *L. sakei, S. carnosus* and *S. xylosus*) greatly reduced the presence of putrescine, cadaverine and tyramine in Spanish sausages to about 90 per cent (Bover-Cid *et al.*, 2000a). Likewise, staphylococci in combination with lactobacilli (amine negative mixed starter cultures) are capable to decreases tyramine, cadaverine, and histamine concentration in sausages (Maijala *et al.*, 1995a).

4.4 Meat Processing

Amine content and profiles may vary depending on various extrinsic and intrinsic factors during meat processing *viz.* redox potential, temperature, additives, curing, the size of the sausage *etc.* (Gardini *et al.*, 2001, Komprda *et al.*, 2004, Latorre-Moratalla *et al.*, 2008). The redox potential of the medium influences biogenic amines production, as conditions resulting in a reduced redox potential stimulate histamine production. Further, histidine decarboxylase activity seems to be inactivated or destroyed in the presence of oxygen (Karovicova and Kohajdova, 2005). Similarly, temperature affects the amine formation in a decisive way. Low temperature restricts or stops the growth of decarboxylases containing bacteria, while the temperature between 20°C and 37°C is optimal for the growth (Karovicova and Kohajdova, 2005).

With respect to additives, potassium sorbate, sodium nitrate, ascorbic acid, sugar *etc.* have been proved as potential barriers for BAs accumulation (Bozkurt and Erkmen, 2004). Sodium nitrites at 45 to 195 ppm level has shown decreased biogenic amine production in sausages (Kurt and Zorba, 2009). The addition of glucono-delta-lactone (GDL) at of 0 to 1 per cent level in the meat decreases histamine and putrescine production through a pH drop in meat (Maijala *et al.*, 1993). The addition of sugar and glycine may also slightly reduce biogenic amine formation (Bover-Cid *et al.*, 2001a; Mah and Hwang, 2009). However, variation in salt level variably affects BAs production. Biogenic amines accumulation decreases markedly with the increase of NaCl concentration, while proteolytic activity is higher for intermediate concentration of salt (Suzzi and Gardini, 2003).

A correlation between contents of BAs in meat samples and use of additives has also been drawn (Jastrzebska *et al.*, 2016). The differences of BAs' (Putricine, Cadaverine, Tyramine, Tryptamin, Histamin) content in meat samples on the fourth and the first days were calculated which clearly indicates the effectiveness of the additives on the inhibition of biogenic formation. All tested compounds reduced the levels of BAs in the following order:

- For poultry: sodium nitrite > disodium diphosphate > sodium chloride > propyl gallate > citric acid; lactic acid; sodium metabisulphite > potassium sorbate > α-tocopherol; ascorbic acid > butylated hydroxyanisole.
- For pork: sodium chloride > disodium diphosphate; sodium nitrite > lactic acid; propyl gallate; ascorbic acid > citric acid > sodium metabisulphite > α-tocopherol > potassium sorbate > butylated hydroxyanisole.
- For beef: sodium nitrite > lactic acid > sodium metabisulphite > propyl gallate > ascorbic acid > disodium diphosphate; sodium chloride > citric acid > potassium sorbate > α-tocopherol > butylated hydroxyanisole.

A relationship has also been found between biogenic amine content and the size of dry fermented sausages. The growth of microorganism is affected by the diameter of the sausage as diameter in turn affects the environment in which microorganisms grow. For example, salt concentration and water activity is different in the sausages with different diameter. Lower salt concentration and higher water activity is a characteristic of the sausages with a larger diameter which make it more suitable to produce certain BAs at high concentration such as tyramine and putrescine. Further, biogenic amines level is higher in the central part of the sausages than in the edge (Suzzi and Gardini, 2003; Ruiz-Capillas and Jimenez-Colmenero, 2004).

4.5 Hygiene and Storage Conditions

Handling of raw materials also predicts the level of biogenic amines in finished products. This is reason that much importance is given to hygienic quality of both raw material and processing units. However, alone hygiene may not be enough to avoid some biogenic amines formation where it acts as a single factor and requires other measures to be applied to control BAs (Latorre-Moratalla *et al.*, 2010).

Besides above, BAs production is greatly affected by storage conditions (temperature and time) too. The quantitative production of BAs depends upon temperature-time complex (Zaman *et al.*, 2009). In general, production rate increases with the temperature increase and vice-versa is true for low temperatures through inhibition of microbial growth and the reduction of enzyme activity. The optimum temperature for the formation of BAs by mesophilic bacteria has been reported to be between 20 to 37°C, while decreases below 5°C or above 40°C. For example, optimum temperature for *Klebsiella pneumonia* to produce cadaverine more extensively is 20°C than at 10°C. Similarly, *Enterobacter cloacae* produce putrescine at 20°C but not at 10°C (Halasz *et al.*, 1994). During storages, many factors are responsible, for instance, two-fold effect have been documented because of unsuitable storage temperatures (*i.e.*, temperatures >5°C), prolonged storage, or temperature abuses during storage period: on proteolysis due to increased microbial growth promoting penetration in the muscle; on amino decarboxylase activity (Maijala *et al.*, 1995b).

4.6 Packaging

The qualitative and quantitative formation of BAs in the food is also determined by the atmosphere (oxygen and CO_2) surrounding it. For example, it has been demonstrated that *E. durans* (Bunkova *et al.*, 2012) and *L. lactis* (Bunkova *et al.*, 2011) tyramine accumulation is favored by anaerobic conditions. A similar trend was reported also by Cid *et al.* (2008) using a strain of *L. curvatus*. However, the main technologies for food preservation based on atmosphere modification are focused on oxygen exclusion. The main aim of oxygen exclusion is not the inactivation of decarboxylase activity but the inhibition of microbial population with decarboxylating properties. Modified atmospheric packaging (MAP) and vacuum packaging (VP) play therefore, an important role to curb decarboxylating bacteria (Curiel *et al.*, 2011).

CO_2 gas, (bacteriostatic) is the main gas used in MAP. Different concentrations of this gas in MAP have been tried to inhibit microbial growth of Enterobacteria and H_2S-producering bacteria (Lopez-Caballero *et al.*, 2002), histamine forming bacteria (Ozogul and Ozogul, 2006), *Pseudomonas* spp. (Li *et al.*, 2014), *Lactobacillus sakei* (Devlieghere *et al.*, 1998) and psychrophilic microorganisms (Arashisar *et al.*, 2004) to prolong shelf-life of foods. A significant reduction in the contents of total BAs in barramundi filets and sardine have been reported at >60 per cent CO_2 concentration in the MAP (Yassoralipour *et al.*, 2012). MAP at 80 per cent CO_2/20 per cent N_2 and VP of rainbow trout resulted in lower production of putrescine and cadaverine (Rodrigues *et al.*, 2016). Curiel *et al.* (2011) demonstrated that under vacuum conditions or MAP (20 per cent CO_2 and 80 per cent N_2) the enterobacteria producing putrescine, cadaverine, and agmatine were inhibited. Study on the effect of air and MAP on the shelf life of chilled chicken showed significantly lower content of putrescine and cadaverine content of gas mixture-packaged samples than that of the air-packaged samples during the storage period. Further, the concentration of BAs (putrescine and cadaverine) was decreasing with increase in CO_2 content (Zhang *et al.*, 2015).

Other than MAP and VP, active packaging technology for food is now trending employing oxygen scavengers which eliminate oxygen in the packaging and in the product during storage (Alvarez, 2000). The use of O_2 scavenger with proper maintenance of chilled storage, reduce the formation of BAs and the risk of *Clostridium botulinum* toxin in seer fish (*Scomberomorus commerson*) (Mohan *et al.*, 2009). In nut shell, active packaging, VP and MAP inhibit formation of BAs more effectively than air packaging, through inhibition of BA forming bacteria or enzyme activity (Naila *et al.*, 2010).

5. Risk Associated with BA in Meat and Meat Products

Biogenic amines play important role in human body such as: regulation of body and stomach pH, gastric acid secretion, the immune response and cell growth and differentiation. In addition, growth, renovation and metabolism of every organ in body are taken care up by amines and also essential for balancing the high metabolic activity of the normal functioning and immunological system of gut. Further, when polyamines are taken up preferentially by tissues with high demands, they could be useful for post-operation patients or during wound healing and for growth and development of the neonate's digestive system (Kalac and Krausova, 2005).

Despite of above mentioned roles, health hazards are associated with BA above tolerance limit due to their psychoactive and vasoactive properties, leading to distinctive pharmacological and toxic effects (Stadnik and Dolatowski, 2010). The most conspicuous symptoms of consumption of high doses of biogenic amines are vomiting, respiratory difficulties, perspiration, palpitation, hypo or hypertension and migraine (Kordiovska *et al.*, 2006).

Psychoactive amines, such as histamine mimics neurotransmitter and cause some neurotransmission disorders. Toxic effects of histamine are exerted via interaction with two types of receptors (H_1 and H_2) on cellular membranes. BAs act either as vasoconstrictor or as vasodilator. Some aromatic amines (tyramine, tryptamine, and β-phenylethylamine) behave as vasoconstrictors while others (histamine and serotonin) do exhibit a vasodilator effect in blood vessels, capillaries and arteries, causing headaches, hypotension, flushing, gastrointestinal distress and oedemas (Onal, 2007, Standarova *et al.*, 2008). The vasoconstrictor amines, specially tyramine work as the initiators of hypertensive crisis in certain patients and are responsible for dietary-induced migraine. The other physiological effects of tyramine include: peripheral vasoconstriction, increased cardiac output, increased respiration, elevated blood glucose, and release of norepinephrine (McCabe-Sellers *et al.*, 2006, Onal, 2007). Tyramine may also induce brain haemorrhage and heart failure (Standarova *et al.*, 2008).

Adverse reactions due to dietary BAs are often termed as intolerance reactions, also called pseudo-allergy or false food allergy. Intolerance is a form of hyper-sensitivity without mediating immune system unlike allergic reactions (Jansen *et al.*, 2003). The most frequent food borne intoxications caused by biogenic amines as mentioned above is named as scombroid poisoning (due to histamine) because this illness is usually related to the consumption of scombroid fish, such as tuna, mackerel, and sardines. The consumption of histamine at higher doses causes life threatening intoxication whereas lower amounts of the same can result in headache, nausea, hot flushes, skin rashes, sweating, respiratory distress, cardiac and intestinal problems (Santos, 1996). The tolerable limit/permissible limit of histamine is considered as 10 mg, however, some sensitive people are able to tolerate 5-10 mg of histamine intake. An intake of 100 mg induce a medium toxicity and 1000 mg intake is considered as highly toxic (Lehane and Olley, 2000).

With respect to polyamines'(cadaverine, putrescine, spermine and spermidine) ill effects, as such they have no adverse health effects, but aggravate the growth of chemically induced aberrant crypt foci in the intestine (Stadnik and Dolatowski, 2010). Formation of N-nitroso compounds is another toxicological risk associated with BAs, especially in cured muscle foods that contain nitrite and nitrate salts as curing agents (Komprda *et al.*, 2004; Karovicova and Kohajdova, 2005; Onal,

2007). The short-lived alkylating agent production due to reaction between nitrosating agents and primary amines, react with other components in the food matrix to generate alcoholic products devoid of toxic activity in the relevant contents. Secondary amines are known to form carcinogenic N-nitrosamines by reaction with nitrosating compounds, while tertiary amines produce a range of labile N-nitroso products. However, as primary biogenic amines can convert to secondary amines such as pyrolidine and piperidine, not only on heating but also during storage at room temperature, and further reaction with nitrite can occur to form carcinogenic nitrosamines. In fatty foods like bacon, N-nitrosopyrrolidine and N-nitrosopiperidine (carcinogen) can be formed from secondary amines such as putrescine or spermidine at high temperature and humid environment (Gonzalez-Fernandez *et al.*, 2003; Karovicova and Kohajdova 2005; Onal, 2007). Therefore, formation of nitrosamines in meat and meat products is a main problem especially in terms of toxicological point of view.

However, in normal circumstances, histamine and tyramine absorbed from foods are rapidly detoxified by the enzymes such as monoamine oxidase (MAO; EC 1.4.3.4), diamine oxidase (DAO; EC 1.4.3.6), and polyamine oxidase (PAO; EC 1.5.3.11). These amines are metabolized into physiologically less active degradation products through acetylation and oxidation in the human body (Bardocz, 1995). However, when detoxification fails either due to high amine intake or the individual is allergic or under treatment with oxidase enzymes (*e.g.*, monoamine oxidase inhibitor, MAOI), biogenic amines may build up in the body and could cause serious toxicological problems (Halasz *et al.*, 1994). Hence, the toxicity of biogenic amines is multifactorial and includes both the factors associated with the food itself (quantitative and qualitative) and the factors associated with the consumer (individual susceptibility and state of health).

5.1 Factors which Intensify the Toxicological Effects

The amount and variety of BAs ingested, individual susceptibility, and the level of detoxification activity in the gut are some factors which enhance the severity of the clinical symptoms. Respiratory and coronary problems, hypertension, vitamin B_{12} deficiency enhance the sensitivity of individuals to lower doses of BAs (Bardocz, 1995). Further, gastrointestinal problems like gastritis, irritable bowel syndrome, Crohn's disease, stomach and colonic ulcers may also pose risk because the activity of oxidases in such cases is usually lower (Stadnik and Dolatowski, 2010). Patients, who are on painkillers, psychopharmaceutics, and drugs used for the treatment of Alzheimer's and Parkinson's diseases can further cause health problem due to the inhibiting effect of these medicines on MAO, DAO and/or PAO which changes the metabolism of biogenic amines (Bouchereau *et al.*, 2000; Moret *et al.*, 2005;Kalac and Krausova, 2005; Latorre-Moratalla *et al.*, 2008). In addition, alcohol intake also aggravates the toxic potential of biogenic amines, as they promote the transportation of these through the intestinal wall (Ruiz-Capillas and Jimenez-Colmenero, 2004; Stadnik and Dolatowski, 2010).

As described before, polyamines do not exert a direct toxic effect, however, in addition to nitrosamines genesis, they inhibit histamine or tyramine detoxifying enzymes and thus act as enhancers of their toxicity. These amines in the intestinal tract compete for the detoxifying enzymes that tend to increase the level of histamine and tyramine in blood (Standarova *et al.*, 2008; Al Bulushi *et al.*, 2009).

6. Biogenic Amine as a Quality Index

A quality index has been proposed by accounting the amounts and ratios of biogenic amines which act as an index of the hygienic conditions of raw material and/or manufacturing practices since their amount increase during microbial fermentation or spoilage. For instance, processing and storage of meat and meat products result in increase in the concentrations of some biogenic amines (tyramine, putrescine, and cadaverine), while others BAs like spermidine and spermine decrease (Ruiz-Capillas and Jiménez-Colmenero 2004; Stadnik and Dolatowski, 2010). A relationship between meat quality and BAs' content (individual or combined) variation have been developed and thereby used as quality indicators of unwanted microbial activity in meat and cooked meat products (Table 15.2). However, the same criteria are difficult to apply in fermented products as amine concentrations vary much more widely in such products as compared to fresh meat and processed meat products (Table 15.1). The wide variation further is multifactorial which includes the type and degree of contamination of raw materials, which are promoted by structural breakdown, manufacturing practices, certain processing stages, and the use of starters. However, these factors may also vary depending upon nature of the product and different phases of treatment and storage (Jairath *et al.*, 2015).

Table 15.2. Index of Biogenic Amines in Meat and Meat Products

Products	*Biogenic Amines*	*References*
Fresh beef meat	Putrescine and cadaverine	Slemr, 1981
Bologna sausages	BAI = Putrecine + cadaverine + histamine + tyramine	Wortberg and Woller (1982)
Pork meat at 6-8°C	BAI = Putrecine + cadaverine + histamine + tyramine	Hernández-Jover *et al.*, 1996b
Raw and cooked ground beef	Tyramine, putrecine	Sayem-El-Daher *et al.*, 1984
Vacuum packed beef at 1°C	Putrecine and cadaverine	Edwards *et al.*, 1985
Wraped and unwrapped fresh meat (pork, beef and rabbit)	Putrecine and cadaverine	Guerrero- Legarreta and Chavez-Gallardo, 1991
Dry sausages	Tyramine, histamine, putrecine and cadaverine	Eerola *et al.*, 1996
Hamburger	Tyramine, histamine, putrecine and cadaverine	Durlu-Ozkaya *et al.*, 2001
Fresh beef meat fresh and packed in aerobic atomospher with biopolymers	Spoilage index: Tyramine and cadaverine	Galgano *et al.*, 2009
Turkey meat under modified atmospheric packaging	Freshness indicators: Cadaverine or putrescine + + cadaverine + tyramine	Fraqueza *et al.*, 2012
Broiler chicken cuts stored in modified atmosphere	Tyramine, putrescine and cadaverine	Rokka *et al.*, 2014

The amines are produced during the end of shelf life and hence their levels can be considered as a spoilage index (amine index) rather than a quality index (Ozogul and Ozogul, 2006). An ideal meat freshness index must include all the biogenic amines related to meat spoilage. For example, tyramine increases considerably during meat storage, this biogenic amine should also be included in a biogenic amine index (BAI). Wortberg and Woller (1982) and Hernandez-Jover *et al.* (1996b) proposed BAI as sum of putrescine, cadaverine, histamine and tyramine. The limits of different categories as per Hernandez- Jover *et al.* (1996b) are: <5 $mg{\cdot}kg^{-1}$ for good quality fresh meat; for acceptable meat but with initial spoilage signs, it should be between 5 and 20 $mg{\cdot}kg^{-1}$; between 20 and 50 $mg{\cdot}kg^{-1}$ for low meat quality and finally, >50 $mg{\cdot}kg^{-1}$ for spoiled meat. The level higher than 20 mg/kg of muscle has been proposed as the upper limits for initiation of spoilage in fresh chicken meat stored aerobically as determined by BAI (putrescine + cadaverine + histamine + tyramine). The results obtained demonstrate the applicability of biogenic amine index (BAI) as indicator for poultry quality and freshness evaluation.

6.1 Recommended/Legal Limits

As mentioned above, the presence of biogenic amines in the food not always causes the ill effects. In different individuals there are different mechanisms for detoxification and the presence of other compounds, these factors vary the toxicity levels of biogenic amines. So, the threshold for toxicity caused by biogenic amines is extremely difficult to establish in a given food product. However, EU, USFDA have established legal limits which is also suggested by many authors.

The histamine level should be below 100 $mg{\cdot}kg^{-1}$ in raw fish, below 200 $mg{\cdot}kg^{-1}$ in salted fish as suggested by European Union regulations for species belonging to the *Scombridae* and *Clupeidae* families and upto 400 $mg{\cdot}kg^{-1}$ in cured products (Karovicova and Kohajdova, 2005). If the histamine level reaches to 50 ppm the food is considered spoiled as per USFDA (FDA, 2011). Netherlands Institute of Dairy Research and Czech Republic (Soufleros *et al.*, 1998) recommended histamine upper limits of 100 to 200 $mg{\cdot}kg^{-1}$ for meat products.

The amine most studied with regard to its toxicological effects is histamine. According to Karovicova and Kohajdova (2005) even about 5-10 mg of histamine intake in sensitive people can be considered as defecting while 10 mg is considered as tolerable limit, medium toxicity is induced by 100 mg and 1000 mg is highly toxic while other authors (Hernandez-Jover *et al.*, 1997; Gardini *et al.*, 2001) state different toxic limits which are 8-40 mg causing slight poisoning, intermediate poisoning caused by 40-100 mg and over 100 mg can lead to intensive poisoning. According to data obtained from food intoxication outbreaks the suggested legal upper limit of histamine is 100 $mg{\cdot}kg^{-1}$ food and for ethanol it is 2 $mg{\cdot}lt.^{-1}$. The toxic doses of other amines is known to a lesser extent. The recommended maximum level of tyramine which has been in the range of 100-800 $mg{\cdot}kg^{-1}$ of food is maximum recommended level. A reported value of 30 $mg{\cdot}kg^{-1}$ for β-phenylethylamine in food has been reported as toxic dose (Gardini *et al.*, 2001).

6.2 Detection of BA in Muscle Foods

Determination or detection of amines is important for two reasons *viz.*, their potential toxicity and useful as food quality markers. The analysis of biogenic amines is required for quality control of raw materials, intermediates and end products, monitoring fermentation processes, process control, research and development (Onal, 2007). Detection of BA comprises of two steps: efficient extraction/separation of amines from food stuff followed by quantitative determination. The extraction of amines is the critical step of the determination process as it affects the recovery of BAs in the samples and thereby negatively affecting the analytical process. The complex matrix sample, the presence of potentially interfering compounds, and the occurrence of several BA simultaneously are typical problems encountered in the analysis.

Different techniques are used to quantify biogenic amines in biological matrices after extraction, such as ion mobility spectrometry (Karpas *et al.*, 2002), thin-layer chromatography (Fadhlaoui-Zid *et al.*, 2012), capillary electrophoresis (Simo *et al.*, 2008), gas chromatography (Almeida *et al.*, 2012), and most commonly high performance liquid chromatography (HPLC) (Jia *et al.*, 2011; Zhai *et al.*, 2012).These methods are well approved. The high performance thin-layer chromatography (HPTLC) technique is the recent improvement in the field of BA detection which has high resolution, more precise quantification and is able to separate BA in one run. An amperometric biosensor has been recently characterized in flow injection analysis (FIA) involving the use of pea seedling amine oxidase (PSAO) as molecular recognition element for the determination of biogenic amines (Telsnig *et al.*, 2012) such as putrescine, cadaverine and tyramine (used as freshness markers). The limit of detection (LOD) of the protocol is 1.1μg/mL for putrescine, 2.2μg/mL for cadaverine and 6.2μg/mL for tyramine.

7. Control of Biogenic Amines

By curbing on the factors responsible for production and enhancement of BAs level, the control is possible. Inhibition of microbial growth and their decarboxylases activity can control the biogenic amines formation. The prevention of biogenic amine formation in food, therefore, can be achieved using temperature control, using high-quality raw material, good manufacturing practices, the use of non-amine forming (amine-negative) or amine oxidizing starter cultures for fermentation (Nieto-Arribas *et al.*, 2009), the use of enzymes to oxidize amines (Dapkevicius *et al.*, 2000), the use of microbial modelling to assess favorable conditions to delay biogenic amine formation (Neumeyer *et al.*, 1997; Emborg and Dalgaard, 2008a, 2008b), packaging techniques (Mohan *et al.*, 2009), high hydrostatic pressure(Bolton *et al.*, 2009), irradiation (Kim *et al.*, 2003; Kim *et al.*, 2005; Min *et al.*, 2007a; Min *et al.*, 2007b), and food additives (Mah and Hwang, 2009).

7.1 Use of Specific Starter Culture

Use of certain specific starter cultures which are either not able to decarboxylate amino acid into biogenic amines (amine-negative) or able to oxidize biogenic amines into aldehyde, hydrogen peroxide, and ammonia (amine oxidizing) in fermentation process can also delay the formation of BAs. Certain specific starters (Bover-Cid *et al.*, 2000a; Suzzi and Gardini, 2003)require optimal growth conditions to dominate over BAs producing (Xu *et al.*, 2010) and other contaminant bacteria (Maijala *et al.*, 1995a; Maijala *et al.*, 1995b; Hu *et al.*, 2008). Starters with inability to form biogenic amines, ability to grow well at the product processing temperature and ability to suppress the growth of wild amine producing microflora are suitable for fermented meat products (Suzzi and Gardini, 2003). Further, bacteriocin (for example curvacin A) producing strains can increase their competitiveness (Hammes and Hertel, 1996). A rapid pH decrease during fermentation is also helpful in preventing biogenic amines accumulation. Moreover, during the later phase of ripening and throughout storage, starter cultures compete with nonstarter bacteria which further avoid excessive biogenic amines production (Suzzi and Gardini, 2003).

Bacteria described as biogenic amine oxidizers include *Micrococcus varians* (Leuschner and Hammes, 1998b), *Natrinema gari* (Tapingkae *et al.*, 2010), *Brevibacterium linen* (Leuschner and Hammes 1998a), *Vergibacillus* spp. SK33 (Yongsawatdigul *et al.*, 2007),*Lactococcus sakei*, *Lactobacillus curvatus* (Dapkevicius *et al.*, 2000), and *S. xylosus* (Mah and Hwang, 2009).

7.2 Microbial Modelling

Microbial modelling can be used to study the growth and inactivation of microorganisms (Zwietering *et al.*, 1990; Xiong *et al.*, 1999; Van Boekel, 2002) with the aim of controlling growth and predicting risk factors (Seo *et al.*, 2007). Modelling microorganisms responsible for biogenic amine formation (Emborg and Dalgaard 2008a, b) has been used to explore options for biogenic amine control.

7.3 High Pressure Processing

High pressure processing is one of the emerging technologies which can be applied at refrigeration, ambient or moderately high temperatures to inactivate pathogenic and spoilage microorganisms, and enzymes in foods with fewer changes in texture, colour and flavour compared to conventional technologies. It has aroused more interest in recent years because it addresses consumer demands for safe better quality products with fewer additives, less processing and prolonged shelf life (Naila *et al.*, 2010). In a fermentation process when high pressure is applied to raw material or the end products, a reduction in the number of bacteria is observed which inhibit biogenic amine formation. Effects of high pressure on fermented sausages have also been studied and it has been observed that treating fermented sausage with high pressure (350 MPa/15 min) reduced lactic acid bacteria and thereby decreased cadaverine, putrescine and tyramine levels during 160 days chilled storage compared to sausage not treated with high pressure. Reduction of lactic acid bacteria together with biogenic amines may cause some problems in fermented foods. Therefore, to control biogenic amine formation in fermented foods, high pressure application could be feasible before starter culture addition (Ruiz-Capillas *et al.*, 2007).

7.4 Application of Additives

Spices and herbs are additives which are generally used in foods for enhancing the flavour or colour attributes. These materials also have antimicrobial and antioxidant activities (Bozkurt, 2006). Many reports show that spices and herbs reduce biogenic amine production. Curcumin (turmeric), capsaicin (red pepper), and piperine (black pepper) are naturally occurring specific inhibitory substances that inhibit biogenic amine formation (Mbarki *et al.*, 2009). The components in spices such as thymol inhibit biogenic amine formation(Yücel and Üren, 2008).The citric acid use in pickled cabbage fermentation and sodium nitrite addition in dry sausage fermentation produced a slight decrease in biogenic amines (Kurt and Zorba, 2009). Use of nitrite in sucuk production has been found to affect the formation of biogenic amines except for spermidine and spermine (Gençcelep *et al.*, 2007). Effect of sodium sulphite on biogenic amine accumulation during the ripening of slightly fermented sausages has been studied. The study reported stimulation of tyramine production in the presence of sulphite, whereas, cadaverine formation was drastically inhibited (Bover-Cid *et al.*, 2001).

7.5 Irradiation

Irradiation of muscle food like pork and beef (Min *et al.*, 2007b), sausage (Kim *et al.*, 2005) and chicken (Min *et al.*, 2007a) have ability to extend the shelf life. Treatment of ground pork and beef inoculated with *Alcaligenes faecalis*, *Bacillus cereus* and *Enterobacter cloacae* with gamma irradiation (2 kGy) showed reduction in histamine, tyramine, spermidine, beta-phenylethylamine, tryptamine, cadaverine, and putrescine formed during 24 h storage at 4°C (Min *et al.*, 2007b). However, some reports of irradiation documented the contradictory results (promoter of formation of other biogenic amines), for instance, irradiation of raw chicken breast and thigh meat at a dose of 2 kGy resulted in reduction of some biogenic amines, but the increment of other biogenic amines level (histamine, spermidine, and spermine), perhaps because irradiation changes the structure and physiological properties of enzymes that form biogenic amines (Min *et al.*, 2007a).

7.6 Restricting Decarboxylase Activity

Biogenic amines are formed by the enzymatic decarboxylation of free amino acids (Figure 15.1). The conducive environments for the production of biogenic amines are: presence of free amino acids and decarboxylase enzymes and appropriate conditions for decarboxylation. Free amino acids are produced by the action of microbial proteolytic enzymes which convert proteins to amino acids and also these amino acids are converted to biogenic amines by the amino decarboxylase enzymes. Biogenic amine formation can be regulated by controlling these rate limiting decarboxylase enzymes. It was reported that the proteolytic enzymes were effective in the reduction of histidine decarboxylase (Yamada *et al.*, 1980). Effect of trypsin and chymotrypsin on the activity of histidine decarboxylase in different rat tissues has been also studied. Use of these enzymes strongly inactivated histidine decarboxylase in rats. Romantsev and Prozorovskii (1984) also reported effect of trypsin on the reduction of histidine decarboxylase from *Micrococcus* sp.; histidine decarboxylase lost 50 per cent of its activity after the first 9 min of hydrolysis. They also studied the thermal stability of this enzyme and found that it preserves 80 per cent activity after 5 min of incubation at 72 °C and at pH 5.55. In the literatures, there is little information about the reduction of biogenic amine formation by inhibition of decarboxylase enzymes.

7.7 Degradation of Biogenic Amines

There are some methods that can be applied to degrade formed biogenic amines such as use of amine oxidizing microorganism and/or enzymes like diamine oxidase and use of amine degrading microorganisms such as *Lactobacillus plantarum* or vineyard ecosystem fungi in wine making process. 44 grapevine and vineyard soil fungi from four locations of Spain were isolated and evaluated for *in vitro* amine degradation in a micro fermentation system. Species of *Pencillium citrinum*, *Alternaria* sp., *Phoma* sp., *Ulocladium chartarum* and *Epicoccum nigrum* showed highest potential for amine degradation (Cueva *et al.*, 2012).

The bacteria with amine oxidizing activity or having oxidizing enzymes may be used to reduce already formed biogenic amine levels in foods. Naila *et al.* (2012) studied the amount of histamine degradation by diamine oxidase in model (buffer) and *in-vivo* (cooked tuna soup used in manufacture of a fish paste product, Rihaakuru) systems. They reported a technique using diamine oxidase to reduce histamine from 500 mg/L to undetectable levels (< 0.5 mg/kg) in Rihaakuru. They found that histamine can be degraded at low salt levels (12 per cent) and pH 6.7, but they also reported that in this condition, product sensory characteristics may vary. Dapkevicius *et al.* (2000) also investigated the degradation of histamine by diamine oxidase in fish silage and fish slurry by using 2 per cent NaCl, 12 per cent saccharose and 0.05 per cent cysteine and found that amine oxidase was only effective in degrading histamine in the presence of cysteine. Use of diamine oxidase can be a potential method for degradation of histamine but there need additional studies to see its effects in degrading other biogenic amines and also on food product characteristics

7.8 External Factor: Temperature

Temperature is an important factor in the formation of biogenic amines. The low temperature slows down the biogenic amine formation by inhibiting microbial growth and reduction of enzyme activity (Duflos, 2009). It was reported that higher fermentation temperature (24°C) favor the growth of lactic acid bacteria which outgrew the amine-positive non-starter microorganisms. Higher temperature can favor proteolytic and decarboxylation reactions resulting in increased amine content after storage (Suzzi and Gardini, 2003). In addition, freezing is an alternative method for stabilization of the level of biogenic amine. Many reports show the efficiency of low temperature application on the reduction of biogenic amines. Yellow fin tuna stored upto 9 days at 0°C and 22°C showed an increase in histamine of 15 mg/kg at 0°C and 4,500 mg/kg at 22°C (Du *et al.*, 2002). The other range *i.e.*, high temperature treatment is also an effective method for the reduction of biogenic amine formation by killing the microorganisms. However, histamine could reform after the application of heat treatment if recontamination occurs and if temperature control could not be supplied (Naila *et al.*, 2010).

Control of the biogenic amine is not always possible with temperature alone because some bacteria can produce biogenic amines even at temperatures below 5°C (Emborg and Dalgaard, 2006). However, maintaining the cold chain in foods that already contain high levels of biogenic amines will generally stabilize the levels of biogenic amines (Bover-Cid *et al.*, 2006).

7.9 Packaging

Packaging (MAP, vacuum and active packaging) either results in inhibition of microorganisms or the enzyme producing biogenic amines leading to delay in production of biogenic amines. There are reports on the successful control of biogenic amines through packaging *e.g.* MAP of chicken (Balamatsia *et al.*, 2006; Patsias *et al.*, 2006), sausage (Kim *et al.*, 2005). Patsias *et al.* (2006) studied precooked chicken meat under air and MAP (30 per cent CO_2/70 per cent N_2) at 4°C for up to 23 days and found reduced level of biogenic amines (putrescine and tyramine) after 23 days of storage under MAP as compared to aerobic packaging.

8. Conclusions

The presence of biogenic amines in meat and meat products depends upon multiple factors and solution for reduction of these amines lies within it by curbing the responsible factors. Normally, human body is capable of detoxify exogenous amines but in certain allergic conditions of medications, BAs could accumulate in the body and could aggravate the situation. Biogenic amines detection is important as they are of public health significance as well as indicators of freshness/quality. New rapid detection techniques have now been evolved to detect their limits. It is the high time for researchers, food technologists, to come together and address the matter in the interest of mankind.

REFERENCES

Al Bulushi, I., Poole, S., Deeth, H. C. and Dykes, G.A. (2009). Biogenic amines in fish: Roles in intoxication, spoilage, and nitrosamine formation-a review. Critical Reviews in Food Science and Nutrition, 49, 369-377.

Almeida, C., Fernandes, J. O. and Cunha, S.C. (2012).A novel dispersive liquid–liquid microextraction (DLLME) gas chromatography-mass spectrometry (GC–MS) method for the determination of eighteen biogenic amines in beer. Food Control, 25, 380-388.

Alvarez, M. F. (2000). Review: active food packaging. Food Science and Technology International, 6, 97–108.

Arashisar, S., Hisar, O., Kaya, M. and Yanik, T. (2004). Effect of modified atmosphere and vacuum packaging on microbiological and chemical properties of rainbow trout (*Oncorynchus mykiss*) fillets. International Journal of Food Microbiology,97, 209– 214.

Ayhan, K., Kolsarici, N. and Ozkan, G.A., 1999. The effects of a starter culture on the formation of biogenic amines in Turkish soudjoucks. International Journal of Food Microbiology, 53, 183 – 188.

Balamatsia, C. C., Paleologos, E. K., Kontominas, M. G. and Savvaidis, I. N. (2006). Correlation between microbial flora, sensory changes and biogenic amines formation in fresh chicken meat stored aerobically or under modified atmosphere packaging at 4 degrees C: possible role of biogenic amines as spoilage indicators. Antonie Van Leeuwenhoek International Journal of General and Molecular Microbiology,89, 9–17.

Banwart, G.J. (1981). Basic Food Microbiology. AVI publishing Company Inc., Westport, CN.

Bardocz S. (1995). Polyamines in food and their consequences for food quality and human health. Trends in Food Science and Technology, 6, 341-346.

Bauer, F., Potzelberge, D., Hellwing, E., and Paulsen, P. (1996). Formation of biogenic amines in fresh meat packed in oxygen permeable foil. In: Proceedings of the 42nd International Congress of Meat Science and Technology, pp. 552–553. Lillehammer, Norway.

Bolton, G. E., Bjornsdottir, K., Nielsen, D., Luna, P. F. and Green, D. P. (2009). Effect of high hydrostatic pressure on histamine forming bacteria in yellow fin tuna and mahi-mahi skinless portions. Institute of Food Technologists (IFT) Conference, Abstract No. 006-05. USA.

Bouchereau, A., Guénot, P. and Larher, F. (2000). Analysis of amines in plant materials. Journal of Chromatography B,747, 49-67.

Bover-Cid, S. and Holzapfel, W.H. (1999). Improved screening procedure for biogenic amine production by lactic acid bacteria. International Journal of Food Microbiology, 59, 391 – 396.

Bover-Cid, S., Izquierdo-Pulido, M. and Vidal-Carou, M. C. (2000). Mixed starter cultures to control biogenic amine production in dry fermented sausages. Journal of Food Protection, 63, 1556 – 1562.

Bover-Cid, S., Izquierdo-Pulido, M. and Vidal-Carou, M.C. (2001b). Effect of the interaction between a low tyramine-producing Lactobacillus and proteolytic staphylococci on biogenic amine production during ripening and storage of dry sausages. International Journal of Food Microbiology,65, 113 – 123.

Bover-Cid, S., Izquierdo-Pulido, M., and Vidal-Carou, M. C. (2001a). Changes in biogenic amines and polyamine contents in slightly fermented sausages manufactured with and without sugar.Meat Science, 57, 215–221.

Bover-Cid, S., Miguelez-Arrizado, J.M. and Vidal-Carou, M.C. (2001). Biogenic amine accumulation in ripened sausages affected by the addition of sodium sulphite. Meat Science, 59, 391-396.

Bover-Cid, S., Miguelez-Arrizado, M. J., Latorre-Moratalla, L. L. and Vidal-Carou, M. C. (2006). Freezing of meat raw materials affects tyramine and diamine accumulation in spontaneously fermented sausages. Meat Science,72, 62-68.

Bozkurt, H. (2006). Utilization of natural antioxidants: Green tea extract and *Thymbra spicata* oil in Turkish dry fermented sausage. Meat Science, 73, 442-450.

Bozkurt, H. and Erkmen, O. (2004). Effects of temperature humidity and additives on the formation of biogenic amines in sucuk during ripening and storage periods. Food Science and Technology International, 10, 21-28.

Bunkova, L., Bunka, F., Drab, V., Kracmar, S., and Kuban, V. (2012). Effects of NaCl, lactose and availability of oxygen on tyramine production by the *Enterococcus durans* CCDM 53. European Food Research and Technology,234, 973–979.

Bunkova, L., Bunka, F., Pollakova, E., Podesvova, T. and Drab, V. (2011). The effect of lactose, NaCl and an aero/anaerobic environment on the tyrosine decarboxylase activity of *Lactococcus lactis* subsp. cremoris and *Lactococcus lactis* subsp. lactis. International Journal of Food Microbiology,147, 112–119. (

Cid, S. B., Miguélez-Arrizado, M. J., Beckerc, B., Holzapfel, W. H., and Vidal- Carou, M. C. (2008). Amino acid decarboxylation by *Lactobacillus* curvatus CTC273 affected by the pH and glucose availability. Food Microbiology,25, 269– 277.

Coton, E. and Coton, M. (2009). Evidence of horizontal transfer as origin of strain to strain variation of the tyramine production trait in *Lactobacillus brevis*. Food Microbiology, 26, 52–57.

Cueva, C., García-Ruiz, A., González-Rompinelli, E., Bartolome, B., Martín-Alvarez, P.J., Salazar, O., Vicente, M.F., Bills, G.F. and Moreno-Arribas, M.V. (2012). Degradation of biogenic amines by vineyard ecosystem, potential use in winemaking, Journal of Applied Microbiology, 112, 672-682.

Curiel, J. A., Ruiz-Capillas, C., de las Rivas, B., Carrascosa, A. V., Jiménez-Colmenero, F., and Munoz, R. (2011). Production of biogenic amines by lactic acid bacteria and enterobacteria isolated from fresh pork sausages packaged in different atmospheres and kept under refrigeration. Meat Science,88, 368–373.

Dapkevicius, M.L.N.E., Nout, M.J.R. and Rombouts, F.M. (2000). Biogenic amine formation and degradation by potential fish silage starter microorganisms, International Journal of Food Microbiology, 57, 107-114.

Devlieghere, F., Debevere, J. and Van Impe, J. (1998). Effect of dissolved carbon dioxide and temperature on the growth of *Lactobacillus* sake in modified atmospheres. International Journal of Food Microbiology, 41, 231–238.

Du, W.X., Lin, C.M., Phu, A.T., Cornell, J.A., Marshall, M.R. and Wei, C.I. (2002). Development of biogenic amines in yellowfin tuna (*Thunnus albacares*): Effect of storage and correlation with decarboxylase positive bacterial flora. Journal of Food Science, 67, 292-301.

Duflos, G. (2009). Histamine risk in fishery products, Bulletin de l'Academie Veterinaire de France, 162, 241-246.

Durlu-Ozkaya, F., Ayhan, K. and Vural, N. (2001). Biogenic amines produced by Enterobacteriaceae isolated from meat products. Meat Science, 58: 163–166.

Edwards, R. A., Daintry, R.H., and Hibard, C. M. (1983). The relationship of bacterial numbers and types to diamine concentration in fresh and aerobically stored beef, pork, and lamb. Journal of Food Technology, 18, 777–788.

Edwards, R. A., Dainty, R. H., Hibbard, C. M. and Ramantanis, S. V. (1987). Amines in fresh beef of normal pH and the role of bacteria in changes in concentration observed during storage in vacuum packs in chilled temperature. Journal of Applied Bacteriology, 63, 427 – 434.

Edwards, R.A., Dainty, R.H., and Hibard, C.M. (1985). Putrescine and cadaverine formation in vacuum packed beef. Journal of Applied Bacteriology, 58, 13–19.

Eerola, S., Maijala, R., Roig Sagúes, A. X., Salminen, M. and Hirvi, T. (1996). Biogenic amines in dry sausages as affected by starter culture and contaminant amine-positive Lactobacillus. Journal of Food Science, 61, 1243–1246.

Eerola, S., Sagues, A. X. R., Lilleberg, L. and Aalto, H. (1997). Biogenic amines in dry sausages during shelf-life storage. Zeitschrift für Le0bensmittel-Untersuchung und –Forschung, 205, 351–355.

El-Habib, R. (2011). Bacteriological quality and biogenic amines determination by HPLC in Bassa fish imported to Saudi Arabia. International Journal of Pharmacy and Pharmaceutical Sciences, 3, 343-347.

Emborg, J. and Dalgaard, P. (2006). Formation of histamine and biogenic amines in colds moked tuna: An investigation of psychro-tolerant bacteria from samples implicated in cases of histamine fish poisoning. Journal of Food Protection, 69, 897-906.

Emborg, J. and Dalgaard, P. (2008b). Modelling the effect of temperature, carbon dioxide, water activity and pH on growth and histamine formation by *Morganella psychrotolerans*. International Journal of Food Microbiology, 128(2), 226–233.

Emborg, J. and Dalgard, P. (2008a). Growth, inactivation and histamine formation of *Morganella psychrotolerans* and *Morganella morganii*– development and evaluation of predictive models. International Journal of Food Microbiology,128(2), 234–243.

European Food Safety Authority Scientific Panel. (2011). Opinion on risk based control of biogenic amine formation in fermented foods.European Food Safety Authority Scientific Journal, 9, 2393.

Fadhlaoui-Zid, K., Curiel, J.A., Landeta, G., Fattouch, S., Reverón, I., de las Rivas, B., Sadok, S. and Muñoz, R. (2012). Biogenic amine production by bacteria isolated from ice-preserved sardine and mackerel. Food Control, 25, 89-95.

Fathi, A.Z., Pooladgar, A.R., Maghami, S.G. and Rahman, I. (2013). The changes evaluation of biogenic amines (Histamine) by HPLC, in shank yellow fin fish (*Acanthopagrus latus*) within 18 days of ice storage. Indian Journal of Science and Technology, 1, 29-34.

FDA, Food and Drug Administration. (2011). Fish and Fishery Products Hazards and Controls Guidance, 4th edn. Department of Health and Human Services, Food and Drug Administration, Center for Food Safety and Applied Nutrition, Washington DC

Fernandez, M., Linares, D. M., Rodriguez, A. and Alvarez, M. A. (2007). Factors affecting tyramine production in *Enterococcus durans* IPLA 655. Applied Microbiology and Biotechnology, 73, 1400–1406.

Foster, J.W. (2004). Escherichia coliacid resistance: tales of an amateur acidophile. Nature Review Microbiology, 2, 898-907.

Fraqueza M.J., Alfaia, C.M. and Barreto, A.S. (2012). Biogenic amine formation in turkey meat under modified atmosphere packaging with extended shelf life: Index of freshness. Poultry Science, 91(6),1465-72.

Galgano, F., Favati, F., Bonadio, M., Lorusso, V. and Romano, P. (2009). Role of biogenic amines as index of freshness in beef meat packed with different biopolymeric materials. Food Research International, 42, 1147–1152.

Gardini, F., Martuscelli, M., Caruso, M. C., Galgano, F., Crudele, M. A., Favati, F., Guerzoni, M. E. and Suzzi, G. (2001). Effects of pH, temperature and NaCl concentration on the growth kinetics, proteolytic activity and biogenic amine production of *Enterococcus faecalis*. International Journal of Food Microbiology, 64, 105 – 117.

Gençcelep, H., Kaban, G. and M. Kaya. (2007). Effects of starter cultures and nitrite levels on formation of biogenic amines in sucuk. Meat Science, 77: 424-430. (

Gonzalez-Fernandez, C., Santos, E. M., Jaime, I. and Rovira, J. (2003). Influence of starter cultures and sugar concentrations on biogenic amine contents in chorizo dry sausage. Food Microbiology, 20, 275-284.

Guerrero-Legarreta, I. and Chavez-Gallardo, A. M. (1991). Detection of biogenic amines as meat spoilage indicators. Journal of Muscle Foods, 2, 263–278.

Halasz, A., Barath, A., Simon-Sarkadi, L. and Holzapfel, W. (1994). Biogenic amines and their production by micro-organisms in food. Trends in Food Science and Technology, 5: 42-49.

Hammes, W. P. and Hertel, C. (1996). Selection and improvement of lactic acid bacteria used in meat and sausage fermentation.Le Lait, 76, 159-168.

Hernandez-Jover, T., Izquierdo-Pulido, M., Veciana-Nogués, M. T., Mariné-Font, A. and Vidal-Carou, M. C. (1997). Biogenic amine and polyamine contents in meat and meatproducts. Journal of Agricultural and Food Chemistry, 45, 2098-2102.

Hernandez-Jover, T., Izquierdo-Pulido, M., Veciana-Nogues, M.T. and Vidal Carou, M.C. (1996a). Ion-pair high performance liquid chromatographic 499 determination of biogenic amines in meat and meat products. Journal of Agricultural and Food Chemistry, 44, 2710–2715.

Hernandez-Jover, T., Izquierdo-Pulido, M., Vecina-Nogués, M. T. and Vidal-Carou, M. C. (1996b). Biogenic amine sources in cooked cured shoulder pork. Journal of Agricultural and Food Chemistry, 44, 3097-3101.

Hu, Y., Xia, W. and Ge, C. (2008). Characterization of fermented silver carp sausages inoculated with mixed starter culture. LWT - Food Science and Technology,41, 730–738.

Huis in't Veld, J. H. J., Hose, H., Schaafsma, G. J., Silla, H. and Smith, J. E. (1990). Health aspects of food biotechnology. In: Processing and Quality of Foods, pp. 2.73-2.97. Zeuthen, P., Cheftel, J. C., Ericksson, C., Gormley, T. R., Link, P. and Paulus, K., Eds., Vol 2. Food Biotechnology: avenues to healthy and nutritious products, Elsevier Applied Science, London and New York.

Ibrahim, H.M., Amin, R.A., Eleiwa, N.Z. and Ahmed, N.M. (2017). Estimation of some biogenic amines on chicken meat products. Benha Veterinary Medical Journal, 32, 23 – 28.

Jairath, G., Singh, P.K.Dabur, R.S., Rani, M. and Chaudhari, M. (2015).Biogenic amines in meat and meat products and its public health significance: a review. Journal Of Food Science and Technology,(DOI 10.1007/s13197-015-1860-x

Jansen, S.C., Dusseldorp, M.V., Bottema, K.C. and Dubois, A.E.J. (2003). Intolerance to dietary biogenic amines: A review. Annals of Allergy, Asthma and Immunology, 91, 233-40.

Jastrzebska, A., Kowalska, S., and Sz³yk, E. (2016). Studies of levels of biogenic amines in meat samples in relation to the content of additives. Food Additives and Contaminants: Part A, 33, 27–40.

Jia, S., Kang, Y. P., Park, J. H., Lee, J. and Kwon, S. W. (2011). Simultaneous determination of 23 amino acids and 7 biogenic amines in fermented food samples by liquid chromatography/quadrupole time-of-flight mass spectrometry. Journal of Chromatography, 1218, 9174-9182.

Kalac, P. and Krausova, P. (2005). A review of dietary polyamines: Formation, implications for growth and health and occurrence in foods. Food Chemistry, 90, 219-230.

Karovicova, J. and Kohajdova, Z. (2005). Biogenic amines in food. Chemical Papers, 59, 70-79.

Karpas, Z., Tilman, B., Gdalevsky, R. and Lorber, A. (2002). Determination of volatile biogenic amines in muscle food products by ion mobility spectrometry. Analytica Chimica Acta, 463: 155–163.

Kebary, K. M. K., El-Sonbaty, A. H. and Badawi, R. M. (1999). Effects of heating milk and accelerating ripening of low fat Ras cheese on biogenic amines and free amino acids development. Food Chemistry, 64, 67–75.

Kim, J. H., Ahn, H. J., Kim, D. H., Jo, C., Yook, H. S., Park, H. J. and Byun, M. W. (2003). Irradiation effects on biogenic amines in Korean fermented soybean paste during fermentation. Journal of Food Science, 68, 80–84.

Kim, J. H., Ahn, H. J., Lee, J. W., Park, H. J., Ryu, G. H., Kang, I. J. and Byun, M. W. (2005). Effects of gamma irradiation on the biogenic amines in pepperoni with different packaging conditions. Food Chemistry, 89, 199–205.

Komprda, T., Burdychová, R., Dohnal, V., Cwiková, O., Sládková, P. and Dvoráčková, H. (2008). Tyramine production in Dutch-type semi-hard cheese from two different producers. Food Microbiology, 25, 219-227.

Komprda, T., Smela, D., Pechova, P., Kalhotka, L., Stencl, J. and Klejdus, B. (2004). Effect of starter culture, spice mix and storage time and temperature on biogenic amine content of dry fermented sausages. Meat Science, 67, 607–616.

Kordiovska, P., Vorlova, L., Borkovcova, I., Karpiskova, R., Buchtova, H., Svobodova, Z., Krizek, M. and Vacha, F. (2006). The dynamics of biogenic amine formation in muscle tissue of carp (*Cyprinus carpio*). Czech Journal of Animal Science, 51, 262-270.

Kurt, S. and Zorba, O. (2009). The effects of ripening period nitrite level and heat treatment on biogenic amine formation of "sucuk"–A Turkish dry fermented sausage, Meat Science, 82, 179-184.

Lange, J., Thomas, K., and Wittmann, C. (2002). Comparison of a capillary electrophoresis method with high-performance liquid chromatography for the determination of biogenic amines in various food samples. Journal of Chromatography B: Analytical Technology Biomedical Life Science, 779, 229–239.

Latorre-Moratalla M.L., Bover-Cid S., Talon R., Garriga M., Zanardi E., Ianieri A., Fraqueza, M. J., Elias, M., Drosinos, E. H. and Vidal-Carou, M. C. (2010). Strategies to reduce biogenic amine accumulation in traditional sausage manufacturing.LWT-Food Science and Technology, 43, 20-25.

Latorre-Moratalla, M. L., Veciana-Nogués, T., Bover-Cid, S., Garriga, M., Aymerich, T., Zanardi, E., Ianieri, A., Fraqueza, M. J., Patarata, L., Drosinos, E. H., Lauková, A., Talon, R. and Vidal-Carou, M. C. (2008). Biogenic amines in traditional fermented sausages produced in selected European countries. Food Chemistry, 107: 912-921.

Lehane, L. and Olley, J. 2000. Histamine fish poisoning revisited. International Journal of Food Microbiology, 58, 1 –37.

Leuschner, R. G. K. and Hammes, W. P. (1998a). Degradation of histamine and tyramine by *Brevibacterium linens* during surface ripening of Munster cheese. Journal of Food Protection, 61, 874–875.

Leuschner, R. G. K. and Hammes, W. P. (1998b). Tyramine degradation by micrococci during ripening of fermented sausage. Meat Science, 49: 289–96.

Li, M., Tian, L., Zhao, G., Zhang, Q., Gao, X., Huang, X., *et al.* (2014). Formation of biogenic amines and growth of spoilage-related microorganisms in pork stored under different packaging conditions. Meat Science,96: 843–848.

Lopez-Caballero, M. E., Goncalves, A., and Nunes, M. L. (2002). Effect of CO2/ O2 -containing modified atmospheres on packed deepwater pink shrimp (*Parapenaus longirostris*). *European Food Research Technology*, 214, 192–197.

Lucas, P. M., Wolken, W. A. M., Claisse, O., Lolkema, J. S. and Lonvaud-Funel, A. (2005). Histamine-producing pathway encoded on an unstable plasmid in *Lactobacillus hilgardii* 0006. Applied Environmental Microbiology, 71, 1417-1424.

Mah, J. H. and Hwang, H. J. (2009). Effects of food additives on biogenic amine formation in Myeolchijeot, a salted and fermented anchovy (*Engraulis japonicus*). Food Chemistry, 114, 168–173.

Maijala, R. L., Eerola, S. H., Aho, M. A. and Hirn, J. A. (1993). The effects of GDL-induced pH decrease on the formation of biogenic amines in meat. Journal of Food Protection, 56,125-129.

Maijala, R., Eerola, S., Lievonen, S., Hill, P. and Hirvi, T. (1995a). Formation of biogenic amines during ripening of dry sausages as affected by starter culture and thawing time of raw materials. Journal of Food Science,60, 1187-1190.

Maijala, R., Nurmi, E. and Fischer, A. (1995b). Influence of processing temperature on the formation of biogenic amines in dry sausages. Meat Science, 39, 9-22.

Masson, F., Talon, R., and Montel, M.C. (1996). Histamine and tyramine production by bacteria from meat products. International Journal of Food Microbiology, 32, 199–207.

Mbarki, R., Miloud, N.B., Semli, S., Dhib, S. and Sadok, S. (2009). Effect of vacuum packaging and low dose irradiation on the microbial chemical and sensory characteristics of chub mackerel (*Scomber japonicus*), Food Microbiology, 26, 821-826.

McCabe-Sellers, B.J., Staggs, C.G. and Bogle, M.L. (2006). Tyramine in foods and monoamine oxidase inhibitor drugs: A crossroad where medicine, nutrition, pharmacy, and food industry converge. Journal of Food Composition and Analysis, 19, S58-S65.

Min, J. S., Lee, S., Jang, A., Jo, C. and Lee, M. (2007b). Irradiation and organic acid treatment for microbial control and the production of biogenic amines in beef and pork. Food Chemistry, 104, 791–799.

Min, J.S., Lee, S., Jang, A., Jo, C. and Lee, M. (2007a). Control of microorganisms and reduction of biogenic amines in chicken breast and thigh by irradiation and organic acids. Poultry Science, 86, 2034–41.

Mohan, C.O., Ravishankar, C.N., Gopal, T.K.S., Kumar, K.A. and Lalitha, K.V. (2009). Biogenic amines formation in seer fish (*Scomberomorus commerson*) steaks packed with O2 scavenger during chilled storage. Food Research International, 42, 411–416.

Moret, S., Smela, D., Populin, T. and Conte, L.S. (2005). A survey on free biogenic amine content of fresh and preserved vegetables. Food Chemistry, 89, 355-361.

Nadon, C.A. Ismond, M.A.H., and Holley, R. (2001). Biogenic amines in vacuumpackaged and carbon dioxide-controlled atmosphere-packaged fresh pork stored at -1.5°C. Journal of Food Protection, 64, 220–227.

Naila, A., Flint, S., Fletcher, G., Bremer, P. and Meerdink, G. (2010). Control of biogenic amines in food: Existing and emerging approaches, Journal of Food Science, 75, 139-150.

Naila, A., Flint, S., Fletcher, G., Bremer, P., Meerdink and Morton, G. R.H. (2012). Prediction of the amount and rate of histamine degradation by diamine oxidase (DAO), Food Chemistry, 135, 2650-2660.

Neumeyer K, Ross T and McMeekin TA. (1997). Development of a predictive model to describe the effects of temperature and water activity on the growth of spoilage pseudomonads. International Journal of Food Microbiology, 38, 45–54.

Nieto-Arribas, P., Poveda, J. M., Seseña, S., Palop, L. and Cabezas, L. (2009). Technological characterization of *Lactobacillus* isolates from traditional Manchego cheese for potential use as adjunct starter cultures. Food Control, 20, 1092–1098.

Ntzimani, A. G., Paleologos, E. K., Savvaidis, I. N. and Kontominas, M. G. (2008). Formation of biogenic amines and relation to microbial flora and sensory changes in smoked turkey breast fillets stored under various packaging conditions at 4°C. Food Microbiology, 25, 509–517.

OECD/FAO (2016), *OECD-FAO Agricultural Outlook 2016-2025*, OECD Publishing, Paris. DOI: http: //dx.doi.org/10.1787/agr_outlook-2016-e. Retrieved on 31st July, 2017.

Onal, A. (2007). A review: Current analytical methods for the determination of biogenic amines in foods. Food Chemistry, 103, 1475-1486.

Ordonez, J.A., De Pablo, B., Perez de Castro, B., Asensio, M.A. and Sanz, B. (1991). Select chemical and microbiological changes in refrigerated pork stored in carbon dioxide and oxygen enriched atmospheres. Journal of Agricultural Food Chemistry, 39, 668–672.

Ozogul, F. and Ozogul, Y. (2006) Biogenic amine content and biogenic amine quality indices of sardines (*Sardina pilchardus*) stored in modified atmosphere packaging and vacuum packaging. Food Chemistry, 99, 574–578.

Papavergou, E. J., Savvaidis, I. N. and Ambrosiadis, I. A. (2012). Levels of biogenic amines in retail market fermented meat products. Food Chemistry, 135, 2750–2755.

Patsias, A., Chouliara, I., Paleologos, E. K., Savvaidis, I. and Kontominas, M. G. (2006). Relation of biogenic amines to microbial and sensory changes of precooked chicken meat stored aerobically and under modified atmosphere packaging at 4°C. European Food Research Technology, 223, 683–689.

Paulsen, P. and Bauer, F. (1997). Biogenic amines in fermented sausages: 2. Factors influencing the formation of biogenic amines in fermented sausages. Fleischswirtschaft International, 77, 32 – 34.

Rodrigues B. L., da Silveira Alvaresb T., Sampaio G. S. L., Cabral C. C., Araujo J. V. A., Franco R. M., Junior C.A.C. (2016). Influence of vacuum and modified atmosphere packaging in combination with UV-C radiation on the shelf life of rainbow trout (*Oncorhynchus mykiss*) fillets. Food Control, 60, 596–605.

Roig-Sagues, A. X. and Eerola, S. (1997). Biogenic amines in meat inoculated with Lactobacillus sake starter strains and an amine-positive lactic acid bacterium. Zeitschrift für Lebensmittel-Untersuchung und –Forschung, 205, 227–231.

Rokka, M., Eerola, S., Smolander, M., Alakomi, H. and Ahvenainen, R. (2014). Monitoring of the quality of modified atmosphere packaged broiler chicken cuts stored in different temperature conditions: B. Biogenic amines as quality-indicating metabolites. Food Control, 15(8), 601–607.

Romantsev, F.E. and Prozorovskii, V.N. (1984). Proteolytic resistance and thermostability of catalase and histidine decarboxylase from *Micrococcus* sp., Bulletin of Experimental Biology and Medicine, 97, 414-415.

Ruiz-Capillas, C. and Jiménez-Colmenero, F. (2004). Biogenic amines in meat and meat products. Critical Review in Food Science, 44, 489-499.

Ruiz-Capillas, C., Colmenero, F.J., Carrascosa, A.V. and Munoz, R. (2007). Biogenic amine production in Spanish drycured "chorizo" sausage treated with high pressure and kept in chilled storage, Meat Science, 77, 365-371.

Santos, M. H. S. (1996). Biogenic amines: their importance in foods. International Journal of Food Microbiology, 29, 213 – 231.

Sayem-El-Daher, N., Simard, R.E. and Fillion, J. (1984). Changes in the amine content of ground beef during storage and processing. Lebensmittel-Wissenschaft and Technologie, 17, 319–323.

Seo, Kyo, Y., Heo S.K., Lee, C., Chung, D.H., Kim, M.G., Lee, K.H., Kim, K.S., Bahk, G.J., Bae, D.H., Kim, K.Y., Kim, C.H., and Ha, S.D. (2007). Development of predictive mathematical model for the growth kinetics of *Staphylococcus aureus* by response surface model. Journal of Microbiology and Biotechnology, 17(9), 1437–44.

Shalaby, A. R. (1996). Significance of biogenic amines to food safety and human health. Food Research International, 29, 675-690.

Simo, C., Moreno-Arribas, M.V. and Cifuentes, A. (2008). Ion trap vs. time of flight MS coupled to CE to analyze biogenic amines in wine. Journal of Chromatography A, 1195, 150-156.

Slemr, T. (1981). Biogenic amine als potentioller chemescher qualitatsindikator fur fleish. Fleischwirtsschaft, 61, 921–924.

Smith, J.S., Kenney, P.B., Kastner, C.L. and Moore, M.M. (1993). Biogenic amine formation in fresh vacuum-packaged beef during storage at 1°C for 120 days. Journal of Food Protection, 56, 497–500, 532.

Smith, T.A. (1980). Amines in food. Food Chemistry, 6, 169–200.

Soufleros, E., Barrios, M. L. and Bertrand, A. (1998). Correlation between the content of biogenic amines and other wine compounds.American Journal of Enology and Viticulture, 49, 266-277.

Stadnik, J. and Dolatowski, Z. J. (2010). Biogenic amines in meat and fermented meat products. Acta Scientiarum Polonorum Technologia Alimentaria, 9, 251-263.

Standarova, E., Borkovcova, I. and Vorlova, L. (2008). The occurrence of biogenic amines in dairy products on the Czech market. Acta Scientiarum Polonorum – Medicina Veterinaria, 7, 35-42.

Stratton, J. E., Hutkins, R. W. and Taylor, S. L. (1991). Biogenic amines in cheese and other fermented foods: a review. Journal of Food Protection, 54, 460-470.

Suzzi G. and Gardini F. (2003). Biogenic amines in dry fermented sausages: a review. International Journal of Food Microbiology, 88, 41-54.

Tapingkae, W., Tanasupawat, S., Parkin, K. L., Benjakul, S. and Visessanguan, W. (2010). Degradation of histamine by extremely halophilic archaea isolated from high salt-fermented fishery products. Enzyme and Microbial Technology, 46, 92–99.

Telsnig, D., Kassarnig, V., Zapf, C., Leitinger, G., Kalcher, K. and Ortner, A. (2012). Characterization of an amperometric biosensor for the determination of biogenic amines in flow injection analysis. International journal of electrochemical science, 7, 10476 – 10486.

Trevino, E., Beil, D. and Steinhart, H. (1997). Formation of biogenic amines during the maturity process of raw meat products, for example of cervelat sausage. Food Chemistry, 60, 521–526.

Van Boekel, M. A. J. S. (2002). On the use of the Weibull model to describe thermal inactivation of microbial vegetative cells. International Journal of Food Microbiology, 74, 139–159.

Vidal-Carou, M. C., Izquierdo, M. L., Matin, M. C. and Marine, A. (1990). Histamina y tiramina en derivados c´ arnico. Revista de Agroquimica y Tecnologia de Alimentos, 30, 102–108.

Vinci, G. and Antonelli, M. L. (2002). Biogenic amines: quality index of freshness in red and white meat. Food Control, 13, 519-524.

Wortberg, B. and Woller, R. (1982). Quality and freshness of meat and meat products as related to their content of biogenic amines. Fleischwirtsch, 62, 1457–1463.

Wortham, B. W., Patell, C. N. and Oliveira, M. A. (2007). Polyamines in bacteria: pleiotropic effects yet specific mechanisms. Advances in Experimental Medicine and Biology, 603, 106-115.

Xiong, R., Xie, G., Edmondson, A. E. and Sheard, M. A. (1999). A mathematical model for bacterial inactivation. International Journal of Food Microbiology, 46, 45–55.

Xu, Y. S., Xia, W. S., Yang, F., Kim, J. M. and Nie, X. H. (2010). Effect of fermentation temperature on the microbial and physicochemical properties of silver carp sausages inoculated with Pediococcus pentosaceus. Food Chemistry, 118, 512–518.

Yamada, M., Watanabe, T., Harino, S., Fukui, H. and Wada, H. (1980). The effects of protease inhibitors on histidine decarboxylase activities and assay of enzyme in various rat tissues. BBA Enzymology, 615, 458-464.

Yassoralipour, A., Bakar, J., Rahman, R. A., and Bakar, F. A. (2012). Biogenic amines formation in barramundi (*Lates calcarifer*) fillets at 8%C kept in modified atmosphere packaging with varied CO2 concentration. LWT - Food Science and Technology,48, 142–146. doi: 10.1016/j.lwt.2012.01.033

Yongsawatdigul, J., Rodtong, S. and Raksakulthai, N. (2007). Acceleration of Thai fish sauce fermentation using proteinases and bacterial starter cultures. Journal of Food Science, 72, M382–90.

Yücel, Ü. and Üren, A. (2008). Biogenic amines in Turkish type pickled cabbage: Effects of salt and citric acid concentration. Acta Alimeteria, 37, 115-122.

Zaman, M. Z., Abdulamir, A. S., Abu Bakar, F., Selamat, J. and Bakar, J. (2009). A review: microbiological, physiological and health impact of high level of biogenic amines in fish sauce. American Journal of Applied Sciences, 6, 1199-1211.

Zhai, H., Yang, X., Xia, L. Li, G., Huang, J. Cen, H. and Hao, S. (2012). Biogenic amines in commercial fish and fish products sold in southern China. Food Control, 25, 303-308.

Zhang, X., Wang, H., Li, N., Li, M. and Xu, X. (2015). High CO2-modified atmosphere packaging for extension of shelf-life of chilled yellow-feather broiler meat: a special breed in Asia. LWT - Food Science and Technology,64, 1123–1129.

Zwietering, M. H., Jongenburger, I., Rombouts, F. M., Van, T. and Riet, K. (1990). Modeling of the bacterial growth curve. Applied and Environmental Microbiology, 56, 1875–1881.

Chapter 16

Technological Development in Meat Processing for Better Quality and Safety

Akhilesh K.Verma[1]*, Pramila Umaraw[1], V.P. Singh[1], Pavan Kumar[2] and Nitin Mehta[2]

[1] Department of Livestock Products Technology, College of Veterinary and Animal Sciences, Sardar Vallabhbhai Patel University of Agriculture and Technology, Meerut – 250 110, Uttar Pradesh

[2]Department of Livestock Products Technology, College of Veterinary Science, Guru Angad Dev Veterinary and Animal Sciences University, Ludhiana – 410 04, Punjab

**e-mail: vetakhilesh@gmail.com*

ABSTRACT

Present chapter deals with the recent advancement in the various meat and meat processing techniques such as high pressure processing (HPP), hydrodyne or hydrodynamic pressure processing or shockwave processing and ultrasonic processing. They describe the advancement in already existing processing techniques; grinding, slicing, tumbling, electrical stimulation and retorting. The chapter discusses the basic principles governing each process, equipments and machineries used along with the effect of the processing technique on meat structure and quality. Most of the techniques have been commercialized (grinding, slicing, tumbling, electrical stimulation and retorting) while few are still under research (hydrodyne or hydrodynamic pressure processing or shockwave processing and ultrasonic processing).

Keywords: *Meat processing, HPP, Shockwaves processing, Ultrasonic wave processing, Tumbling.*

1. Introduction

Meat and meat products are an important part of diet globally. They have even occupied their space in the fast food sector; fast food sector and ready to eat and ready to cook meals have boosted the meat processing sector. The processed meat market value is expected to increase up-to 1.5 trillion dollars by 2022 from 714 billion dollars in 2016 (https://www.statista.com/2019). The major driving forces for increased production and processing are urbanization, busy-lifestyle, dual income and convenience. Nowadays utilization and consumption pattern of meat has gone to a higher level than just nourishing, relishing and preserving. Today's fast moving and busy life has shrunken the cooking time and efforts. Ready-to-eat meat or ready-to-cook meat with convenient one or two diet meal packs has overtaken longer home cooking processes. Consumer demands have also shifted from normal diet to nutritious and healthy diet with increased functionality. Efforts are being made for developing and enhancing the functionality of meat products. Such products require a better and sophisticated processing so as to prevent undesirable effects on functional components. The processing techniques earlier were more time, energy and labour intensive which sometimes becomes inefficient during large scale production. The processing process should be economical with minimum waste, microbiologically safe with minimum chances of contamination and recontamination as well as should retain and maintain the nutritional value of meat. Thus, the processing industry must embrace novel and advanced processing equipments, tools and methods:

1. To produce value added products and provide variety of meat products.
2. To increase demand and marketability and meet life style requirements.
3. To utilize different carcasses effectively and to utilize different by-products.
4. To combine and compliment different meat with advantage.
5. To incorporate non-meat ingredients for quality and economy.
6. To preserve, transport and distribute to larger populations.
7. To facilitate export of meat products and compete with imports.
8. To promote entrepreneur ventures and employment.

2. High-pressure Processing (HPP)

High pressure processing is a non-thermal processing technique that involves use of high pressure to destroy microorganisms in food without affecting the sensorial or nutritional characteristics of meat or meat products and causes no physico-chemical alterations in them. It is a safe and consumer friendly, potential preservation method used for inactivation of pathogenic and spoilage microbes. Today's consumer is looking for nutritional, healthy, minimally processed and chemical free foods and is also willing to pay higher prices for it. Thus, the meat industry has employed high pressure processing to develop meat products with good eating and keeping properties. High pressure processing was first applied for inactivation of microorganisms in milk by Hite (1899). Today, it is widely being used for microbial inactivation of foods and has emerged as an alternative method of processing and preservation of meat and meat products. However Yagiz *et al.* (2009) reported that high pressure processing treatment may enhance the lipid oxidation and persuade colour and texture profile changes in meat and meat products. Commercial application of HPP requires standardization of conditions to inactivate selected microbes for every product, novel packaging, innovative processing combinations as well as proper marketing.

2.1 Principle

The basic principle of high pressure processing is application and even distribution of high pressure leading to desirable changes in food. The changes in food are governed by following principles (Balasubramaniam *et al.*, 2015).

Isostatic Principle: states that uniform application of pressure (hydrostatic pressure) acts equally in all directions which is independent of time and space. This principle governs the functionality and applicability of HPP at commercial scales. Generally fluids are used to create such a condition. The pressure created is distributed uniformly and instantaneously throughout the food that is independent of its shape and size.

Le Chatelier's Principle or Equilibrium Law: this principle governs the physical and chemical changes that occur in food due to hydrostatic pressure. It states that in an equilibrium state any changes in constrains such as pressure, temperature or concentration of reactants will shift the equilibrium towards counteracting the effect of constrains.

2.2 Microscopic Ordering

Pressure and temperature exert an antagonistic effect on degree of molecular ordering in a substance *i.e.* at constant temperature, increase in pressure would result in further ordering of molecules. The changes in food system due to high pressure are accompanied by various thermal changes also. Thus, the combined pressure-thermal effect on food can be additive, synergistic or antagonistic.

Meat is a nutritional food with high protein, fat, mineral, oligonutrients and vitamins at a pH range of 5.5-7 that favours growth of microbes if left unattended. All preservative techniques are designed towards minimizing factors favouring growth of microorganisms. The HPP does not alter the environmental and/or micro-environmental conditions rather it directly acts on the microorganisms to destroy or inhibit their growth. Microbial proliferation of growth is arrested by various physico-chemical changes such as alteration in membrane lipids, internal pH and/or irreversible changes in physiological processes (Molina-Guitierrez *et al.*, 2002). Efficacy of the inactivation of microbes present in meat and meat products depends on various factors such as type of microbial population, number of microbes, pressure, temperature, time of application and meat product itself. Another distinctive effect of HPP is on the textural qualities of meat. Among all foods, muscle food and within meat the myoproteins are most responsive to pressure due to their molecular configuration and due to pressure sensitivity of glycolytic processes (Sun and Holley, 2010). High pressure causes change in volume of food which in turn affects the protein structure. Cheftel (1995) reported that the covalent bonds are unbroken but the meager energy bonds such as hydrogen bonds and the hydrophobic bonds can be irreversibly modified. The electrostatic and hydrophobic bonds are disrupted under pressure and on de-pressurization new intra or inter molecular bonds are formed within or among proteins (Sun and Holley, 2010). Thus, membranes, enzymes and other large molecules are affected but smaller compounds like vitamins and flavourants are preserved which retains nutritional qualities of foods (Linton and Patterson, 2000). The ultimate effect on muscle protein is hegemonized by pressure and temperature changes. The temperature changes in a HPP system is due to adiabatic heating. Work of compression, induces temperature increase in the medium and food. Both thermal and HPP heating causes denaturation of proteins but the mechanism of action is different and thus the eating quality is also different. Denaturation in HPP system is due to reduction of volume causing changes in electrostatic and hydrophobic bonds while the same in thermal heating is due to violent agitation leading to destruction of hydrogen and covalent bonds (Okamoto *et al.*, 1990). Iwasaki *et al.* (2006) reported that heat induced gels (chicken myofibrillar protein gels) were stranded consisting of bundles of protein but that of HPP treated was a fine network gel. With chemical point of view, high pressure processing is softer as compared to thermal processing.

2.3 Chemical Changes in Meat and Meat Products during HPP

HPP involves application of high pressure 100-1000 MPa on vacuum packaged meat or meat products in a flexible high barrier packaging material. Vacuum packaging is required for minimizing the detrimental effect of oxygen present in head space. Under pressure the oxygen solubility would increase and in-turn this dissolved oxygen would initiate and propagate free radical reactions leading to enhanced lipid and protein oxidation. The other major concern regarding head space is that the compressibility of gases (present in head space) and water (present in food) are different and thus the product and head space would react differently to the pressure applied, hampering the process.

Application of high pressure processing on water molecules decreases the melting point of water under pressure and raises ionization, reduction in pH. At low pressure processing these differences are reversible however; they can produce marked effects by transforming processed products properties at high pressure processing. At high pressure ionic dissociation of water molecules occur, decreasing pH. This pH drop affects the microbial inactivation kinetics (Balasubramaniam *et al.*, 2015).

Processing of meat at high pressure (>400 MPa) converts ferrous to ferric myoglobin, denatures global protein and degrades calpains. Texture is improved due to degradation of lysosomal membranes enhancing autolytic activity and by inhibition of calpastatin at pressure < 200 MPa, some unacceptable changes also occur in colour of fresh meat. Cathepsin-H and amino peptidases are inactivated at > 200 MPa while cathepsin D is inactivated at 500 MPa (Montero and Gomez-Guillen, 2002). High pressure processing at much higher pressure might bring some undesirable effect on quality of meat. Such conditions bring about changes in molecular configuration of proteins and also damage the lipid membranes making it prone to lipid oxidation during storage.

3. Tumbling

Tumbling can be defined as a process carried out in a revolving, stainless steel, cylindrical vessel (tumbler) with or without baffles that works (tumble, drops, revolves) on fresh or cured uncooked meat chunks to enhance brine uptake and to produce sticky solubilised protein extract that enhances coherency and hold chunks together (Pearson and Gillett, 1996). It is a physical process that involves transfer of kinetic energy to meat proteins by mechanical action such as revolving and impaction. The process along with added salt and phosphates solubilises various meat proteins that forms binding agent for meat as well as improves texture, firmness and cooking yield (Feiner, 2006). It also accelerates the curing process.

3.1 Principle

The mechanical action of tumblers causes disruption of muscle cells at the edges of meat chunks or pieces. This disruption along with added salt extracts the salt soluble proteins *i.e.* myofibrillar proteins which form tacky fluid exudates that binds the meat chunks or pieces together.

3.2 Chemical Changes in Meat during Tumbling

Myofibrillar proteins or the salt soluble proteins are the major proteins extracted during tumbling. Sarcoplasmic proteins released due to disruption of muscle cells play a minor role in forming exudate. Myosin is the key protein responsible for strong bonding between adjacent meat chunks in formed and sectioned meat products. Other myofibrillar proteins such as actomyosin and actin form comparatively weaker bonds (Pearson and Gillet, 1996). Heterogenous nature of meat and its low porosity hampers brine uptake. The hydrodynamic mechanism improves the meat quality during tumbling. During vacuum tumbling meat degases by reduced pressure, this draws brine into the meat structure and gets locked when the pressure is restored. Massaging tempers muscle structure causing infiltration of brine as result more water is bound and thus cooking yield is increased.

3.3 Equipment and Design

Tumbler or massager basically consists of a rotating single or double jacketed cylindrical vat with or without baffles. The baffles ensure even mixing of meat chunks with salt, marinade or protein exudates. The shape, size and direction of baffles play important role in achieving proper mixing. It may or may not have adjunct vacuum conditions. Continuous vacuum and pulsed vacuum are the variation in vacuum tumblers. These vats are attached with temperature and pressure probes, weight sensors and microprocessors. Tumbling or massaging causes destruction of meat chunks during process, extent depends on the processing conditions.

3.4 Vacuum Tumbling

Vacuum tumbling is an extension of conventional tumbling where vacuum tumblers are used to remove air and thus prevent frothing of protein exudates.

Vacuum also facilitates rapid and consistent brine uptake, as the air is gradually removed, meat tissues are exposed causing cellular absorption of brine into meat. Later when vacuum is released meat regains its original state locking marinade into the structure. The expansion and pounding of meat during tumbling improves the texture of meat making it more tender and juicier. The whole process of marination is accelerated. Vacuum tumbling also reduces bacterial load as high salt concentration in brine along with vacuum causes bacterial cells to rupture. The amount of vacuum required depends on the age, sex, species, type and size of meat. Tougher meat, beef, lamb and pork require higher vacuum and longer time to achieve desired texture while poultry meat requires lower vacuum and lesser time.

Tumbling duration plays an important role in development of desired meat characteristics. Tumbling duration upto 0-9 hr increases hydration quality and cooking yield in pork with 30 and 40 per cent innoculation (Patrascu *et al.*, 2013). It also improved the textural characteristics by reducing the shear force value and hardness of meat chunks. In a study to develop marketable processing technique for preparation of boneless pork chops continuous vacuum tumbling marination was observed to be better than conventional static marination and vacuum intermittent tumbling marination. Continuous vacuum marination improved the product yield, tenderness, flavour and juiciness of pork chops. The cooking loss was appreciably reduced (Gao *et al.*, 2015).

4. Meat Grinders

Comminution is the subdivision of meat into smaller size. Various forms of comminution of meat are grinding, chopping, flaking, slicing, homogenization *etc.* Meat grinders are used for particle reduction of meat for further processing and depending on the particle size these grinders can be classified as coarse meat grinders and fine meat grinders (better known as homogenisers). These can also be classified according to the type of meat handled *i.e.* normal grinders, mixture grinders and frozen meat grinders. The holes are generally made at 90° to the plate but recent advanced plates have holes at less than 90° to the plates. It has been suggested that these changes have increased the efficiency and functionality of grinders (Weiss *et al.*, 2010).

In a continuous meat processing plant pumping grinders are used. Pump grinders consists of a vacuum pumping machine attached to meat grinders. Vacuum causes a pumping action and pushes meat through the grinders. Vacuum pumping has enabled grinding of meat as small as 0.25 mm with a continuous and consistent supply (Haack and Schnackel, 2008). It has been shown in various studies that keeping quality of pump grinder meat is better than conventional grinding owing to lesser air pockets and damaging effect of vacuum on microbial load (Honikel, 2004).

4.1 Principle

The working principle of grinders is that meat cuts are pushed by pitched screw augur through grinding plates that have standard hole size and number and rotating knives (Rust, 2004).

4.2 Chemical Changes in Structure during Grinding

Coarse grinding reduces the particle size of meat structure with the pushing action of augur the meat chunks are pressed towards plates with holes and cutting knives. The size reduction ensures even distribution of other ingredients and parts consistency to the products. Fine grinding causes much more size reduction.

Bowl choppers have been traditionally used for preparation of fine meat processing especially for emulsion type products. Emulsion is a mixture of two immiscible liquids. Meat emulsion is not a true emulsion as its particle size is greater than 50 micron. Meat emulsion consists of fat droplets dispersed in a proteins solution. The emulsion is stabilized by the myofibrillar proteins. These proteins, with their hydrophobic ends towards fat and their hydrophilic ends towards solution, surround oil droplets and prevent its coherence, imparting stability to the emulsion.

4.3 Equipments and Design

The basic conventional grinders consists of inlet meat feeding tray, a pumping system for continuous system or a pusher for manual feeding, pitched screw or augur and set of knives and perforated plates. The major advances made are in the pumping system and the perforated plates and knife combination. The pumping systems have evolved to feed meat through vacuum. This vacuum feeding system has higher efficiency, productivity and is more hygienic than conventional system. The developments in perforated plate have been less. Few advances have been made in the hole structure and placements. In a newer version, the holes are in a multi-start spiral fashion with less than 90° angle from the plate surface. These changes have extended the working-life of tools and have improved product quality (Weiss, 2010).

Traditionally fine meat homogenization was done by bowl choppers that consisted of a rotating bowl with set of knives for reducing meat particles. With advancement and automation of system large bowl choppers were used with upto 1200L capacity (Seydelmann, 2009). But such large choppers require higher energy for maintaining the efficiency and also have higher non-working zone of meat batter that further reduces the efficiency. Nowadays, for emulsion type meat preparations are done in fine meat homogenizers. These fine meat homogenizers are fully automated and can be used for both batch or continuous processing with lower production cost. These produce consistent and fine meat batters with lesser air incorporation which prevents oxidative degradation of meat batters.

5. Slicers

Meat slicers are widely used for portioning of meat and meat products. With increased urbanization, commercialization of convenience foods, ready to eat meat *etc.* slicing is becoming integral part of meat processing. Continuous slicing lines are increasingly been installed in meat processing plants (Holac, 2009). Traditional slicers have been used to accurately portion meat and meat products of specific desired dimensions. Nowadays slicers with varying designs are available. Spiral slicers have become quite popular for preparing thin sheet of meat for gourmet dressing. Automated slicer lines are now equipped with processors for feeding instructions to produce decorative designs while slicing. Heavy duty meat slicers can handle large volume of meat with high efficiency and continuous supply. An essential requirement for slicing is that the product must be frozen for even finishing, accurate and consistent weight and structure but recent advancements in knife design and structure has eliminated the need of freezing surfaces of soft products (Weiss, 2010).

6. Retorting

Retorting is the process of cooking and/or commercially sterilizing the meat products in a package (retort pouches, cans *etc.*) such that it receives sufficient thermal heat for specified predetermined time so as to achieve process lethality. The temperature and time for process lethality has to be predetermined for specific product depending on its size, composition, retorting medium and packaging. Uniform and constant temperature controls are the key to successful retorting. Processing is generally carried out at high temperatures (>100°C; 110-130°C) and high pressure (2-3 atmosphere). Heating medium used are water, steam and steam-air combination. Steam heated water with forced circulation is widely used for meat and meat products packed in glass containers, tin cans or retorting pouches. Sterility of retort products can be attributed to:

1. Hermetic sealing that ensures sterile condition
2. Process lethality that ensures destruction of harmful microorganisms with standardized temperature-time combination reaching to the most inaccessible part of containers *i.e.* cold spot
3. Hygienic post process handling of containers

6.1 Principle

Meat and/or meat products sealed in containers are subjected to high temperature at increased pressure for sufficient time followed by immediate cooling (shock) so as to inactivate/destroy harmful microorganisms and to achieve commercial sterility. Absolute sterility is practically unachievable for food items such as meat because subjecting meat to extreme conditions or absolute sterility would adversely affect its eating quality thus commercial sterility is followed. Commercial sterility is a condition where microorganisms capable of growing in foods stored at ambient temperatures are destroyed. According to FAO, "Commercial sterility means the absence of microorganisms capable of growing in the food at normal non-refrigerated conditions at which the food is likely to be held during manufacture, distribution and storage." (Codex Alimentarius Commission (WHO/FAO) CAC/RCP 40-1993).

6.2 Equipments and Design

Retorting was initially done for food products stored in glass containers, later with the advent of tin manufacturing, hermetically sealed tin cans were widely used for packaging of meat and meat products *i.e.* canning. Recently with the advancement of laminate packaging retortable pouches is *en-vogue*. Retort pouches are hermetically sealed, flexible consisting of three to four layers of plastic/foil. The number and type of material used for layering depends on the requirement of food, its processing conditions and functionality of layer. Generally, there is a food contact layer *i.e.* the layer which comes in direct contact with the food, an outer layer and a middle layer. The outer layer provides the required strength, toughness and fulfils the labeling requirements. PET (polyethylene terephthalate) is the widely used material (material of choice) for outer layer as it provides desired strength along with good printing characteristics. Pouches are reverse printed. The middle layer is responsible for impeding characteristics of the retort pouch. This layer acts as a barrier for moisture, gas and light. Aluminum foils are mostly used for this purpose. The innermost layer of cast polypropylene (CPP) is used for providing heat sealability, strength and compatibility with food surfaces.

Retorting is done in a closed vessel known as retorts. The processing can be done in batches or in a continuous line. Batch retorts are mostly used as they are more economical. Continuous retorting is done on large scale for huge volume of products and thus is more expensive but saves labour and energy and have higher efficiency. Batch retorts are steam retorts, steam/air retorts, water spray retorts, water immersion retorts, crate less retorts and agitating retorts. Continuous retorts are hydrostatic cookers, continuous rotary atmospheric sterilizer and continuous rotary pressure sterilizers and pallet hydrostatic sterilizers.

6.3 Chemical Changes in Meat during Retorting

Retorting exposes meat to high temperatures that causes changes in protein structure primarily in secondary, tertiary and quaternary structures. The bonds are broken and proteins are unfolded. Such proteins become chemically more bioavailable for absorption after digestion. The myofibrillar proteins undergo denaturation and chemical modification while the intramuscular collagen tissues convert into gelatine causing tenderization of meat. The major changes observed depend on the type of meat, species, size and processing conditions (temperature, time). Retorting at high temperatures, generates volatile compounds such as amino acids, organic acids, peptides *etc.*, which impart characteristic flavour while thermal processing of lipid components develop the characteristic aroma of the final cooked product.

7. Electrical Stimulation

Electrical stimulation (ES) is a process of passing electrical current through the carcass for accelerating the bio-chemical processes and thus reaching a desired pH level in relatively less time. It is the process of hastening rigor mortis before chilling. It is widely used for preventing cold shortening and increasing the tenderness of meat. Efficient electrical stimulation lowers the muscle pH below 6 before the carcass temperature decreases below 10°C thus prevents cold shortening and reduces carcass processing time (from slaughter to chilling). It also improves muscle texture by increasing proteolytic degradation and some physical disruption of muscle structure (Lang *et al.*, 2016).

7.1 Principle

Electrical stimulation of carcasses causes change in conductivity of membranes especially the sarcoplasmic membranes. These alterations expose the calsequestrins (calcium binding proteins) which causes releases of calcium ions. Increased calcium level has two fold effect, firstly it activates contractile actomyosin ATPase and it also enhances the *phosphorylase a* titre. ATPase increases the rate of ATP breakdown and the *phosphorylase a* increases the rate of post-mortem glycolysis. Both factors accelerate post-mortem pH fall. The increased calcium levels also activate proteolytic enzymes such as calpains. Activated proteolytic enzymes along with relatively high temperature (body temperature) contributes to the observed tenderness in electrically stimulated carcasses.

7.2 Chemical Changes in Meat

Freshly slaughtered animals are electrically stimulated by passing low or high voltage current through it. As an effect accelerated glycolysis causes muscle contractions and quick fall in muscle pH. The rate of pH fall is both temperature (of carcass) and method (of electrical stimulation) dependent. The rate of pH fall is higher at higher temperatures. Tenderness induced by electrical stimulation can be attributed partly to avoidance of cold shortening and partly to proteolytic degradation. Evidence suggests 50-60 Hz frequency causes disruption of myofibrillar matrix at lower pH. Tenderness in beef carcasses was optimal when post-mortem glycolysis being intermediate type (Marsh, 1987). Higher rate of glycolysis ensures earlier rigour development. This increased muscle stiffness facilitates deboning. Apart from tenderization ES has positive effect on colour and flavour of meat.

7.3 Equipments and Design

Traditionally for electrical stimulation, the rail is deployed as ground and a live electrode is placed on carcass. The current with high or low frequency is passed for specific time through the carcass. Generally the live electrode is placed on head or neck of the freshly slaughtered animal. Nowadays modernized batch and continuous electrical stimulation systems are available. The batch system consists of a cabinet for stimulation. The carcasses are manually inserted with the electrode bars. The continuous system consists of stationary rubbing electrodes that are sterilized in between two applications. The operating variations occur in voltage, current, wave form and frequency. Largely high voltage and low voltage equipments are used. A voltage range of 300 to 1000 V is used for more consistent and greater pH decline (Troy and Kerry, 2010). High voltage application requires shorter duration exposure but have their safety concerns. Application of 50 -120 V for more than 40 s exposure time is done in low voltage stimulation for highest efficiency (Gursansky *et al.*, 2010)).

8. Ultrasonic Wave Processing

Ultrasound waves are sound waves with frequencies greater than 20 kHz or above the audible range of human ear. It is a form of energy which when passes through a medium; it induces high compression and depression of particles imparting energy. Acoustic cavitation is the characteristic of ultrasound waves which depends on wave frequency, intensity, products characteristics, temperature and pressure (Dolatowski *et al.*, 2007). Application of ultrasonics in food industry can be classified on the basis of energy generated (sound power (W), intensity (W/m^2)

and sound/energy density (Ws/m^3)) into low-energy and high -energy or power-ultrasound (Knorr *et al.*, 2004). Low energy (Intensity< 1 W/cm^2 or frequency 10-100 kHz) ultrasounds are used for non-invasive applications in food such as analytical observations, physico-chemical evaluation, composition *etc.* (Jayasooriya *et al.*, 2004). Simal *et al.* (2003) evaluated composition of fish tissues, chicken and raw meat mixtures using ultrasonic waves and semi-empirical equations. High energy (Intensity > 1 W/cm^2) ultrasound waves in the range of 10-1000W/cm^2 intensity with frequency in the range of 20-100 KHZ are used for causing structural, physical or chemical changes. Application of ultrasound waves is a recent advancement in food technology. It is considered a sustainable, energy efficient technique. In meat industry ultrasound waves have been applied for meat tenderization, curing and marination.

8.1 Principle

The ultrasound waves produce various changes by acoustic cavitation. During sonication under the effect of acoustic field cavitation occurs *i.e.* gas bubbles are formed in the medium. Fluctuating pressure (expansion and compression) causes rectified diffusion. Rectified diffusion is governed by various factors such as intensity of waves, energy, medium (presence of surface active agents) and is influenced by "area effect" and "shell effect". Area effect relates to the mass diffusion into gas bubbles during expansion and compression while shell effect refers to the changes in thickness of interface boundary during fluctuations. During oscillation due to rectified diffusion a stage comes when the bubbles reach maximum resonance size and they violently collapse generating chemical reactions (reactive radicals) and physical forces such as shock waves, turbulence, microjets or microstreams and shear force along with very high temperature (upto 5500°C) and pressure (upto 50 MPa). These effects can accelerate the transport of substances to and from enzymes, and enhance mass transfer in enzymes, increasing thereby enzymes' catalytic efficiency. Ultrasonic cavitation creates intense shear forces. This means, ultrasonic tenderizing and the physicochemical change of meat structure is obtained due to mechanical effects only.

8.2 Changes in Meat

Sonication induced acoustic cavitation causes physico-chemical changes in meat. The technique has been applied to improve tenderness of meat, brine uptake in cured products, facilitate marination and drying (Jayasooriya *et al.*, 2007; Stadnik and Dolatowski, 2011; Ozuna *et al.*, 2013). Tenderness is improved by both physical disruption and due to enzymatic degradation (Alarcon-Rojo *et al.*, 2015). Improved tenderness can be accounted to the myofibrillar degradation, changes in Z-line (Got *et al.*, 1999), degradation of troponin T and design proteins, activation of calpains (Wang *et al.*, 2018), activation of apoptosis cascade (Chen *et al.*, 2015), reduced thermal stability of collagen proteins (Chang *et al.*, 2012). Collagen is a connective tissue that is responsible for background toughness of meat. Collagen is very un-reactive and does not produce changes until high temperature exposure. Collagen based toughness is difficult to overcome but with application of power sonication collagen is solubilised. It produces conformational changes in collagen protein structure, such that its denaturation temperature decreases and also causes fragmentation of collagen macromolecules.

Ultrasonic application also affects meat pH. The pH value is increased and thus moisture retention capacity of meat is enhanced leading to higher cooking yield. Sonication also promotes brine uptake during marination. Disruption of muscle opens-up its structure permitting deeper penetration of marinade which is held tightly within imparting intense flavour and better eating quality.

9. Hydrodynamic Pressure Processing or Hydrodyne or Shock Wave Processing

Hydrodynamic pressure processing is a novel technique for tenderisation of meat which involves generation of shock waves that causes physical disruption of muscle structure as well as intracellular organelles' membrane damage, inducing enzymatic degradation. Mechanical pressure pulses propagating through meat causes instantaneous tearing of muscle structure (Bolumar *et al.*, 2014; Bolumar *et al.*, 2013). The tenderizing effect is dependent on the condition (fresh, frozen) and type of sample (size, shape, species, age, toughness of cut,) and treatment conditions (number of detonations, amount of explosive, voltage) (Solomon, *et al.*, 2004; Zuckerman *et al.*, 2013).

9.1 Equipment and Design

The meat to be treated is packed and placed in a container containing water. The high pressure shock waves are generated by detonation of small quantities of high explosives or by high voltage electrical discharge through the water travel through water in fractions of milliseconds. At intersection of water, package and meat, pressure is increased causing physical disruption of enclosed meat. A prototype developed consists of a steel tank filled with water and covered with a steel dome (2.3 tonne; 2.7 m diameter). Meat is packaged in a pressure and water resistant package and placed in the tank. This tank is placed about 3 m in ground. An explosive is detonated at 1 meter distance from the meat. Explosion causes shockwaves that are restrained within the dome.

10. Conclusions

Recent advances in meat processing technology are aimed towards minimizing nutritional loss during processing, reducing production wastage, shortening the time of processing with better efficiency, effectiveness, larger production, continuous processing, extending shelf-life and improving microbial safety. Mechanization reduces the human intervention. Along with computer operated systems these processing technologies can be better monitored, controlled and operated. Although these techniques are very promising for large scale production research and development measures are required for developing small scale processing units for entrepreneurs, farmers and small business owners. Further developments can be made towards developing strategies and methodology for technological advancements in preparation and processing of functional foods such that the functionality of products is not affected by the processing technique.

REFERENCES

Alarcon-Rojo, A. D., Janacua, H., Rodriguez, J. C., Paniwnyk, L., and Mason, T. J. (2015). Power ultrasound in meat processing. Meat Science, 107, 86–93.

Marsh, B.B., Ringkob, T.P., Russell, R.L., Swartz, D.R., and Pagel, L.A. (1987). Effects of early-postmortem glycolytic rate on beef tenderness. Meat Science,21(4),241-248.

Balasubramaniam, V. M., Martinez-Monteagudo, S. I., and Gupta, R. (2015). Principles and application of high pressure–based technologies in the food industry. Annual review of food science and technology, 6, 435-462.

Bolumar, T., Bindricha, U., Toepfl, S., Toldra, F., and Heinz, V. (2014). Effect of electrohydraulic shockwave treatment on tenderness, muscle cathepsin and peptidase activities and microstructure of beef loin steaks from Holstein young bulls. Meat Science, 98, 759–765.

Bolumar, T., Enneking, M., Toepfl, S., and Heinz, V. (2013). New developments in shockwave technology intended for meat tenderization: Opportunities and challenges. A review. Meat Science, 95, 931–939.

Chang, H. J., Xu, X. L., Zhou, G. H., Li, C. B., and Huang, M. (2012). Effects of characteristics changes of collagen on meat physico-chemical properties of beef semitendinosus muscle during ultrasonic processing. Food Bioprocess Technology, 5, 285–297.

Cheftel J. C. (1995). High-pressure, microbial inactivation and food preservation. Food Science Technology International, 1, 75–90.

Chen, L., Feng, X., Zhang, Y., Liu, X., Zhang, W., Li, C., and Zhou, G. (2015). Effects of ultrasonic processing on caspase-3, calpain expression and myofibrillar structure of chicken during post-mortem aging. Food Chemistry, 177, 280–87.

Codex Alimentarius Commission (WHO/FAO).Code of Hygienic Practice for Aseptically Processed and Packaged Low-Acid Foods. CAC/RCP 40-1993.

Dolatowski, Z. J., Stadnik, J., and Stasiak, D. (2007). Applications of ultrasound in food technology. Acta Scientiarum Polonorum, Technologia Alimentaria, 6, 89–99.

Feiner, G. (2006) Meat products handbook - Practical science and technology. Woodhead Publishing Limited, Cambridge, England.

Gao, T., Li, J., Zhang, L., Jiang, Y., Ma, R., Song, L., Gao, F., and Zhou, G. (2015). Effect of Different Tumbling Marination Treatments on the Quality Characteristics of Prepared Pork Chops. Asian-Australasian Journal of Animal Sciences, 28, 260-267.

Got, F., Culioli, J., Berge, P., Vignon, X., Astruc, T., Quideau, J. M., and Lethiecq, M. (1999). Effects of high-intensity high-frequency ultrasound on aging rate, ultrastructure and some physicochemical properties of beef. Meat Science, 51, 35–42.

Gursansky, B., O'Halloran, J. M., Egan, A., and Devine, C. E. (2010). Tenderness enhancement of beef from *Bos indicus* and *Bos taurus* cattle following electrical stimulation. Meat Science, 86, 635–641.

Haack, E., and Schnäckel, W. (2008). From meat to emulsion — a single operation/ Separation systems for upgrading material properties of meat — Part 2. Fleischwirtschaft International, 23(5), 23–28.

Hite BH. (1899). The effect of pressure in the preservation of milk. Bull West Virginia University Agriculture Experiment Station, 58,15–35.

Holac (2009). Holac/continuous high-volume slicing. Fleischwirtschaft International, 89(12), 60.

Honikel, K. O. (2004). Minced meats. In C. Devine, M. Dikeman, and W. K. Jensen (Eds.), Encyclopedia of meat sciences (pp. 854–856). Oxford: Elsevier.

https://www.statista.com/topics/4880/global-meat-industry/(2019) accessed on dated 06/03/2019.

Iwasaki, T., Noshiroya, K., Saitoh, N., Okano, K., and Yamamoto, K. (2006). Studies of the effect of hydrostatic pressure pretreatment on thermal gelation of chicken myofibrils and pork meat patty. Food Chemistry, 95,474–483.

Jayasooriya, S. D., Bhandari, B. R., Torley, P., and D'Arey, B. R. (2004). Effect of high-power ultrasound waves on properties of meat: A review. International Journal of Food Properties, 7, 301–319.

Jayasooriya, S. D., Torley, P. J., D'Arcy, B. R., Bhandari, B. R., 2007. Effect of high power ultrasound and ageing on the physical properties of bovine *Semitendinous* and *Longissimus muscles*. Meat Science 75 (4), 628–639.

Knorr, D., M., Heinz, V., and Lee, D.-U. (2004). Applications and potential of ultrasonics in food processing. Trends in Food Science and Technology, 15, 261-266.

Lang, Y., Sha, K., Zhang, R., Xie, P., Luo, X., Sun, B., and Liu, X. (2016). Effect of electrical stimulation and hot boning on the eating quality of Gannan yak *longissimus lumborum*. Meat Science, 112, 3–8.

Linton, M., and Patterson, M. F. (2000). High pressure processing of foods for microbiological safety and quality (a short review). Acta Microbiologica et Immunologica Hungarica, 47,175-82.

Molina-Guitierrez, A., Stippl, V., Delgado, A., Ganzle, M. G., and Vogel, R. (2002). *In-situ* determination of the intercel- lular pH of *Lactococcus lactis* and *Lactobacillus plantarum* during pressure treatment. Applied and Environmental Microbiology. 68, 4399–406.

Montero, P. and Gómez-Guillén, C. (2002). High pressure applications on miosystems, In Symposium on Emerging Technologies for the Food Industry (pp. 29), 11–13 March 2002, Madrid, Spain.

Okamoto, M., Kawamura, Y., and Hayashi, R. (1990). Application of high pressure to food processing: textural comparison of pressure- and heat-induced gels of food proteins. Agricultural and Biological Chemistry, 54,183-190.

Ozuna, C., Puig, A., Garcýa-Perez, J. V., Mulet, A., and Carcel, J. A. (2013). Influence of high intensity ultrasound application on mass transport, microstructure and textural properties of pork meat (*Longissimus dorsi*) brined at different NaCl concentrations. Journal of Food Engineering, 119, 84–93.

Patra°cu, L., Dobre, I., and Alexe, P. (2013). Effect of tumbling time, injection rate and k-carrageenan addition on processing, textural and colour characteristics of pork biceps femoris muscle. The Annals of the University Dunarea de Jos of GalatiFascicle VI –Food Technology, 37(1), 69-84.

Pearson, A.M. and T.A. Gillett, 1996. Processed Meats. 3rd Edn., Chapman and Hall, London, Pages, 150-152.

Rust, R. E. (2004). Processing equipment | mixing and cutting equipment. In C. Devine, M. Dikeman, and W. K. Jensen (Eds.), Encyclopedia of meat sciences (pp. 1057–1061). Oxford: Elsevier.

Seydelmann. (2009). Seydelmann/Zerkleinerung in mehreren vertikal angeordneten Ebenen. Fleischwirtschaft, 89(12), 59.

Simal, S., Benedito, J., Clemente, G., Femenia, A., and Rossello, C. (2003). Ultrasonic determination of the composition of a meat-based product. Journal of Food Engineering, 58(3), 253–257.

Solomon, M. B., Liu, M., Patel, J., Paroczay, E., and Eastridge, J. (2004). Tenderness improvement in fresh and frozen/thawed beef strip loins treated with hydrodynamic-pressure processing. Journal of Animal Science, 82, 18–22.

Stadnik, J., and Dolatowski, Z. J. (2011). Influence of sonication on Warner-Bratzler shear force, colour, and myoglobin of beef (*M. semimembranosus*). European Food Research and Technology, 233, 553–559.

Sun, X. D., and Holley, R. A. (2010). High hydrostatic pressure effects on the texture of meat and meat products. Journal of Food Science, 75,R17–R23.

Troy, D. J., and Kerry, J. P. (2010). Consumer perception and the role of science in the meat industry. Meat Science, 86, 214–226.

Wang, A., Kang, D., Zhang, W., Zhang, C., Zou, Y., and Zhou, G. (2018). Changes in calpain activity, protein degradation and microstructure of beef *M. semitendinosus* by the application of ultrasound. Food Chemistry, 245, 724–730.

Weiss, J., Gibis, M., Schuh, V., and Salminen, H. (2010). Advances in ingredient and processing systems for meat and meat products. Meat Science, 86, 196–213.

Yagiz, Y., Kristinsson, H. G., Balaban, M. O., Welt, B. A., Ralat, M., and Marshall, M. R. (2009). Effect of high pressure processing and cooking treatment on the quality of Atlantic salmon. Food Chemistry, 116, 828–835.

Zuckerman, H., Bowker, B. C., Eastridge, J. S., and Solomon, M. B. (2013). Microstructure alterations in beef intramuscular connective tissue caused by hydrodynamic pressure processing. Meat Science, 95, 603–607.

Chapter 17

Functional Meat Products: A Future of Meat Industry

P.K. Singh[1]*, Shivesh Singh[2], Ashok Pathera[3] and Sanjay Yadav[4]

[1]Assistant Professor, Livestock Products Technology, College of Veterinary Science and Animal Husbandry, Rewa – 486 001, M.P.
[2]Assistant Professor, Department of Livestock Products Technology, College of Veterinary Science and Animal Husbandry, NDUA&T, Kumarganj, Ayodhya, U.P.
[3]Assistant Professor, School of Bioengineeirng and Food Technology Shoolini University, Solan – 173 229, H.P.
[4]Associate Professor, Department of Livestock Products Technology, College of Veterinary and Animal Science, LUVAS, Hissar, Haryana
**e-mail: drpradeeplpt@yahoo.co.in*

ABSTRACT

With the change in lifestyle there will be change in the nutritional requirement. Similarly different age group, like children, adolescent, old age people, pregnant mother etc. don't have uniform nutrient requirement. Hence the coming age will be the age of designer and functional foods to meet the specific requirement. The products not tailored to meet the requirement of population would become an obsolete phenomenon in the time to come. Functionality of a food item can be enhanced by unmasking the functional ingredients or by incorporating it as an additive and another approach is to remove or reduce the unwanted substances that may be present in food product. The food like meat is going to play a crucial role in production of functional meat products, as is highly demanded to provide essential nutrient to population and is also a source of several functional elements.

***Keywords**: Meat products, Functional, Designer, Probiotics.*

1. Introduction

With the change in life style of people and the increase in awareness, the products not tailored to meet out the requirement would not find their place in future food platter. Even in current scenario people have shown their concern towards life style and food related ailments and the only answer to this problem is the functional food. Functional food products are going to be the future of food industry and there are several strategies for its development. Goldberg (1994) suggested that the most common way to increase functionality in food products is incorporation of nutraceuticals like dietary fiber, oligosaccharides, sugar alcohols, amino acids, peptides and proteins *etc.* Another approach of designing functional food is reduction of undesirable food components such as fat, sugar, cholesterol, sodium, nitrites. Proper designing and formulation may reduce the deleterious effects and at the same time it may enhance the desired features in a food product. In this perspective it is pertinent to state here that it has been found that there are several health benefits associated with different food constituents. And enhancement of these beneficent substances inherently present in the food items at the production stage could be the more recent approach to produce a functional food product. In this format the bioactive chemicals already present in food like the L-carnitine in meat, polyphenols in tea, long chain polyunsaturated fatty acids in fish are of great importance.

The meat being an animal origin product is of great concern as it has been associated with several ill health effects, in recent years. However both the production and processing practices could be targeted to enhance the functionality of product and to reduce the undesired properties. Hence the strategies for achieving healthier meat products involve modification at the farm level and manipulation of meat raw materials. The farm level modifications include genetic selection, controlled feeding, growth-promoting and nutrient partitioning agents, in addition to the genetic manipulation techniques which is still the most researched area. Hoffmann *et al.* (2010) observed that the production practices result in change of meat constituents like protein, fat content, fatty acid composition and level of antioxidants like vitamin E and selenium. However it has been found that nutritional strategies dealing with fatty acids profile are easier to be applied in mono-gastric animals like pig than in ruminants as bio-hydrogenation is inherent in ruminants (Scollan *et al.*, 2006).

The other approach to enhance the functionality in meat based products is the processing practices as it is possible to reformulate it by reducing some compounds normally present in a specific product to appropriate amounts (fat, salt, nitrites, *etc.*), or by incorporating ingredients which are health-enhancing like fibre, antioxidants, MUFA, PUFA, vegetable proteins and probiotics (Zhang *et al.*, 2010). The processing techniques here involve the reduction of fat, calories, sodium, cholesterol, nitrites, modification of the fatty acid profile by replacing part of the animal fat with another more suitable to health needs, and incorporation of functional ingredients (Arihara, 2006). Hence the functional meat products either contain lesser quantity of harmful compounds like cholesterol, fat *etc.* or possess nutritional ingredients that improve health (Diplock *et al.*, 1999; Yue, 2001). Meat products containing dietary fibers are considered as an excellent one due to enhancement in their functionality (Hur *et al.*, 2009; Kumar *et al.*, 2010).

In this present discussion an attempt is made to produce an elaborate view on the issue of functionality of meat food products. This is about the different approaches that could be adopted to produce a healthier and safe meat product.

2. Production Practices

The meat production is the first step in designing the need based meat products. There are several practices applicable at farm level for the desirable modification in meat based composition. Velasco *et al.* (2004) observed that weaning in lambs had more prominent effect than the feed variation on the quality characteristics of the meat. In another study it was found that a decrease in the intramuscular fat content results in decreased sensory characteristics like juiciness and flavour (Chizzolini *et al.*, 1999). It has been found that the possible modification in meat composition at production level includes variation in fat profile; a higher percentage of unsaturated fatty acids, especially n-3 fatty acids result in improvement of n-3: n-6 ratio, PUFA: SFA ratio and reduction in fat content. While reduction of fat content in meat is related with nutrition and genetic aspects both, however the changes in fatty acids profile depends mainly on nutrition (Jiménez-Colmenero *et al.*, 2001; Scollan *et al.*, 2006). Hence for the meat designing, the factors considered are selection of breeds, alteration in animal feeding practices and intervention in animal metabolism (Chizzolini *et al.*, 1999). Enser *et al.* (2000) found that feeding practices directly affected the pork composition; pigs fed on linseed showed an increase in the n-3 PUFA in meat. To enrich chicken, pork, beef and lamb with CLA, dietary supplementation has been found to be an effective mode (Raes *et al.*, 2004). De La Torre *et al.* (2006b) observed an increased CLA level in beef from cattle fed with linseed meal.

Inhibition of oxidative deterioration could be achieved by increasing muscle antioxidants through diet, as increase in unsaturated fatty acids may increase the oxidation risk in meat based food products. Nowadays, farm animal dietary supplementation with vitamin E has increased the concentration of this compound in animal tissue to the extent that meat became a moderate vitamin E source. Similarly in another finding Dal Bosco *et al.* (2004) reported that enrichment of n-3 PUFA increased the oxidative stability of rabbit meat; the α-linolenic acid-vitamin E diet favoured the accumulation of long chain polyunsaturated n-3 in the meat and it inhibited the oxidation in meat and improved its nutritional value. It has been found that lambs fed with concentrates having a rich source of carotenoids and fat soluble vitamins like retinol and tocopherol contained significantly higher retinol and tocopherol levels in their body fat and meat products from these lambs could contribute to health benefits (Alvarez *et al.*, 2014).To reduce lipid and myoglobin oxidation in fresh meat and meat products vitamin E fortification of animal feed has been found to be very effective (Houben *et al.*, 2000). Dunshea *et al.* (2005) reviewed the effects of dietary factors on quality and nutritional value of meat and stated the positive effect of a supplementation of vitamin E at levels of 100–200 mg/kg feed on the quality of meat and meat products.

It has been observed that dietary supplementation of animals enhances the level of minerals naturally occurring in meat, and prominent of them are iron and selenium, which are deficient in diets of population in many countries (Jiménez-Colmenero, 2007a). Supplementation with selenium, magnesium or iron has been employed to increase the concentration of these healthy elements in meat (Olmedilla-Alonso *et al.*, 2013).Taking into consideration the protective role of selenium in chronic degenerative diseases and low or diminishing selenium status in some parts of the world, enrichment of meat and meat products in selenium should be considered (Biesalski, 2005).

3. Processing Practices

Manipulation of meat raw materials could be practiced during the slaughter and dressing operations which involve mainly the reduction of fat content in meat by removing external and internal fat from the carcass. The other processing practices target the product formulation to manipulate the content of a product, so that the health favouring ingredients could be enhanced while those having unwanted effect could be reduced.

Meat is nutrient rich food but is found to be deficient in certain important features like fiber, hence its enrichment is practiced with fibers from vegetable origin. Similarly the fat and nitrite content in meat products have been associated with ill health effects and they can be manipulated during the product formulations. In preparation of the product, it is possible to reformulate it by reducing some compounds present in a product to appropriate amounts (fat, salt, nitrites, *etc.*), or by incorporating ingredients which are health-enhancing like fibre, antioxidants, MUFA, PUFA, vegetable proteins and probiotics (Zhang, *et al.*, 2010). There are several approaches for the possible modification in meat formulation to reduce the health risk but the most common strategies are reduction of fat, calories, sodium, cholesterol, nitrites, modification of the fatty acid profile by replacing part of the animal fat with another more suitable to health needs, and incorporation of functional ingredients (Arihara, 2006). As the modification of fat with MUFA and PUFA may increase the susceptibility of product for oxidation hence the control of oxidation in meat products is another concern. Consumer concerns over safety and toxicity of synthetic antioxidants forced the food industry to explore natural sources (Coronado *et al.*, 2002). There are several natural antioxidants of plant origin which are still being explored for their antioxidant potential.

3.1 Dietary Modifications

3.1.1 Lipid

Meat reformulation strategies with respect to lipid content targets the - reduction of total fat and energy, reduction of cholesterol and modification of fatty acid profiles (Jiménez Colmenero *et al.*, 2012). Kerry and Kerry (2006) observed that fat can be reduced by utilization of leaner meat raw materials and using the fat replacers based on carbohydrate or protein or by adding water. Hur *et al.* (2004) reported that substitution of fat in beef patties with CLA sources improved the colour stability by inhibition of lipid oxidation and oxymyoglobin oxidation, and an improvement in anti-carcinogenic, anti-diabetic, anti-obesity, anti-atherogenic, and anti-oxidative properties was observed (Khanal and Olson, 2004). Fat already present in the meat is replaced with the health favourable fats like MUFAs and PUFAs by using that which is naturally available in vegetable sources like olive, corn, soybean, peanut *etc.* and marine lipids (Ansorena and Astiasaran, 2013). Contrary to plant oils, fish and algae oils are one of the dietary sources that are rich in long chain ω-3 PUFAs. In particular oily cold water fish such as salmon, herring, mackerel, anchovies and sardines are major sources of ω-3 PUFAs with mackerel containing the highest amount with 2.2 g per 100 g of fish (Weiss *et al.,* 2010). A dietary supplementation of food products with ω-3 PUFAs, such as eicosapentaenoic acid (EPA; 20: 5) and docosahexaenoic acid (DHA; C22: 6) has been suggested to replace saturated, monounsaturated and ω-6 polyunsaturated fatty acids in foods (Jimenez-Colmenero, 2007b) to enhance the health benefits. It should be noted though, that in order to achieve the same physiological benefits as the higher molecular weight DHA and EPA, approximately five times the amount of ALA (Alpha-Linolenic Acid) compared to DHA and EPA needs to be consumed. This is because the body is extremely inefficient in converting ALA into the physiologically active DHA and EPA forms (Weiss *et al.,* 2010).

The joint statements by WHO and FAO states that the recommended ratio of polyunsaturated fatty acids (PUFAs) and saturated fatty acids (SFAs) in diets should be between 0.4 and 1.0 while ω-6/ω-3 PUFA ratio should be between 1 and 4, respectively (WHO, 2003; Wood *et al.,* 2003). However, the current trend denotes, that they are not only deficient in ω-3 PUFAs (especially long chain fatty acids) but also contain excessive amounts of ω-6 PUFAs, with an ω-6/ω-3 PUFA ratio of 15–20 (Simopoulos, 2002).

In addition, some of the oxidation products generated from meat fat have shown to possess mutagenic and carcinogenic potential (Gao *et al.,* 1987), making an extensive oxidation of meat and meat products a health problem. Compounds such as vitamin E, lycopene or lutein have also shown to not only reduced lipid oxidation but also to have health benefits. For example, lutein together with zeaxanthin has been associated with the maintenance of good eye health due to its ability to reduce the risk of Age Related Macular Degeneration (AMD). Hence the fortification of meat products with some of these substances having health benefits may thus be highly desirable to create a new class of functional foods (Granado-Lorencio *et al.,* 2010).

3.1.2 Fiber

Meat is considered nutrient rich food but for the fiber, it has been associated with a number of modern day malaise. The emergence of a range of chronic diseases, including colon cancer, obesity, cardiovascular diseases, and several other disorders have been associated with a diet containing an excess of energy-dense foods rich in fat and sugar (Best, 1991; Kaefersteins and Clugston 1995; Beecher, 1999). In accordance to the findings, the dietary fiber in the daily diet needs to be increased (Eastwood, 1992; Johnson and Southgate, 1994). Fiber in foods has an effect on the abatement of caloric content. Cofrades *et al.* (2000) found that fiber is suitable for incorporation in meat products as it enhances the cooking yield due to its water-binding and fat-binding properties. Fibers have been studied alone or in combination for formulations of reduced-fat meat products (Chang and Carpenter, 1997; Desmond *et al.*, 1998; Grigelmo-Miguel *et al.*, 1999; Mansour and Khalil, 1999). Yilmaz (2004) utilized the rye bran in the production of meatballs as rye consumption has been reported for several health beneficial effects; inhibit breast and colon tumor growth in animal models, lower glucose response in diabetics, lower the risk of death from coronary heart diseases. As the incorporation of rye bran in meatballs resulted in lowering of total trans-fatty acid content and an increase in the ratio of total unsaturated fatty acids to total saturated fatty acids.

Oat, another source of fiber has a good water-absorption capacity and claimed health benefits making it a potential ingredient to be used in meat products. Steenblock *et al.* (2001) determined its effect on the quality characteristics of meat products like low fat bologna and fat-free frankfurters where it was observed that oat fiber at levels up to 3 per cent resulted in greater yields as purge from the product was reduced. In a similar study it was recorded that oat bran and oat fiber provided the flavour, texture, and mouth feel of fat in ground beef and pork sausages (García *et al.*, 2002). The effect of a short-chain fructo-oligosaccharides (FOS), a component of dietary fiber, was studied on cooked sausages (Caceres *et al.*, 2004) and it was found that the addition did not affect the pH, a_w or weight losses because the presence of soluble dietary fiber (SDF) leads to a compact gel structure. The energy values of product recorded a decrease from 279 kcal/100 g in the conventional control to 187 kcal/100 g in the reduced-fat sausages having 12 per cent added fiber, however the hardness of the samples with SDF was also lower, and the acceptability of product was higher. Mendoza *et al.* (2001) prepared low-fat, dry-fermented sausages supplemented with 7.5 and 12.5 per cent of inulin and having a fat content reduced by 25 to 50 per cent of the original amount and the results indicated a softer texture and a tenderness, springiness, and adhesiveness very similar to that of conventional sausages, hence a low-calorie product can be obtained with approximately 10 per cent inulin where the energy value reduced by 30 per cent of the original product.

In addition to the cereals and grains, vegetables and fruits could be good source of dietary fiber, however they contain health benefits too that are mainly attributed to organic micronutrients such as carotenoids, polyphenolics, tocopherols, vitamin C *etc.* (Schieber *et al.*, 2001). Inner pea fiber may improve the sensory properties of lower fat ground beef by retaining substantial amounts of both the moisture and fat that are normally lost during cooking. At 10 per cent and 14 per cent when added in a dry form to the lower-fat beef patties, it improved tenderness and cooking yield without having negative effects on sensory features like juiciness and flavour (Anderson and Berry, 2000).

3.1.3 Protein

Although the meat is a protein rich food but sometimes other ingredients used as a product formulation could impart their specific characteristics to improve the products' functional quality. Whey proteins found to have improved colour properties, emulsion stability, and resulted in lower chewiness and elasticity in sausages (Yetim *et al.*, 2001). Whey proteins showed excellent nutritional and functional properties in low-fat meat products (Perez-Gago and Krochta, 2001).

The pre-heated whey protein when used in poultry meat batter, lead to increased water holding capacity, improved rheological properties, and reduced cooking loss as the isolates formed gel at low protein concentrations and low temperature in the presence of added salt (Hongsprabhas and Barbut, 1999).

The soy products like soy flour, and soy protein concentrate and isolate rich in protein content when incorporated in meat products, improve water and fat binding ability, enhance emulsion stability, improve nutritional content, and increase yields (Chin *et al.*, 2000). However, soy flour produced some beany flavour and soy protein concentrates provided some undesirable palatability in soy-added meat products (Smith *et al.*, 1973). But the recent processing techniques have some advantages in overcoming the beany flavour and other undesirable effects of soy incorporation. Incorporation of tofu powder resulted in lower fat and higher protein and moisture content, but did not affect sensory parameters in lean pork sausages (Ho *et al.*, 1997). Soy protein isolates are very hydrophilic and can reduce cooking loss; addition of 2.5 per cent soy protein isolate in Argentina sausage, decreased drip loss during refrigerated storage and did not affect the flavour, aroma, juiciness characteristics (Porcella *et al.*, 2001).

Wheat proteins form visco-elastic mass of gluten through the interaction with water and could be a great additive in food products (Pritchard and Brock, 1994). Gluten produced from wheat flour can be used as a binder or extender in meat products (Janssen *et al.*, 1994). Chymotrypsin-hydrolyzed wheat gluten resulted in lower microbial transglutaminase activity and improved thermal gelation and emulsifying properties of myofibrillar protein isolates (Xiong *et al.*, 2008). Wheat proteins at 3 per cent and 6 per cent added to sausages had hardness of the product increased but springiness decreased (Li *et al.*, 1998). Addition of 3.5 per cent wheat protein flour increased water holding capacity and decreased cooking loss, and at the same time the textural and sensory properties of frankfurters were also improved (Gnanasambandam and Zayas, 1992).

3.1.4 Salt

Excessive dietary sodium intake is associated with development of hypertension and consequently with increased risk of CVD. A direct relationship between an excessive intake of sodium and an increased incidence of hypertension has been demonstrated in various scientific publications (Dahl, 1972). The total amount of dietary salt should be maintained at about 5-6 g per day, as it has been found that the consumption of more than 6 g NaCl per day by an individual results in hypertension (Ruusunen and Puolanne, 2005). At present the daily sodium intake by an adult is almost three times of the recommended daily allowance. The prominent source of sodium in the diet is sodium chloride, common part of food formulation, as it imparts the essential attributes of food including meat products; flavour, texture and shelf life. With only 50 to 90 mg of sodium per 100 g of meat, it is relatively poor in sodium content however the derivatives or the products have much higher because of the salt added as an ingredient. The consumption of meat and meat products contributes about 16– 25 per cent to the total daily intake of sodium chloride and thus is second only to bread with respect to salt levels (WHO, 2003). Although the meat product like other food products contain 2 per cent salt, but in products like cured one, salt can reach a level of 4-5 per cent (Desmond, 2006). Basic strategies to reduce the sodium content in processed foods include the use of salt substitutes, specially potassium chloride (KCl) in combination with masking agents like yeast extracts, lactates, monosodium glutamate and nucleotides, which do not have a salty taste, but improve the saltiness of products when used in combination as they act as flavour enhancers (Desmond, 2006). Potassium chloride has a slightly bitter taste and to prevent the product from having unacceptable sensory properties, masking substances have to additionally be added to the products (Desmond, 2007). It has been found that partial replacement of NaCl by calcium ascorbate to be a viable way of decreasing sodium content in fermented sausages. Carboxy-methyl cellulose and carrageenan in combination with sodium citrate have been found to improve salty flavour in frankfurters (Ruusunen *et al.*, 2003). Recent approach which is not fully explored is that the physical structure of sodium chloride could be altered and a change in particle size of salt crystals could lead to a more rapid dissolution behavior in the mouth thereby yielding a more pronounced salty taste of the product.

4. Other Strategies

4.1 Bioactive Components

The Foundation for Innovation in Medicine in 1991 defined bioactives as "... any substance that may be considered a food or part of a food and provides medical or health benefits, including the prevention and treatment of disease" (IFIC, 2006).The most common type of bioactive components associated with the meat are omega-3 fatty acids, CLA, carnitin and bioactive peptides. A value is always added to the commodities from which bioactives are derived or extracted and most of them are naturally occurring compounds and can be obtained from plant/animal source.

In terms of omega-3 fatty acids, fish is considered as a significantly better source than the red meat where beef and lamb have more omega-3fatty acids than chicken and pork. Large-scale population (epidemiologic) studies have shown that recommended amounts of eicosa-penta-enoic acid (EPA) and docosa-hexa-enoic acid (DHA) from fish oil leads to lower triglycerides, heart attack, abnormal heart rhythms, and strokes in people with known cardiovascular diseases. Incidence of cardiovascular diseases can be reduced by regular consumption of ω-3 fatty acid-enriched pork which leads to decrease in the content of serum triglycerides and increase in the production of serum thromboxane (Coates *et al.*, 2009).

4.1.1 Conjugated Linoleic Acid (CLA)

CLA is a collective term to describe a mixture of positional and geometric isomers of linoleic acid, having double bonds at positions 7 and 9, 8 and 10, 9 and 11, 10 and 12, and 11 and 13 in the fatty acid chain (Eulitz *et al.*, 1999). In accordance to fermented milk products, probiotic bacteria may lead to enhanced formation of CLA in meat products (Arihara, 2006).The two main isomers of conjugated linoleic acid (CLA); cis-9,trans-11 and trans-10,cis-12, show marked biological activities. The most common CLA isomer found in beef is cis-9 and trans-11 (Schmid *et al.*, 2006). Conjugated linoleic acid (CLA), found in extracts of grilled beef, is a positional and geometric isomer of octa-deca-dienoic acid and observed to have anticarcinogenic property (Nagao and Yanagita, 2005). As rumen bacteria leads to conversion of linoleic acid to CLA by their isomerase activity, it is most common in fat of ruminants. The concentration of CLA significantly increased by substitution of fat (Hur *et al.*, 2004). Its content in meat is variable and affected by a number of factors like breed, age, feed composition *etc.* (Dhiman *et al.*, 2005). Supplementation of diet with feed ingredients, high in linolenic acid result in rise of conjugated linoleic acid (CLA) content in meat (Scollan *et al.*, 2006). The most common isomer is the cis-9, trans-11,which has been linked to health benefits like antitumor, antiatherogenic, antioxidative and anticarcinogenic properties (Azain, 2003; Jose *et al.*, 2005; De La Torre *et al.*, 2006a). In addition to these properties, CLA may also play an important role in control of obesity (trans-10, cis-12 isomer reduces rate of fat deposition), reduction of the diabetes risk and modulation of bone metabolism owing to its antioxidative and immunomodulative effect (Arihara, 2006; Schmid *et al.*, 2006). Beef contains 1.2-14 mg/g of CLA in fat and up to 130 mg per 100 g of lean tissue (Hoffmann, 2010). Realini *et al.* (2004) found that intramuscular fat from Hereford steers fed on pasture had twice the CLA content than the concentrate/ stall fed. *Longissimus* muscle-lipid from grass-fed and concentrate-supplemented beef contained CLA as 10.8 mg/g and 3.7mg/g respectively (French *et al.*, 2000). In conclusion, dietary supplementation of synthesized CLA can increase the content of CLA and change the fatty acid profile in non-ruminant animal fat and muscle, therefore, dietary supplementation of CLA is a reasonable way of developing a value-added meat product (Zhang, 2010).

4.1.2 Carnitine

Meat also provides L-carnitine (beta-hydroxy-gamma-trimethyl amino butyric acid), which transports long chain fatty acids across the inner mitochondrial membranes as L-carnitine ester where they are processed by β-oxidation to produce biological energy during exercise. It is not considered as an essential nutrient but is required during extra energy need like pregnancy and after heavy exercise and a recommended intake of 24-81mg/d has been suggested (Tanpaichitr and Leelahagul, 1993). L-carnitine has been associated with biological activities like absorption of calcium to improve skeletal strength and chromium picolinate to help in building of lean muscle tissue. Vescovo *et al.* (2002) reported that L-carnitine can block apoptosis and prevent skeletal muscle myopathy in conditions like heart failure. L-carnitine is frequently recommended as a food supplement for boosting fat combustion, chiefly within the context of weight reduction diets and in sports' world to enhance performance and to produce energy during exercise.

4.1.3 Bioactive Peptides

Bioactive peptides are either absorbed in the intestine to reach the different parts of body through circulatory system or they may produce local effects in the digestive tract itself (Erdmann *et al.*, 2008). As meat is a source of high quality proteins hence it presents one of the most important sources for isolation of bioactive peptides in recent years (Ryan *et al.*, 2011). In addition to myosin and actin other proteins from thick and thin filaments and connective tissues like fibrillar collagen are well used for peptide generation (Udenigwe and Ashton, 2013). Bioactive peptides displaying different favourable effects like antihypertensive, antioxidant, and antiproliferative have been found in the hydrolysates of meat and fish proteins (Matsui *et al.*, 2006; Kim *et al.*, 2009). A number of peptides with angiotensin-converting enzyme (ACE)-inhibitory activity were isolated from pork, beef and chicken. These peptides act differently than hypotensive drugs. ACE (angiotensin-converting enzyme) is directly blocked by these substances hence the conversion of angiotensin-I into angiotensin-II does not occurs, and the major product of the renin–angiotensin system *i.e.* angiotensin-II which is a powerful vasoconstrictor is not formed. Other actions of ACE result in hydrolysis of bradykinin, a potent vasodilator, and it also induces the release of aldosterone, leading to increase of sodium concentration and blood pressure (Escudero *et al.*, 2012; He *et al.*, 2013; Udenigwe and Ashton, 2013). Udenigwe and Ashton (2013) found that some peptides obtained from meat sources have antithrombotic properties and they can be used in future to prevent or control vascular ailments. A peptide with molecular weight of 2.5 kDa from porcine muscle (*longissimus dorsi*) was isolated and investigated for its effect on conditions like thrombosis (Shimizu *et al.*, 2009). Peptides isolated from meat and other organisms of marine sources are found to have anti-cancer activity like inhibition of cell proliferation and cytotoxic effect against cancerous cells (Udenigwe and Aluko, 2012). In a study, apoptosis in a human lymphoma cell line (U937) was found to be induced by hydrophobic peptide isolated from anchovy sauce (Ryan *et al.*, 2011).

4.2 Probiotic

The probiotics are microorganisms which can exert some health benefits to the host when ingested in adequate levels in live and are mainly the strains from species of *Bifidobacterium* and *Lactobacillus* (FAO, 2006). It can impart beneficial effect on the health of host when ingested in appropriate amount by inhibiting the growth of harmful bacteria via competitive exclusion and by generating substances like organic acids and antimicrobial agents in the colon (Salminen *et al.*, 1996).) The ideal probiotic should have desirable properties: adherence and colonization in human guts; production of antimicrobial substances; resistance to acid and bile released in gut; antagonism against pathogenic bacteria; and immune modulation properties (Brassart and Schiffrin, 2000; Stanton *et al.*, 2003; Agrawal, 2005)

In 2000, Erkkila and Petaja explored the survival of lactic acid bacteria in several meat starter cultures and found that strains of *Lactobacillus sakei* and *Pediococcus acidilactici* have the best survival capacities under acidic conditions and high levels of bile salt. Although the awareness and market for meat products with probiotic effect is not in good shape, however some probiotic meat products have been found in countries like Germany and Japan (Arihara, 2006). Arihara *et al.* (1998) have shown that *L. gasseri* JCM1131 is applicable for meat fermentation as a potentially probiotic strain. However, development of fermented meat products with probiotic seems challenging since the viability of those bacteria is affected by high content of curing salt and low pH due to acidification and low water activity due to drying (De Vuyst *et al.*, 2008). Erkkila *et al.* (2000, 2001a, b) found that strains like GG and E-97800 are suitable for use as probiotic starter cultures in fermenting dry sausage. It has been found that dry sausages processed without heat treatment could be a most suitable carrier for probiotic organisms like *Bifidobacterium lactis, Lactobacillus casei, L. paracasei* and *L. rhamnosus* (Tyoppone *et al.*, 2003; Arihara, 2006).

According to Lucke (2000), a probiotic fermented sausage produced with *Bifidobacterium* in Germany resulted in poor survival of *Bifidobacterium* during the sausage ripening suggesting that a very high inoculums is required for achieving the minimum level of probiotic bacterial population (6 log cfu/g) in the final product. Muthukumarasamy and Holley (2006) studied the effectiveness of a microencapsulation technique for protecting probiotic bacteria during sausage processing and it has been suggested as a promising method to increase the survival ability of probitics during the meat fermentation (Audet *et al.*, 1988; Sheu and Marshall, 1993).

Probiotics may also play some role in inhibition of pathogens such as *Helicobacter pylori* and *Salmonella* (Mcnaught and Macfie, 2001, Santosa *et al.*, 2006). Jahreis *et al.* (2002) reported that the consumption of probiotic sausage increased the antibodies against oxidized low density lipoprotein without introducing significant effects on the serum concentration of different cholesterol fractions and triglycerides in human. There is also an expectation that in meat products, analogically to fermented milk products, probiotic bacteria may increase the formation of CLA (Arihara, 2006). Because of their health benefits probiotics could promote, market for the meat industry. Meat products, processed by fermentation without heating have been considered an excellent vehicle for probiotics.

4.3 Antioxidant

4.3.1 Meat Origin

Various endogenous antioxidants, including tocopherols, ubiquinone, cartenoids, ascorbic acid, glutathione, lipoic acid, uric acid, spermine, carnosine, and anserine, have been found in skeletal muscle (Decker *et al.*, 2000). Neutralization and reduced release of free radicals by antioxidants prevent oxidative damage in our body. Histidine-containing dipeptides, carnosine (β-alanyl-L-histidine) and anserine (N-β-alanyl-1-methyl-L-histidine) are the best-known antioxidants found in meat based food items (Sarmadi and Ismail, 2010; Young *et al.*, 2013). These peptides attribute their antioxidant activity to their ability of chelating transition metals like cobalt, zinc and copper (Arihara, 2006; Young *et al.*, 2013).

The concentrations of carnosine in meat range from 500 mg per kg of chicken thigh to 2700 mg per kg of pork shoulder. On the other hand, anserine is especially abundant in chicken muscle (Brown, 1981). These antioxidative peptides have been reported to play many roles, such as prevention of diseases and aging related to oxidative stress (Hipkiss and Brownson, 2000; Hipkiss *et al.*, 1998). Study demonstrated the bioavailability of carnosine by determining its concentration in human plasma after ingestion of beef (Park *et al.*, 2005).The hydrolysates have been found to have these peptides in ample amount. Kim *et al.* (2009) obtained two antioxidative peptides from venison protein hydrolysates, APVPH I (MQIFVKTLTG) and APVPH II (DLSDGEQGVL), respectively, where their free radical scavenging activity was higher than that of vitamin C. Park and

Chin (2011) assessed the antioxidant activity of pepsin-digested protein extracted from pork ham and observed that low molecular weight peptides (<7 kDa) have activity to inhibit lipid peroxidation developing in linoleic acid emulsion system. Saiga *et al.* (2003) tested porcine myofibrillar protein hydrolysates (obtained by treatment with papain and actinase E) as a source of antioxidant peptides and found that these hydrolysates inhibited peroxidation of linoleic acid and possessed 1,1-diphenyl-2-picrylhydrazyl (DPPH) radical scavenging activity and metal ion chelating activity. Arihara *et al.* (2005) found three antioxidative peptides isolated from a papain-treated hydrolysate of pork actomyosin and sequenced them as Asp-Leu-Tyr-Ala, Ser-Leu- Tyr-Ala, and Val-Trp. Di Bernardini *et al.* (2012) investigated the antioxidant activity of sarcoplasmic proteins isolated from bovine brisket muscle (*Pectoralis profundus*) by enzymatic hydrolysis with papain and found that the 10-kDa and the 3-kDa peptidic fractions demonstrated antioxidant activities using *in vitro* assays [DPPH free radical scavenging activity, the ferric ion reducing antioxidant power (FRAP) and the Fe2+ metal chelating ability]. A number of strategies are being used to enhance antioxidant activity in meat systems and to reduce the formation of oxidation products with their subsequent impact on ageing, cancer and cardiovascular disease (Decker *et al.*, 2000).

4.3.2 Plant Origin

Another approach is the incorporation of antioxidant from the plant sources to enhance the oxidative stability of product and impart the health effect of antioxidants in meat food products. The application of synthetic antioxidant is limited due to its association with potential health hazards (Mendis *et al.*, 2005). When incorporated in food products, they can improve flavour by retarding the lipid oxidation, inhibit the growth of microorganisms, and may play roles in reducing the disease risk (Tanabe *et al.*, 2002). Lipid oxidation is one of the causes for the deterioration of meat and derivatives because their appearance determines the onset of a large number of undesirable changes in flavour, texture, and nutritional value (Gil *et al.*, 2001). Natural antioxidants extracted from plants such as rosemary, sage, tea, soybean, citrus peel, sesame seed, olives, carob pod, and grapes can be used as alternatives to the synthetic antioxidants because of their equivalent or greater effect on the inhibition of lipid oxidation (Tang *et al.*, 2001). The protective effect of fruit and vegetable fibers has generally been attributed to their antioxidant constituent, including vitamin C and E, carotenoid, glutathione, flavonoids and phenolic acids, as well as other unidentified compounds (Eberhardt *et al.*, 2000). Flavonoids, a group of more than 4000 polyphenolics, are products of plant metabolism and naturally occurring polyphenolics present in fruit and vegetables, and play an integral part in human diet (Hollman *et al.*, 1997).

The functional property of pork patties with added green tea leaf extract was explored by Jo *et al.* (2003). This extract did not have negative effects on the physical and sensory properties and had beneficial biochemical properties; the researchers concluded that irradiated green tea extract powder can be used to add functional properties to pork patties. The addition of tea catechins to cooked meat and poultry significantly inhibited the pro-oxidative effect of NaCl (Tang *et al.*, 2001). Catechins composed of four compounds epicatechin, epicatechin gallate, epigallocatechin, and epigallocatechin gallate, is a predominant group of polyphenols present in green tea leaves (Zhong *et al.*, 2009).

In their studies into "salami" products, Severini *et al.* (2003) found that the partial substitution of pork back-fat by extra virgin olive oil did not affect the chemical, physical, and sensory characteristics of the products, with the exception of water activity and firmness. Moreover the addition of the extra virgin olive oil, which is rich in unsaturated fatty acids, did not reduce the shelf life in terms of lipid oxidation, probably due to the antioxidant effect of both polyphenols and tocopherols. Another extract used in meat products is rosemary, which is a source of large number of phenolic compounds with antioxidant activities; carnosol, carnosic acid, rosmanol, epirosmanol, isorosmanol, rosmarinic acid, rosmaridiphenol, and rosmariquinone (Gil *et al.*, 2001; Fernandez-Lopez *et al.*, 2003). Coronado *et al.* (2002) manufactured wiener sausages with its extract, and low lipid oxidation was observed in the product during long-term frozen storage.

Sage is the dried leaf of a mint family and major antioxidant compounds in it include carnosol, carnosic acid, rosmadial, rosmanol, epirosmanol, and methyl carnosate (Cuvelier *et al.*, 1994). Similarly oregano is a traditional Mediterranean spice and its essential oil contains compounds like carvacrol and thymol with major antioxidant activity (Vekiari *et al.*, 1993).

Green tea extract is known to decrease total cholesterol, increase the HDL fraction, and decrease lipoprotein oxidation (Tang *et al.*, 2001). The tea extract promote health by preventing lipid oxidation and providing antibacterial, anti-carcinogenic and antiviral ability (Yang *et al.*, 2000). Carotenoids are related to CVD risk reduction, cancer prevention and immune-stimulation (Torrisen, 2000). Tocopherols are known for their efficient antioxidant capacity in foods and biological systems and epidemiological studies provided the inverse relationship between coronary artery disease and vitamin E supplementation (Pryor, 2000). Flavonoids are effective antioxidants because of their scavenging properties, chelators of metal ions (Kandaswami and Middleton, 1994) and may protect tissues against free oxygen radicals and lipid peroxidation.

4.4 Reduction of Nitrites

There has been some concern over the use of nitrites and nitrate in foods, especially in processed meat products. In these products nitrite participates in the formation of nitrosamine (Fine, 1982). Nitrosamine has been associated with the ill health effects and is known to be a cancerous agent. However in recent years residual nitrite level has been substantially reduced with the addition of ascorbic acid/sodium ascorbate that accelerates curing and allows for lower nitrate addition. Ascorbate and erythorbate help in reducing residual nitrite and limiting the formation of nitrosamine.

There are two basic strategies for reducing the potential health risks of nitrites in meat products; where one focuses in reduced addition of nitrite, the other is to use N-nitrosamine inhibitors (Jiménez-Colmenero *et al.*, 2001). Use of erythrosine, betanin as colouring agents and sorbic acid, fumaric acid esters and lactic acid bacteria have been studied as potential components for replacing nitrites. The second strategy of inhibition of nitrosamines formation – is possible with addition of ascorbates and erythorbate. It has been also shown that vitamin E added to meat products decreases the production of nitrosamines (Hoffmann *et al.*, 2010). Other strategies which allow lowering in level of residual nitrite include cover curing in modified atmosphere and application of starter cultures (Cierach, 2007). Processing and handling factors may hold more potential and may be more effective in reducing nitrite than chemical reductants.

The idea of utilizing a low pH to reduce residual nitrite and potential nitrosamine formation has been applied to retard the nitrosamine formation upon frying and provide additional protection against *C. botulinum*. As an alternative for nitrite curing, mixtures based on meat starter cultures, enzymes and vegetable juices and spices have been tested (Hoffmann *et al.*, 2010).

4.5 Enzymes

The meat structure is a vital concern in deciding the acceptability of a product. It has been found that meat from aged animals may be tough enough to be considered a premium product. However on the other side sometimes we need to restructure a meat product to give it a proper texture and shape and enhance its acceptability among the consumers. These structural changes in meat could be done with the help of enzymes. The endogenous enzymes present in meat play an important role in postmortem changes and conversion of muscle to meat. Many muscle endogenous proteases are possibly involved in meat protein hydrolysis including calpains, cathepsins, dipeptidyl peptidases, and aminopeptidases. Among these enzymes, cathepsins and calpains are the most important endopeptidases for muscle proteolysis (Di Luccia *et al.*, 2005). Instead at times we resort to external sources for the desirable changes in meat and meat products. Traditionally, proteases derived from plants have been used. Examples of cystein proteases from plants are papain present in papaya and bromelain from plants of Bromeliaceae family; pineapple belongs to this category. It has been found that the plant-derived proteases are non-specific in their action and thus do not only tenderize but also degrade the texture of meat, which may result in over-tenderization. Newer, more specific proteases have been identified by Benito *et al.* (2003) where they reported the use of fungal extracellular protease for the texture changes in pork loin and found that EPg222 protease isolated from *Penicillium chrysogenum* had desired effect on dry-cured ham. It was observed that incubation of pork loins with 0.012 mg/ml of the enzyme for up to 32 days under sterile conditions significantly decreased hardness as measured by instrumental texture analysis by a factor of 3 (from 54.5 N in control loins to 14.2 N in treated pork loins after 32 days) and conclusion of the finding was that the externally applied enzyme was able to counteract increases in hardness resulting from protein denaturation in preparation of dry-cured hams. Similarly, Qihe *et al.* (2006) reported the use of elastase from *Bacillus sp.* EL31410 in beef meat where they observed that a 4 hour treatment of freeze-dried meats that had been dipped in 1 per cent elastase solutions led to a 30 per cent decrease in relative hardness after storage for 96 h. Elastases could be found in aquatic species too and the enzymes may have very different specificities compared to those found in mammalian muscles, the gastric elastases recovered from carp, catfish *etc.* with the ability to rapidly degrade elastin (Shahidi and Kamil, 2001).

In contrast to the above described "structure breakers", enzymes may also be used to form covalent bonds between proteins. The most prominent enzymes that belong to this class are transglutaminases (TGase). Generally, transglutaminases are calcium-dependent enzymes that catalyze acyl transfer reactions with the ε-amino group of lysine and lysyl residues acting as acyl acceptors and γ-carboxamide groups of glutamine acting as acyl donors (Kumazawa *et al.*, 1996). In the process, ε-(γ-glutamyl)-lysine cross-links are formed. Microbial transglutaminase has been successfully used to improve the functional properties of meat gels by catalyzing formation of glutamyl-lysine bonds in myosin, myosin and actin, myosin and fribonectin and fibrin and actin (Kahn and Cohen, 1981).Tseng *et al.* (2000) similarly suggested that high quality, low salt chicken meat balls could be formulated with the help of TGase. Lantto *et al.* (2006) investigated the effect of addition of tyrosinase and freeze dried apple pomace powder that contained both TGase and tyrosinase to an industrial pork meat homogenate. They reported an increase of gel strength with the addition of the powder, but little effect of tyrosinase likely to due to the presence of cystein, which is a known tyrosinase inhibitor. Similarly laccase, a copper-containing polyphenol-oxidase (p-diphenol oxidase) that oxidizes polyphenols, methoxy substituted phenols, and diamines has not yet been investigated for use in meat products. However, with addition of suitable substrates such as fibers or milk proteins to meat products, laccase may become a candidate for texture improvements (Minussi *et al.*, 2002).

The enzymes play an important role in the flavour enhancement of meat products and are an important feature when the meat is cured or aged. As meat is protein and fat dense food hence the activity of proteases and lipases has major effect on the flavour changes. The free amino acids and fatty acids are the prime player of flavour enhancement. The levels of free amino acids depend on aminopeptidase activity and the type of meat products (Toldra *et al.*, 2000). They not only directly attribute to flavour characteristics (Mottram, 1998) and taste properties (Koutsidis *et al.*, 2007) of meat products, but also serve as water-soluble flavour precursors. Large increases in free amino acid contents also occur during the curing of meat products, and glutamate is the major free amino acid found in the final product (Cordoba *et al.*, 1994). Similarly during curing, lipolysis and auto-oxidation are responsible for the changes in lipids (Toldra, 1998). Phospholipids (PLs) and triglycerids (TGs) degraded by phospholipases and lipases release free fatty acids and fatty acids could undergo oxidation to form peroxides which further react with peptides, amino acids leading to secondary oxidation products to form aroma compounds (Zhou and Zhao, 2007). Although oxidation have been associated with deterioration of meat quality during storage and processing but it is a crucial reaction to develop typical flavour of meat products, especially in dry-cured meat products (Chizzolini *et al.*, 1998).

5. Conclusion

With the advancement at the economic front, meat and meat products are expected to have additional functions and not only the necessary nutrients.

Increasing attention has been paid to the physiological functions inherent in meat and development of functional meat products in recent years (Jimenez – Colmenero, 2007a). The strategies to fortify foods with functional compounds to increase micronutrients and limit undesirable constituents can be done at animal production level or by dietary modification of meat raw materials and reformulation of meat products. The new meat products must have not only high nutritional value and health enhancing properties, but also have to be safe, tasty and convenient. However the consumer acceptance is the key for the success of functional foods in the market. Sensory attributes, especially taste, are among the most important factors that affect consumers' choice of functional foods (Krystallis *et al.*, 2008). There are very few comprehensive studies on the consumer acceptance and the market size for functional meat and meat products. Most conclusions are drawn from the fact that functional ingredients itself may be beneficial to human hence further studies are needed to provide strong scientific evidences for the human health benefits of functional meat and meat products. With increased scientific data, meat scientists and industry have to spend more efforts in informing and educating consumers about the health benefits of functional meat and meat products. The bio-availability of added functional ingredients should be maintained during the processing and commercial storage, is another sphere to be evidenced by research findings. Hence a holistic approach is required in the coming days for further study involving the functionality of meat products.

REFERENCES

Agrawal, R. (2005). Probiotics: an emerging food supplement with health benefits. Food Biotechnology, 19, 227–246.

Alvarez, R.A., Mele´ndez-Martý´nez, A.J., Vicario, I.M., and Alcalde, M.J. (2014). Effect of pasture and concentrate diets on concentrations of carotenoids, vitamin A and vitamin E in plasma and adipose tissue of lambs. J. Food Comp. Anal., 36, 59–65.

Anderson, E.T., and Berry, B.W. (2000). Sensory, shear and cooking properties of lower fat beef patties made with inner pea fiber. J. Food Sci., 65(5), 805–810.

Ansorena, D., and Astiasarán, I. (2013). Enrichment of meat products with omega-3 fatty acids by methods other than modification of animal diet. Woodhead Publishing Series in Food Science, Technology and Nutrition. Woodhead Publishers, pp.299-318.

Arihara, K. (2006). Strategies for designing novel functional meat products. Meat Science, 74, 219-229.

Arihara, K., Ota, H., Itoh, M., Kondo, Y., Sameshima, T., Yamanaka, H., Akimoto, M., Kanai, S., and Miki, T. (1998). *Lactobacillus acidophilus* group lactic acid bacteria applied to meat fermentation. Journal of Food Science, 63, 544-547.

Arihara, K., Tomita, K., Ishikawa, S., Itoh, M., Akimoto, M., and Sameshima, T. (2005). Anti-fatigue peptides derived from meat proteins. Japan patent, submitted to government.

Audet, P., Paquin, C., and Lacroix, C. (1988). Immobilized growing lactic acid bacteriawith κ-carrageenan–locust bean gumgel. Applied Microbiology and Biotechnology, 29, 11-18.

Azain, M. J. (2003). Conjugated linoleic acid and its effects on animal products and health in single-stomached animals. Proceedings of the Nutrition Society, 62, 319-328.

Beecher, G.R. (1999). Phytonutrients role in metabolism: effects on resistance to degenerative processes. Nutr. Rev., 57,3-6.

Benito, M. J., Rodriguez, M., Acosta, R., and Cordoba, J. J. (2003). Effect of the fungal extracellular protease EPg222 on texture of whole pieces of pork loin. Meat Science, 65, 877-884.

Best, D. (1991). Whatever happened to fiber. Prep Foods, 160,54–56.

Biesalski, H. K. (2005). Meat as a component of a healthy diet – are there any risks or benefits if meat is avoided in the diet? Meat Sci., 70, 509-524.

Brassart, D., and Schiffrin, E. J. (2000). Pre- and probiotics. In M. K. Schmidl, and T.P.Labuza (Eds.), Essential of functional foods (pp. 205-216). Gaithersburg MD, Aspen Publication.

Brown, C. E. (1981). Interactions among carnosine, anserine, ophidine and copper in biochemical adaptation. Journal of Theoretical Biology, 88, 245-256.

Cáceres, E., García, M.L., Toro, J., and Selgas, M.D. (2004). The effect of fructo-oligosacharides on the sensory characteristics of cooked sausages. Meat Sci., 68, 87-96.

Chang, H.C., and Carpenter, J.A. (1997). Optimizing quality of frankfurters containing oat bran and added water. J. Food Sci., 62,194-202.

Chin, K. B., Keeton, J. T., Miller, R. K., Longnecker, M. T., and Lamkey, J.W. (2000). Evaluation of konjac blends and soy protein isolate as fat replacements in low-fat bologna. Journal of Food Science, 65, 756-763.

Chizzolini, R., Zanardi, E., Dorigoni, V., and Ghidini, S. (1999). Calorific value and cholesterol content of normal and low-fat meat and meat products. Trends Food Sci. Technol., 10, 119-128.

Chizzolini, R., Novelli, E., and Zanardi, E. (1998). Oxidation in traditional Mediterranean meat products. Meat Science, 49, S87-S99.

Cierach M. (2007). Azotyny w procesie peklowania miêsa funkcje, aspekty zdrowotne, peklowanie bezazotynowe. Cz. II. Gosp. Miêsna, 5, 18-21.

Coates, A. M., Sioutis, S., Buckley, J. D., and Howe, P. R. C. (2009). Regular consumption of n–3 fatty acid-enriched pork modifies cardiovascular risk factors. British Journal of Nutrition, 101, 592-597.

Cofrades, S., Guerra, M.A., Carballo, J., Fernández-Martín, F., and Jiménez-Colmenero, F. (2000). Plasma protein and soy fiber content effect on bologna sausage properties as influenced by fat level. J. Food Sci., 65, 281-287.

Córdoba, J. J., Rojas, T. A., González, C. G., and Barroso, J. V. (1994). Evolution of free amino acids and amines during ripening of Iberian cured ham. Journal of Agriculture and Food Chemistry, 42, 2296-2301.

Coronado, S.A., Trout, G.T., Dunshea, F.R., and Shah N.P. (2002). Antioxidant effects of rosemary extract and whey powder on the oxidative stability of wiener sausages during 10 months frozen storage. Meat Sci., 62, 217-224.

Cuvelier, M., Berset, C., and Richard, H. (1994). Antioxidant constituents in sage (Salvia officinalis). Journal of Agricultural and Food Chemistry, 42, 665-669.

Dahl, L. K. (1972). Salt and hypertension. The American Journal of Clinical Nutrition, 25 (2), 231-244.

Dal Bosco, A., Castellini, C., Bianchi, L., and Mugnai, C. (2004). Effect of dietary a-inolenic acid and vitamin E on the fatty acid composition, storage stability and sensory traits of rabbit meat. Meat Sci., 66, 407-413.

De La Torre, A., Debiton, E., Juaneda, P., Durand, D., Chardigny, J., Barthomeuf, C., Bauchart, D., and Gruffat D. (2006a). Beef conjugated linoleic acid isomers reduce human cancer cell growth even associated with other beef fatty acids. Br. J. Nutr., 95, 346-352.

De La Torre, A., Gruffat, D., Durand, D., Micol, D., Peyron, A., Scislowski, V., and Bauchart, D. (2006b). Factors influencing proportion and composition of CLA in beef. Meat Sci.,73,258-268.

De Vuyst, L., Falony, G., and Leroy, F. (2008). Review: Probiotics in fermented sausages. Meat Science, 80, 75-78.

Decker, E. A., Livisay, S. A., and Zhou, S. (2000). Mechanisms of endogenous skeletal muscle antioxidants: chemical and physical aspects. In E. A. Decker, C. Faustman, and C. J. Lopez-Bote (Eds.), Antioxidants in muscle foods (pp. 25–60). New York: Wiley- Interscience.

Desmond, E., Troy, D.J., and Bucley, J. (1998). Comparative studies on non-meat ingredients used in the manufacture of low-fat burgers. J. Muscle Foods, 9, 221-224.

Desmond, E. (2006). Reducing salt: A challenge for the meat industry. Meat Science, 74 (1), 188-196.

Desmond, E. (2007). Reducing salt in meat and poultry products. In D. Kilcast, and F. Angus (Eds.), Reducing salt foods (pp. 233–255). Cambridge, UK: Woodhead Publishing Ltd.

Dhiman, T. R., Nam, S. H., and Ure, A. L. (2005). Factors affecting conjugated linoleic acid content in milk and meat. Critical Reviews of Food Science and Nutrition, 45, 463-482.

Di Bernardini, R., Mullen, A. M., Bolton, D., Kerry, J., O'Neill, E., and Hayes, M. (2012). Assessment of the angiotensin-I-converting enzyme (ACE-I) inhibitory and antioxidant activities of hydrolysates of bovine brisket sarcoplasmic proteins produced by papain and characterization of associated bioactive peptidic fractions. Meat Sci., 90, 226-235.

Diplock, A.T., Aggett, P.J., Ashwell, M., Bornet, F., Fern, E.B., and Roberfroid, M.B. (1999). Scientific concepts of functional food in Europe: Consensus document. Brit. J. Nutr., 81, 1-27.

Dunshea, F. R., D'Souza, D. N., Pethick, D. W., Harper, G. S., and Warner, R. D. (2005). Effects of dietary factors and other metabolic modifiers on quality and nutritional value of meat. Meat Science, 71(1), 8-38.

Eastwood, M.A. (1992). The physiological effect of dietary fiber: an update. Ann. Rev. Nutr., 12, 19-35.

Eberhardt, M.V., Lee, C.Y., and Liu, R.H. (2000). Nutrition-Antioxidant activity of fresh apples. Nature, 405, 903-904.

Enser, M., Richardson, R.I., Wood, J.D., Gill, B.P., and Sheard, P.R. (2000). Feeding linseed to increase the n-3 PUFA of pork: Fatty acid composition of muscle, adipose tissue, liver and sausages. Meat Sci., 55, 201-212.

Erdmann, K, Cheung, B.W.Y., and Schroder, H. (2008). The possible roles of food-derived bioactive peptides in reducing the risk of cardiovascular disease. J. Nutr. Biochem., 19, 643-654.

Erkkila, S., Petaja, E., Eerola, S., Lilleberg, L., Mattila-Sandholm, T., and Suihko, M. L. (2001a). Flavour profiles of dry sausages fermented by selected novel meat starter cultures. Meat Science, 58, 111-116.

Erkkila, S., Suihko, M. L., Eerola, S., Petaja, E., and Mattila-Sandholm, T. (2001b). Dry sausages fermented by *Lactobacillus rhamnosus* strains. International Journal of Food Microbiology, 64, 205-210.

Erkkila, S., and Petäjä, E. (2000). Screening of commercial meat starter cultures at low pH and in the presence of bile salts for potential probiotic use. Meat Science, 55, 297-300.

Escudero, E., Aristoy, M. C., Nishimura, H., Arihara, K., and Toldra, F. (2012). Antihypertensive effect and antioxidant activity of peptide fractions isolated from Spanish dry-cured ham. Meat Science, 91, 306-311.

Eulitz, K., Yurawecz, M. P., Sehat, N., Fritsche, J., Roach, J. A., Mossoba, M. M., Kramer, J.K., Adlof, R.O., and Ku, Y. (1999).Preparation, separation, and conformation of the eight geometrical cis/trans conjugated linoleic acid isomers 8, 10 through 11, 13–18: 2. Lipids, 34, 873-877.

Fernández-López, J., Sayas-Barberá, M.E., Navarro, C., Marín, F., and Pérez-Alvarez, J.A. (2003). Evaluation of the antioxidant potential of hyssop (*Hyssopus officinalis* L.) and rosemary (*Rosmarinus officinalis L.*) extract in cooked pork meat. J. Food Sci., 68,660-664.

Fine, D.H., Challis, B.C., Hartman, P., and Van Ryzin, J. (1982). Endogenous synthesis of volatile nitrosamines: model calculations and risk assessment. IARC Sci. Publ., 41, 379-396.

Food and Agriculture Organization/World Health Organization of the United Nation (2006). FAO Food and Nutrition Paper 85. Probiotics in food health and nutritional properties and guidelines for evaluation, Rome.

French, P., O'Riordan, E. G., Monahan, F. J., Caffrey, P. J., Vidal, M., Mooney, M. T., Troy, D.J., and Moloney, A.P. (2000). Meat quality of steers finished on autumn grass, grass silage or concentrate based diets. Meat Science, 56, 173-180.

Gao, Y. T., Blot, W. J., Zheng, W., Ershow, A. G., Hsu, C. W., Levin, L. I., Zhang, R., and Fraumeni, J.F. Jr. (1987). Lung cancer among Chinese women. International Journal of Cancer, 40(5), 604-609.

García, M.L., Domínguez, R., Galvez, M.D., Casas, C., and Selgas, M.D. (2002). Utilization of cereal and fruit fibres in low fat dry fermented sausages. Meat Sci., 60,227-236.

Gil, M.D., Bañón, S.J., Cayuela, J.M., Laencina, J., and Garrido, M.D. (2001). Utilización de extractos de plantas como antioxidantes naturales en carney productos cárnicos: revisión. Eurocarne, 101,1-10.

Gnanasambandam, R., and Zayas, J. F. (1992). Functionality of wheat germ protein in comminuted meat products as compared with corn germ and soy proteins. Journal of Food Science, 57, 829-833.

Goldberg, I. (1994). Functional foods: designer foods, pharmafoods, nutraceuticals. Chapman and Hall, New York.

Granado-Lorencio, F., Lopez-Lopez, I., Herrero-Barbudo, C., Blanco-Navarro, I., Cofrades, S., Perez-Sacristan, B., Delagado-Pando G. *et al.* (2010). Lutein-enriched frankfurter-type products: Physicochemical characteristics and lutein in vitro bioaccessibility. Food Chemistry, 120(3), 741-748.

Grigelmo-Miguel, N., Abadía-Serós, M.I., and Martín-Belloso, O.A. (1999). Characterisation of low-fat high-dietary fiber frankfurters. Meat Sci, 52,247-256.

He, R., Alashi, A., Malomo, S. A., Girgih, A. T., Chao, D., Ju, X., and Aluko, R. E. (2013). Antihypertensive and free radical scavenging properties of enzymatic rapeseed protein hydrolysates. Food Chemistry, 141(1), 153-159.

Hipkiss, A. R., and Brownson, C. A. (2000). A possible new role for the anti-aging peptide carnosine. Cell and Molecular Life Science, 57, 747-753.

Hipkiss, A. R., Preston, J. E., Himsworth, D. T., Allende, L., Abbott, N. J., and Michaelis, J. (1998). Pluripotent protective effects of carnosine, a naturally occurring dipeptide. Annals of the New York Academy of Sciences, 854, 37-53.

Ho, K. G., Wilson, L. A., and Sebranek, J. G. (1997). Dried soy tofu powder effects on frankfurters and pork sausage patties. Journal of Food Science, 62, 434-437.

Hoffmann, M., Waszkiewicz-Robak, B., and Œwiderski, F. (2010). Functional food of animal origin. Meat and meat products. Nauka Przyr. Technol., 4, (5) 63.

Hollman, P.C. (1997). Relative bioavailability of the antioxidant flavonoid quercetin from various foods in man. FEBS Lett., 418, 152-156.

Hongsprabhas, P., and Barbut, S. (1999). Effect of pre-heated whey protein level and salt on texture development of poultry meat batters. Food Research International, 32, 145-149.

Houben, J. H., Van Dijk, A., Eikelenboom, G., and Hoving-Bolink, A. H. (2000). Effect of dietary vitamin E supplementation, fat level and packaging on colour stability and lipid oxidation in minced beef. Meat Science, 55(3), 331-336.

Hur, S.J., Lim, B.O., Park, G.B., and Joo, S.T. (2009). Effect of various fiber additions on lipid digestion during in *vitro* digestion of beef patties. J. Food Sci., 74(9), C653-C657.

Hur, S.J., Ye, B.W., Lee, J.L., Ha, Y.L., Park, G.B., and Joo, S.T. (2004). Effects of conjugated linoleicacid on colour and lipid oxidation of beef patties during cold storage. Meat Sci., 66,771-775.

IFIC. (2006). International food information Council Qualified health claims consumer research projects executive summary. http: //ific.org/research/qualhealthclaimsres.cfm

Jahreis, G., Vogelsang, H., Kiessling, G., Schubert, R., Bunte, C., and Hammes, W. P. (2002). Influence of probiotic sausage (*Lactobacillus paracasei*) on blood lipids and immunological parameters of healthy volunteers. Food Research International, 35, 133-138.

Janssen, F. W., de-Baaij, J. A., and Hagele, G. H. (1994). Heat-treated meat-products: Detection of modified gluten by SDS-electrophoresis, western-blotting and immunochemical staining. Fleischwirtschaft, 74(168-170), 176–178.

Jiménez-Colmenero, F., Herrero, A., Cofrades, S., and Ruiz-capillas, C. (2012). Meat and Functional foods. In: Y.H.Hui (ed), Hand book of meat and meat processing, 2nd ed. (pp.225-248), Boca Raton: CRC Press, Taylor and Francis group.

Jiménez-Colmenero, F. (2007a). Functional foods based on meat products. In: Handbook of food products manufacturing. Ed. Y. Hui. Wiley, New York, 989-1116.

Jiménez-Colmenero, F., Carballo, J., Cofrades, S. (2001). Healthier meat and meat products: their role as functional foods. Meat Sci. 59, 5-13.

Jimenez-Colmenero, F. (2007b). Healthier lipid formulation approaches in meat-based functional foods. Technological options for replacement of meat fats by non-meat fats. Trends in Food Science and Technology, 18(11), 567-578.

Jo, C., Ho Son, J., Bae Son, C., and Woo Byun, M. (2003). Functional properties of raw and cooked pork patties with added irradiated, freeze-dried green tea leaf extract powder during storage at 4°C. Meat Sci., 64,13-17.

Johnson, I.T., and Southgate, D.A.T. (1994). Dietary fiber and related substances. In: J. Edelman, S. Miller (edt), Food safety series (pp 39-65). London: Chapman and Hall.

Jose, M., Fernández-ginés, Juana Fernández-López, Estrella Sayas-Barberá, Jose, A., and Pérez-alvarez (2005). meat products as functional foods: a review. Journal of Food Science, 70(2), R**37-R43.**

Kaeferstein, F.K., and Clugston, G.A. (1995). Human health problems related to meat production and consumption. Fleisch. Technol., 75, 889-892.

Kahn, D. R., and Cohen, I. (1981). Factor XIIIa-catalyzed coupling of structural proteins. Biochimica Biophysica Acta, 668, 490-494.

Kandaswami, C., and Middleton, E. Jr. (1994). Free radical scavenging and antioxidant activity of plant flavonoids. Adv. Exp. Med. Biol., 366, 351-376.

Kerry, J.F., and Kerry, J.P. (2006). Producing low–fat meat products. In: Williams, C. and Buttriss, J. (ed.) Improving the fat content of foods (pp.336-379). Woodhead Publishing Limited and CRC Press, LLC University College Cork, Cambridge.

Khanal, R.C., and and Olson, K.C. (2004). Factors affecting conjugated linoleic acid content in milk, meat and egg-a review. Pak. J. Natr., 3,82-98.

Kim, E. K., Lee, S. J., Jeon, B. T., Moon, S. H., Kim, B. K., Park, T. K., Han, J. S. (2009). Purification and characterization of antioxidative peptides from enzymatic hydrolysates of venison protein. Food Chem., 114, 1365-1370.

Koutsidis, G., Koutsidis, J. S., Elmore, M. J., Oruna-Concha, M. M., Campo, J. D. W., and Mottram, D. S. (2007). Water-soluble precursors of beef flavour: II. Effect of postmortem conditioning. Meat Science, 79, 270-277.

Krystallis, A., Maglaras, G., and Mamalis, S. (20080. Motivations and cognitive structures of consumers in their purchasing of functional foods. *Food Qual. Prefer.*, 19, 6, 525-538.

Kumar, V., Biswas, A.K., Chatli, M.K., and Sahoo, J. (2010). Effect of banana and soybean hull flours on vacuum packaged chicken nuggets during refrigeration storage. Int. J. Food Sci. Technol., https: //doi.org/10.1111/ j.1365-2621.2010.02461.x.

Kumazawa, Y., Nakanishi, K., Yasueda, H., and Motoki, M. (1996). Purification and characterization of transglutaminase from walleye pollock liver. Fisheries Science, 62, 959-964.

Lantto, R., Plathin, P., Niemisto, M., Buchert, J., and Autio, K. (2006). Effects of transglutaminase, tyrosinase, and freeze dried apple pomace powder on gel forming and structure of pork meat. LWT- Food Science and Technology, 39, 1117–1124.

Li, R., Carpenter, J. A., and Cheney, R. (1998). Sensory and instrumental properties of smoked sausage made with mechanically separated poultry (MSP) meat what protein. Journal of Food Science, 63, 1–7.

Di Luccia, A., Picariello, G., Cacace, G., Scaloni, A., Faccia, M., Liuzzi, V., Alviti G, and Spagna Musso, S. (2005). Proteomic analysis of water soluble and myofibrillar protein changes occurring in dry-cured hams. Meat Science, 69, 479-491.

Lucke, F. K. (2000). Utilization of microbes to process and preserve meat. Meat Science, 56, 105-115.

Mansour, E.H., and Khalil, A.H. (1999). Characteristics of low-fat beef burgers as influenced by various types of wheat fibers. J. Sci. Food Agric., 79,493-498.

Matsui, T., Matsumoto, K., Mahmud, T.H.K., and Arjumand, A. (2006). Antihypertensive peptides from natural resources. In: Advances in Phytomedicine (pp255-271), Oxford: Elsevier, UK, Volume 2.

Mcnaught, C., and Macfie, I. (2001). Probiotics in clinical practice: a critical review of the evidence. Nutr. Res., 21, 343-353.

Mendis, E., Rajapakse, N., and Kim, S. (2005). Antioxidant properties of a radical scavenging peptide purified from enzymatically prepared fish skin gelatin hydrolysate. Journal of Agricultural and Food Chemistry, 53, 581-587.

Mendoza, E., García, M.L., Casas, C., Selgas, M.D. (2001). Inulin as fat substitute in low fat, dry fermented sausages. Meat Sci., 57,387-393.

Minussi, R. C., Pastore, G. M., and Duran, N. (2002). Potential applications of laccase in the food industry. Trends in Food Science and Technology, 13, 205-216.

Mottram, D. S. (1998). Flavour formation in meat and meat products: A review. Food Chemistry, 62, 415-424.

Muthukumarasamy, P., and Holley, R. A. (2006). Microbiological and sensory quality of dry fermented sausages containing alginate-microencapsulated *Lactobacillus reuteri*. International Journal of Food Microbiology, 111, 164-169.

Nagao, K., and Yanagita, T. (2005). Conjugated fatty acids in food and their health benefits. Journal of Bioscience and Bioengineering, 100, 152-157.

Olmedilla-Alonso, B., Jiménez-Colmenero., F., and Sánchez-Muniz, J. (2013). Development and assessment of healthy properties of meat and meat products designed as functional foods. Meat Sci., 95, 919-930.

Park, S. Y., and Chin, K. B. (2011). Antioxidant activities of pepsin hydrolysates of water-and salt-soluble protein extracted from pork hams. Int. J. Food Sci. Technol., 46, 229-235.

Park, Y. J., Volpe, S. L., and Decker, E. A. (2005). Quantitation of carnosine in humans plasma after dietary consumption of beef. Journal of Agricultural and Food Chemistry, 53, 4736-4739.

Perez-Gago, M. B., and Krochta, J. M. (2001). Denaturation time and temperature effects on solubility, tensile properties and oxygen permeability of whey protein edible films. Journal of Food Science, 66, 705-710.

Porcella, M. I., Sanchez, G., Vaudagna, S. R., Zanelli, M. L., Descalzo, A. M., Meichtri, L. H., Gallinger, M.M., and Lasta, J.A. (2001). Soy protein isolate added to vacuum-packaged chorizos: Effect on drip loss, quality characteristics and stability during refrigerated storage. Meat Science, 57, 437-443.

Pritchard, P. E., and Brock, C. J. (1994). The glutenin fraction of wheat protein: The importance of genetic background on its quantity and quality. Journal of Science and Food Agriculture, 65, 401-406.

Pryor, W. (2000). Vitamin E and heart disease: basic science to clinical intervention trials. Free Radic. Biol. Med., 28 (1), 141-164.

Qihe, C., Guoqing, H., Yingchun, J., and Hui, N. (2006). Effects of elastase from a Bacillus strain on the tenderization of beef meat. Food Chemistry, 98, 624-629.

Raes, K., De Smet, S. Demeyer, D. (2004). Effect of dietary fatty acids on incorporation of long chain polyunsaturated fatty acids and conjugated linoleic acid in lamb, beef and pork meat: a review. Animal Feed Science and Technology, 113, 199-221.

Realini, C. E., Duckett, S. K., Brito, G. W., Dalla Rizza, M., and De Mattos, D. (2004). Effect of pasture vs. concentrate feeding with or without antioxidants on carcass characteristics, fatty acid composition, and quality of Uruguayan beef. Meat Science, 66, 567-577.

Ruusunen, M., and Puolanne E. (2005). Reducing sodium intake from meat products. Meat Sci., 70(3), 531-541. doi: 10.1016/j.meatsci.

Ruusunen, M., Vainionpaa, J., Puolanne, E., Lyly, M., Lahteenmaki, L., Niemisto, M., and Ahvenainen, R. (2003). Effect of sodium citrate, carboxymethyl cellulose and carrageenan levels on quality characteristics of low-salt and low-fat bologna type sausages. Meat Science, 64(4), 371–381.

Ryan, J. T., Ross, R. P., Bolton, D., Fitzgerald, G. F., and Stanton, C. (2011).Bioactive peptides from muscle sources: meat and fish. Nutrients, 3(9), 765-791.

Saiga, A., Tanabe, S., and Nishimura, T. (2003). Antioxidant activity of peptides obtained from porcine myofibrillar proteins by protease treatment. Journal of Agricultural and Food Chemistry, 51, 3661–3667.

Salminen, S., Laine, M., von Wright, A., Vuopio-Varkila, J., Korhonen, T., and Mattila- Sandholm, T. (1996). Development of selection criteria for probiotic strains to assess their potential in functional foods: A Nordic and European approach. Bioscience Microflora, 15, 61-67.

Santosa S., Farnworth E., and Jones P., 2006. Probiotics and their potential health claims. Nutr. Rev. 64, 6: 265-274.

Sarmadi, B., and Ismail, A. (2010). Antioxidative peptides from food proteins: A review. Peptides, 31, 1949-1956.

Schieber A., Stintzing, F.C., and Carle, R. (2001). By-products of plant food processing as a source of functional compounds–recent developments. Trends in Food Science and Technology, 12,401-413.

Schmid, A., Collomb, M., Sieber, R., and Bee, G. (2006). Conjugated linoleic acid in meat and meat products: A review. Meat Science, 73(1), 29-41.

Scollan, N., Hocquette, J., Nuernberg, K., Dannenberger, D., Richardson, I., Moloney, A. (2006). Innovations in beef production systems that enhance the nutritional and health value of beef lipids and their relationship with meat quality. Meat Sci., 74, 17-33.

Severini, C., De Pilli, T., and Baiano, A. (2003). Partial substitution of pork back fat with extra-virgin olive oil in "salami" products: effects on chemical, physical and sensorial quality. Meat Sci., 64,323-331.

Shahidi, F., and Kamil, Y. V. A. J. (2001). Enzymes from fish and aquatic invertebrates and their application in the food industry. Trends in Food Science and Technology, 12, 435-464.

Sheu, T. Y., and Marshall, R. T. (1993). Micro-encapsulation of lactobacilli in calcium alginate gels. Journal of Food Science, 54, 557–561.

Shimizu, M., Sawashita, N., Morimatsu, F., Ichikawa, J., Taguchi, Y., and Ijiri, Y. (2009). Antithrombotic papain hydrolyzed peptides isolated from pork meat. Thrombosis Research, 123, 753-757.

Simopoulos, A.P. (2002).The importance of the ratio of omega-6/omega-3 essential fatty acids. Biomedicine and Pharmacotherapy, 56(8), 365-379.

Smith, G. C., Hynunil, J., Carpenter, Z. L., Mattil, K. F., and Cater, C.M. (1973). Efficacy of protein additives as emulsion stabilizers in frankfurters. Journal of Food Science, 38, 849-855.

Stanton, C., Desmond, C., Coakley, M., Collins, J. K., Fitzgerald, G., and Ross, P. (2003). Challenges facing development of probiotic-containing functional foods. In E. R. Farnworth (Ed.), Handbook of fermented functional foods (pp. 27–58). Boca Raton, FL: CRC Press.

Steenblock, R.L., Sebranek, J.G., Olson, D.G., and Love, J.A. (2001). The effects of oat fiber on the properties of light bologna and fat-free frankfurters. J. Food Sci., 66(9),1409–1415.

Tanabe, H., Yoshida, M., and Tomita, N. (2002). Comparison of the antioxidant activities of 22 commonly used culinary herbs and spices on the lipid oxidation of pork meat. Animal Science Journal, 73, 389-393.

Tang, S., Kerry, J.P., Sheehan, D., Joe Bukley, D., Morrissey, P.A. (2001). Antioxidative effect of added tea catechins on susceptibility of cooked red meat, poultry and fish patties to lipid oxidation. Food Res. Int. 34,651-657.

Tanpaichitr, V., and Leelahagul, P. (1993). Carnitine metabolism and human carnitine metabolism. Nutriton, 9, 246-254.

Toldrá, F. (1998). Proteolysis and lipolysis in flavour development of dry-cured meat products. Meat Science, 49, S101–S110.

Toldrá, F., Aristoy, M. C., and Flores, M. (2000). Contribution of muscle aminopeptidases to flavour development in dry-cured ham. Food Research International, 33, 181-185.

Torrisen, O. (2000). Dietary delivery of caretonoids. In: E. Decker, C. Faustman, C. Lopez-Bote (Eds.), Antioxidants in muscle foods – nutritional strategies to improve quality (pp.289-314). Wiley, New York.

Tseng, T. F., Liu, D. C., and Chen, M. T. (2000). Evaluation of transglutaminase on the quality of low-salt chicken meat-balls. Meat Science, 55, 427-431.

Tyoppone, S., Petaja, E., and Mattila-Sandholm, T. (2003). Bioprotectives and probiotics for dry sausages. International Journal of Food Microbiology, 83, 233-244.

Udenigwe, C. C., Ashton, H. (2013).Meat proteome as source of functional biopeptides, *Food Research International*, 54, 1021–1032.

Udenigwe, C. C., and Aluko, R. E. (2012). Food Protein-Derived Bioactive Peptides: Production, Processing, and Potential Health Benefits. Journal of Food Science, 77(1), R11-R24.

Vekiari, S. A., Oreopoulou, V., Tzia, C., and Thomopoulos, C. D. (1993). Oregano flavonoids as lipid antioxidants. Journal of American Oil Chemistry Society, 70, 483-487.

Velasco, S., Caneque, V., Lauzurica, S., Perez, C., and Huidobro, F. (2004). Effect of different feeds on meta quality and fatty acid composition of lambs fattened at pasture. Meat Sci., 66, 457-465.

Vescovo, G., Ravara, B., Gobbo, V., Sandri, M., Angelini, A., Della Barbera M., Dona, M., Peluso, G., Calvani, M., Mosconi, L., and Dalla Libera, L. (2002). Carnitine: a potential treatment for blocking apotosis and preventing skeletal muscle myopathy in heart failure. American Journal of Physiology, 283, C802–C810.

Weiss, J., Monika Gibis, Valerie Schuh, Hanna Salminen Jochen Weiss, Monika Gibis, Valerie Schuh, Hanna Salminen. (2010). Advances in ingredient and processing systems for meat and meat products. Meat Science, 86, 196-213.

WHO. (2003). Diet, nutrition and the prevention of chronic disease, Report of a Joint WHO/FAO Expert Consultation. WHO Technical Report Series: World Health Organization, Geneva.

Wood, J. D., Richardson, R. I., Nute, G. R., Fisher, A. V., Campo, M. M., Kasapidou, E., Sheard, P. R., and Enser, M. (2003). Effects of fatty acids on meat quality: A review. Meat Science, 66(1), 21-32.

Xiong, Y. L., Agyare, K. K., and Addo, K. (2008). Hydrolyzed wheat gluten suppresses transglutaminase mediated gelation but improves emulsification of pork myofibrillar protein. Meat Science, 80, 535-544.

Yang, C. S., Chung, J. Y., Yang, G. Y., Chhabra, S. K., and Lee, M. J. (2000). Tea and tea polyphenols in cancer prevention. Journal of Nutrition, 130, S472-S478.

Yetim, H., Muller, W. D., and Eber, M. (2001). Using fluid whey in comminuted meat products: effects on technological, chemical and sensory properties of frankfurter type sausages. Food Research International, 34, 97-101.

Yilmaz, I. (2004). Effects of rye bran addition on fatty acid composition and quality characteristics of low-fat meatballs. Meat Sci., 67,245–249.

Young, J. F., Therkildsen, M., Ekstrand, B., Che, B. N., Larsen, M. K., Oksbjerg, N., Stagsted, J. (2013). Novel aspects of health promoting compounds in meat. Meat Sci., 95, 904-911.

Yue X. (2001). Perspectives on the 21st century development of functional foods: Bridging Chinese medicated diet and functional foods. Int. J. Food Sci. Technol., 36, 229-242.

Zhang, W., Xiao, S., Samaraweera, H., Lee, E. J., and Ahn, D. U. (2010). Improving functional value of meat products. Meat Science, 86, 15-31.

Zhong, R. Z., Tan, C. Y., Han, X. F., Tang, S. X., Tan, Z. L., and Zeng, B. (2009). Effect of dietary tea catechins supplementation in goats on the quality of meat kept under refrigeration. Small Ruminant Research, 87, 122-125.

Zhou, G. H., and Zhao, G. M. (2007). Biochemical changes during processing of traditional Jinhua ham. Meat Science, 77, 114-120.

Chapter 18

Sensory Quality and its Improvement in Meat and Meat Products

G.V. Bhaskar Reddy[1]*, Param Debberma[2] and Pradeep Kumar Singh[3]

[1]Department of Livestock Products Technology, College of Veterinary Science, Sri Venkateswara Veterinary University, Tirupathi – 517 502, A.P.
[2]Department of Livestock Products Technology, College of Veterinary Science and Animal Husbandry, Bodhjung Nagar, Agartala – 799 008, Tripura
[3]Department of Livestock Products Technology, College of Veterinary Science and Animal Husbandry, Rewa, M.P.
**e-mail: vbreddylpt@gmail.com*

ABSTRACT

Sensory evaluation of meat includes the use of either trained sensory panellists to rate eating quality differences or untrained, consumer sensory evaluation to determine consumer preference or acceptability of fresh meat and processed meat products. This chapter will discuss the fundamental science regarding the human senses, perceptions of sensory evaluation parameters, importance of sensory quality of meat and meat products, various intrinsic, extrinsic and technological factors affecting the sensory quality of meat products, description of sensory evaluation parameters, consumer's preferences regarding various species of fresh meat and processed meat products. The discussion here will also focus on different strategies for improvement of sensory quality of meat products, sensory testing facilities and environment, preparation of samples for sensory evaluation, various sensory evaluation methods, lexicons of different sensory parameters, consumer sensory panel evaluation, instrumental measures of textural properties of meat products and analysis of sensory data.

Keywords: *Sensory quality, Meat and meat products, Quality improvement, Sensory characteristics.*

1. Introduction

Sensory evaluation is a common and very useful tool in quality assessment of meat and meat products. It makes use of the senses to evaluate the general acceptability and quality attributes of the products. "Sensory evaluation" refers to scientific evaluation of food products through the application of the human senses and it is defined as a scientific method used to evoke, measure, analyse and interpret those response to products as perceived through the senses of sight, smell, touch, taste, and hearing. The instruments can analyse only a single component at a time, whereas overall impression and taste of meat products can be assessed only by human sensory perception. The different components involved in the sensory perception of the meat and meat products are appearance and colour, flavour, texture and tenderness and juiciness.

- ☆ *Sense of sight* is used to evaluate the "general appearance" of the product such as "colour", "size", "shape" *etc.*
- ☆ *Sense of smell* for the "odour".
- ☆ *Sense of taste* for the "flavour" which includes the four basic tastes "sour", "sweet", "bitter" and "salty".
- ☆ *Sense of touch* for the "texture" either by "mouth feel" or "finger feel".

2. Importance of Sensory Evaluation of Meat and Meat Products

Meat and meat products have properties that are related to the five senses of taste, smell, sight, feel and sound. The four basic tastes of salty, sweet, bitter, and sour can be readily identified in different meat products. Smell is a very important sensory property, as the detection of aromatics and/or odours by the olfactory nerve comprises the major components of muscle food flavour. Texture or feel of food influences perceptions of acceptability. The tenderness of meat products has been shown to affect consumer acceptability just as the detection of mouth-coating or a residual substances, usually fat, in the mouth and throat after the consumption of high fat meat products can influence human perception of acceptability or unacceptability.

Sensory evaluation is very essential for development of new product; to bring up an opportunity for improving the lacunae in the product; performance of the product against its competitor in the market can be evaluated; it can be used as a consumer acceptability index to determine the consumer's preference for a particular product; it assists in determining the shelf life of the product and suitability of packaging material; it gives an idea about raw material incorporation and its suitability in the formulation; it helps to visualize parameters of the product which are better than any other product; it is useful for research and product analysis evaluating the eating quality of foods and is nowadays frequently used both within the food industry and research; it can be used in quality control (of *e.g.* raw material, process and product), product development (copying competing products, product improvement) and shelf life evaluation.

3. Factors Affecting the Sensory Quality

Factors affecting the overall sensory quality of meat can be grouped into: internal or genetic factors (species, breed, genotype, sex, age and individual, *etc.*) and external factors or technological (conditions of animal slaughter and cooling systems of carcasses, the fattening systems, water-holding capacity, hydration capacity, rate of maturation and storage, rate of loss by boiling and roasting, meat resistance and pH value *etc.*).

3.1 Internal or Genetic Factors

Species of Animal: Meat quality and characteristics differ among animal species, even within more similar or homologous groups. These differences are mainly due to the species-dependent adipose tissues.

Breed of Animal: Breed is a clear source of variation in carcass morphology related to fat quantity or meat quality. However, the effect of breed on sensory meat quality, such as pH, colour, texture and sensory characteristics is slight.

Genotype: The genotype has an adequate bearing on the sensory qualities of meat irrespective of the management conditions in pre and post slaughter.

Sex: Gender effects are mainly related to the quantity of fat deposited, deposition site, growth rate and carcass yield. Meat from female is more tender and juicy and there is strong typical flavour in meat of un-castrated animals due higher level of testosterone.

Age: Age plays an important role in defining the degree of succulence of the meat, young animals have juicy meat than the adults, due to fine fibre and higher water content.

Individual: Individual plays significant differences in sensory quality of meat and meat products.

3.2 External Factors or Technological Factors

Conditions and Systems of Cooling of Carcasses: The chilling system employed has a significant impact on several key aspects of animal carcasses, such as microbiology, meat safety, eating quality and even production yield.

The Fattening Systems: Male are superior to female in terms of daily weight gain, hot and cold carcass weights, intra-muscular fat ratios and dripping loss of semitendinosus muscle. At the end of fattening period, generally female lambs reach a lower body weight than the male lambs, but internal fat, pelvic fat and kidney fat proportions of female lambs are higher than that of male lambs.

Water-Holding Capacity: This property affects the retention of vitamins, minerals and salts, as well as the volume of water retained. Muscles that lose water easily are drier and lose more weight during refrigeration, storage, transport and marketing.

Hydration Capacity: Hydration capacity of meat is the property of meat to absorb liquid when submerged in it. The hydration, meat bulk and weight, improves tenderness and juiciness, and weakening of the cohesion force component of muscle fibres. The capacity of hydration is influenced by the same factors as binding capacity or water retention.

Rates of Maturation and Storage: Losses are acquired in meat to lose a certain amount of water and its juice during maturation or storage. In general, meat looses a larger amount of water in aging or storage, have a low coefficient of loss through training. The same attribute depends on the species, breed, sex, age of slaughter, individual, being fattening, like aging and preserving meat (carcass or portions maturing cut) *etc.*

Rates of Loss by Boiling and Roasting: It is a criterion for the expression of water retention capacity for processed meat. Thicker muscle fibers (≥ 67.3 µ) have a delinquency rate by boiling or roasting more than the thin (≤ 64.9 µ). Factors of variation for the rate loss by boiling or roasting is considered to be: species, breed, body region, like muscle, the state of fattening, keeping processes, processing of meat, *etc.*

pH Value: It can be, as a valuable indicator of its quality assessment. Living muscle has a pH value 7.0-7.1 and more, after slaughter, muscle pH changes and after about 12 to 24 hours it reaches at a level of 5.4 to 5.6, the ultimate pH, any variation affects the sensory quality.

Freezing and Storage: Neither the industrial freezing method nor time of frozen storage causes enough change in sensory quality to be detected by a panel test or rejected by consumers. There are only slight differences in tenderness and undesirable odours and flavours caused by freezing, although lipid oxidation increases as freezing rate is decreased and/or storage duration increased.

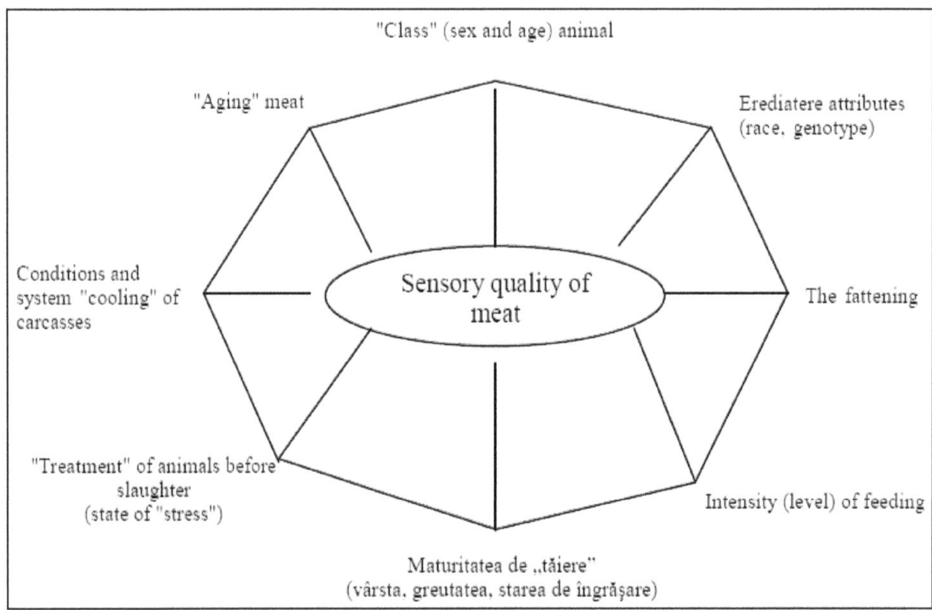

Figure 18.1. Factors Influencing Sensory Characteristics of Meat (Georgescu and Banu, 2000).

Ageing: Meat ageing shows that it has a significant effect on sensory parameters when evaluated by a trained sensory panel. Most sensory attributes are affected by ageing, which makes this factor of utmost importance in terms of quality.

Type of Storage: Diverse packaging techniques (vacuum, modified atmosphere or film) have been developed to maintain the product´s optimum characteristics. Studies on lamb meat recommended early vacuum packed products (1-2 days post-slaughter) to maintain meat properties during long periods (until 29 days) without modifying negatively its textural and sensory characteristics.

Manner and Time of Cooking: Sensory appraisal of meat is conditioned by cooking method and time. When sensory analyses are developed by a trained panel or by consumers under standardized situations (laboratory or hall test), ruminants' meat is usually cooked on a grill, without any condiments, using *Longissimus dorsi* as the muscle selected for the test.

4. The Description of Sensory Parameters

Tenderness: Tenderness of meat is an important indicator of quality, and therefore important to the meat industry as well as consumers. Consumer preferences show a strong demand for tender meats. Tenderness is a measure of force required to compress a substance between molar teeth, or the force necessary to attain a given deformation (Heymann *et al.*, 1990). Tenderness can be either initial or sustained. Initial tenderness is considered the 'first bite' impression and sustained tenderness includes the continued chewing of the sample. The procedure by which panellists evaluate hardness during the first down stroke of mastication appears to change from compression to biting as hardness increases (Boyd and Sherman, 1975). Cooked meat is considered to rank midway between 'hard' and 'soft' foods. Tenderness of meat is subject to species, breed, line, age, state of fattening, meat chilling methodology, heat treatment, *etc.*

Texture: Texture is a complex sensory manifestation of the structure or inner makeup of the product. It can be explained on the basis of textural qualities of meat products. Texture of meat is the relationship between meat components, namely the actual meat, fat, bones, tendons, *etc.* As such, the definition of meat texture, involves muscle tissue composition, thickness, density and structure of muscle fibers, structure and amount of connective tissue (fat, fibrous, cartilage, bone), *etc.* Texture of meat is positively correlated with tenderness, consistency and amount of fat in muscle tissue. In birds, the texture is smooth and sarcolemma of muscle fibers are smooth and thin.

Consistency of Meat: According to Georgescu and Banu (2000), the meat consistency is the resistance by which it opposes the deformation on pressing a finger on its surface. Many factors influence this trait, including: species, age (young animals/birds have meat less consistent than adults), chemical modification post-slaughter stage (chilled meat has the consistency harder than matured), the state of fattening (inter and intra muscular fat imparts more consistency than that of subcutaneous), sex (males have more consistent meat than females), the degree of freshness of meat. Normally, the best meat for consumption should have an elastic consistency and firm.

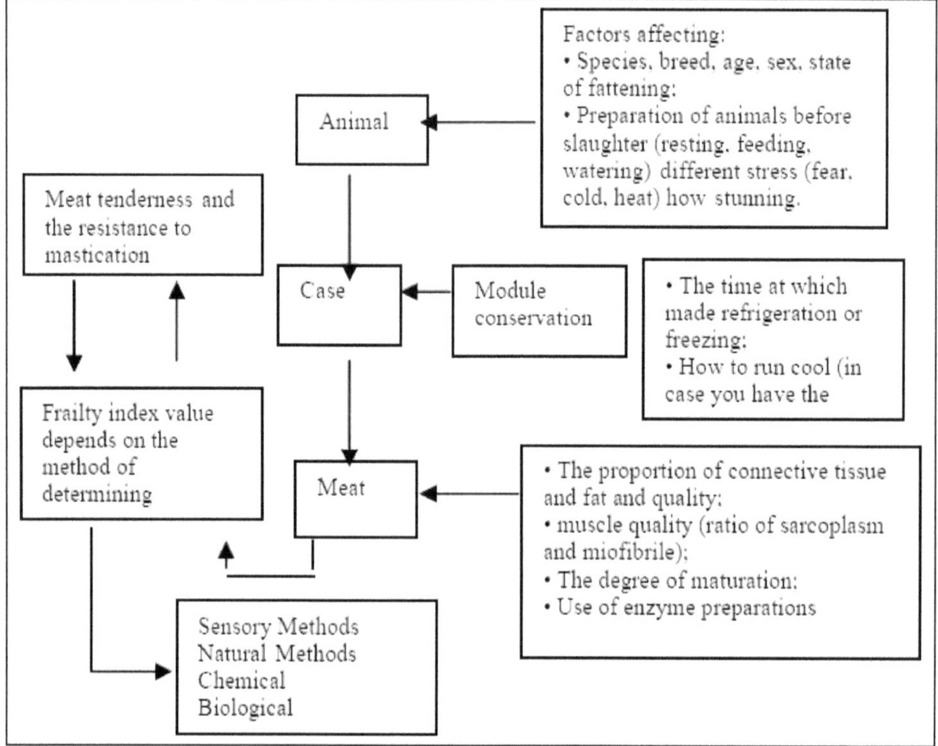

Figure 18.2. Factors Influencing the Tenderness of Meat.

Juiciness: Juiciness of a sample in sensory terms is defined as the degree and amount to which moisture inside the sample is released upon chewing (Heymann *et al.*, 1990). It is often measured as the initial impression of moistness or release of juice during the first bite of the sample. The level of meat succulent is determined by the capacity of the meat to retain a certain amount of intracellular and intercellular juices, this level is influenced by water and fat content in meat. Age plays an important role in defining the degree of succulence of the meat, as always; young animal/bird meat is juicier than old animal/bird meat due to the higher water content in former.

Meat Colour: Colour and other aspects of appearance influence meat and meat product's appreciation and quality, especially by the consumer. Man has subjective standards for the acceptable range and preferred optima for these qualities for almost every food. Intensity of meat colour is given by the quantities of hemoglobin and myoglobin present in meat. Brightness of meat is the power of reflectance of light by the meat, being influenced by bleeding techniques used, ie, the amount of blood removed from the body of the animal. The ratio of muscle tissue and fat, the ratio of the pigment; its reduced and oxidized state, tone colour, *etc.* Colour is expressed in terms of hue, chroma and value in the meat products whereas; appearance is the overall impression of meat size and shape, surface texture and clarity of the product. Colour also indicates the state of freshness of meat and meat products. In general, the colour of fresh meat, except poultry meat, is in the shades of red colour. Remarkable changes in the meat colour occur when fresh meat has been boiled or cooked. It looses its red colour almost entirely and turns to grey (in pressure cooking and boiling) or brown (in roasting, broiling or canning). The reason for this is the destruction of the myoglobin through heat treatment.

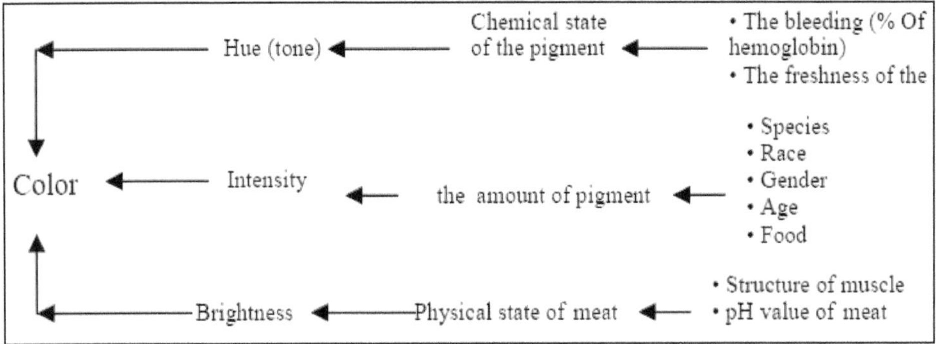

Figure 18.3. Factors Affecting the Colour of Meat.

Appearance: The appearance of meat is the presentation of meat on the outside, is a trait-dependent process, subjective to the degree of freshness. Thus, well chilled meat has a dry film surface, while the un-chilled meat surfaces are wet and poorly presented. Frozen meat is clean, covered with a thin layer of fine ice crystals and the defrosted have wet surface.

Flavour or Aroma of Meat: Flavour has been defined as the sum of perceptions resulting from stimulation of the sensory ends that are grouped together at the entrance of the alimentary and respiratory tracts. For practical sensory analysis the term is restricted to the impressions perceived *via* the chemical senses from a product in the mouth. Flavour is an attribute that embodies the taste and smell of meat, as determined by a number of variables such as sex, age, heredity, growth conditions provided to animals for slaughter, slaughter conditions, cooling and storage of meat obtained, type and amount of additives added to the processed meat, the degree of consumer perceptions and the meaning of the concept of quality meat, *etc.*

Off Flavour: Numerous off flavours have been reported in meat and meat products' sensory tests. Aromas and flavours such as metallic, sour, sweet, musky, and several others have been included in profiles of processed meat products. Jeremiah *et al.* (1990) reported a predominance of sour notes from pale, soft and exudative (PSE) condition in pork loins that may be due to the increased accumulation of lactic acid during post mortem glycolysis. Off flavours were also detected in extreme dark, firm and dry (DFD) condition in pork loins.

Chewiness: Chewiness is defined as the energy required to masticate a solid food product to a state ready for swallowing. It is related to the primary parameters of hardness and cohesiveness. Chewiness can be divided into first chew, chew down, and rate of melt (Meilgaard *et al.*, 1991). The number of chews to disintegrate the sample is determined by counting the number of chews necessary before the sample is swallowed.

Mouth Feel Or Mouth Coating: It is overall impression of meat residue remains in the mouth after swallowing and chewing of meat products. It is determined by binding ability of meat proteins with fat globulins in the meat products. It is a good quality attribute especially in frankfurters, sausages, patties *etc.*

Overall Acceptability: It is an overall acceptability status of any meat food product but it is not an average of all sensory attributes, it may be affected by certain attributes and may not be affected by others. In this attribute we see how a particular meat product impressed taste panellists in totality.

5. Consumer Preferences

Sensory enjoyment of meat is related to several traits as described below: visual appearance and in-mouth perception of both texture and flavour, although preferences for these are not homogenous among consumers. These quality traits depend on several intrinsic and extrinsic factors such as species, genotype, nutrition, age, ante mortem and post mortem treatment, slaughter procedure, storage conditions and ageing time, although not all of the factors affect all of the cues. Thus, sometimes it can be difficult to improve one meat characteristic because it depends on various stakeholders in the meat production chain. Moreover, modification in the production chain to produce a desired characteristic can negatively affect other characteristics. These properties affect the consumer's acceptance and meat preferences and consequently their intention to purchase and willingness to pay.

Visual appearance: Visual appearance characteristics (*i.e.*, colour, fat content, marbling, drip loss) are intrinsic quality cues highly related with consumers' expectations of meat quality and their choice at the point of purchase because these characteristics are used to assess food quality. Meat colour is related with the different forms of the sarcoplasmatic protein myoglobin and it depends on ante mortem (*i.e.* diet, housing, genetic) and post mortem factors (*i.e.* cold chain management and packaging). Colour has been reported to be one of the most important fresh meat characteristics at the point of purchase probably because consumers use inadequate colour as an indicator of spoilage and wholesomeness. Consumers relate red–purple colour with freshness and brown colour with lack of freshness.

Texture: It is a multi-parameter sensory attribute and consumer perception and the acceptability of tenderness and juiciness are the most studied. Tenderness and juiciness are eating quality attributes that positively influence (to a greater or lesser degree) most consumers' preferences in pork, beef and lamb meat. Furthermore, experienced characteristics of the meat such as tenderness and juiciness, as well as taste, are highly correlated with the overall experienced quality, intention to purchase and willingness to pay.

Flavour: Meat flavour is very complex, and it is created mainly when meat is treated thermally because raw meat has only a bloody taste and very little aroma. When cooked, lipids and water-soluble components form several volatile compounds, mainly by means of lipid degradation and Maillard reactions or through reactions between their products. These volatile compounds are the main contributors to meat flavour.

Marketing: Much of the information that consumers receive regarding meat and meat quality is provided through adverts, information campaigns, labels or brands. This information is used by consumers, together with other factors, to create their quality expectations, which in turn influence the choice of the product, purchasing decisions and willingness to pay.

Price: Price is an important extrinsic quality cue related with consumers' purchasing decisions, but though it has a positive effect on expected quality, its relationship with actual eating quality is not clear and it is affected by demographic characteristics. A positive association between low beef prices and low quality because of the discounts offered for beef close to the expiration date has been found. Compared with other cues, price has been reported to affect meat perception after quality, taste, being hormone free and healthiness.

Quality Labelling: Certification is an important attribute that can affect consumer preferences in one way or another, depending on the country. In some countries, most of the consumers prefer to buy meat from known butchers without veterinary stamps over meat that was certified by government veterinarians, probably because of distrust of government food safety enforcement. Additionally, lamb quality through certification was the most preferred followed by quality through labelling and branding and finally by quality through origin. Another important aspect related to quality labels for Muslim consumers that live in non-Muslim countries in Western society is the certified halal label, which assures the halal authenticity of meat and meat products, as well as meat wholesomeness. The importance of the certified halal label is higher in young and more acculturated female Muslims, who are also more willing to pay for it.

6. Strategies for Improvement of Sensory Quality

Improving Colour of the Meat: Thus, it is very important to improve colour stability because it will increase the shelf life of meat and meat products by increasing the time that meat will be visually accepted by consumers at the point of purchase. It can be increased by feeding antioxidants, packing in an atmosphere modified with different gases depending on national legislation (McMillin, 2008) or, in sausages or ground meat, by adding antioxidants that minimize and delay the transformation of myoglobin into metmyogloblin, which is responsible for the brownish colour.

Reducing Fat of the Meat: In pork, one option to reduce fat and increase performance is the use of genotypes with a mutation in the ryanodine (Ryr1) gene (known as halothane gene), but that usually decreases meat quality by increasing drip losses and the incidence of PSE meat (Huff-Lonergan, 2009). Marbling content is dependent on several factors such as breed, diet, sex and weight. Marbling is a visual appearance characteristic of the meat that is considered less important than colour and fat content in pork, although some studies showed marbling to be the most important cue in consumers' perception of quality and in intention to buy (Papanagiotou *et al.*, 2013). In general, marbling has been reported to negatively affect consumer preferences and the acceptability of red meat (Moeller *et al.*, 2010).

Improving Tenderness of the Meat: It is possible to improve the tenderness of meat by optimizing production system factors such as feeding or genetics. However, post mortem factors are crucial in determining tenderness and, if those factors are not optimized, the *in vivo* strategies are useless. Carcass refrigeration after slaughter, hot carcass hanging, ageing time, as well as cooking procedure and temperature are post mortem factors that strongly affect the tenderness of meat (Ngapo *et al.*, 2013). Thus to satisfy consumers' demands it is important to optimize all of these parameters to obtain improved tenderness depending on species and muscle (Hunt *et al.*, 2014).

Improving Flavour of the Meat: Alterations during storage and serving conditions affect the flavour. Flavour depends on intrinsic and extrinsic factors (*i.e.*, species, genetics, sex, feeding regimen, and management practices). A meat's flavour is characteristic of the species. Lamb is differentiated from beef and pork because of its flavour intensity and aroma (Matsuishi *et al.*, 2004), and those people who differentiated among the types of meat scored lamb's flavour stronger and with lower palatability. Lamb flavour differs with age of the animal, feeding and genetics, among other factors. For instance, the odour and flavour of lamb from grass-fed or grain-fed animals is related to the incorporation of n-3 and n-6 polyunsaturated fatty acids into muscle. Meat from lambs fed on pasture or older lambs usually has an intense mutton odour and flavour, probably due to its higher content of α-linolenic acid and its oxidation products, and may also have a strange and rancid flavour that some consumers probably dislike because it is not familiar. Similarly beef odour and flavour, and its acceptability or preference by consumers, are affected by genetics and feeding.

7. Sensory Testing Facilities

7.1 Sensory Testing Environment

The methods used for preparing and presenting samples to the panel require decisions about the testing environment, number of sessions, and physical condition of the samples. The validity of sensory panel results is partially dependent on control of various factors within the testing environment. The sensory testing facilities should be in an accessible location with sufficient space, temperature and humidity control, and freedom from noise and odours. There should be a slight positive pressure within the sensory room to prevent cooking odours from entering the room. Sensory panel booths are commonly used for product evaluations as they maximize the ability to control the testing environment (lighting, temperature, food odours, noise, *etc.* Incandescent, fluorescent, or both incandescent and fluorescent lighting can be used in the testing area. Whichever approach is chosen, the lighting should be uniform. The use of a dimmer switch is desirable because it allows for a variety of intensities ranging from 70 to 80 foot candles, which is typical in an office area, to higher levels of 110 foot candles.

7.2 Preparation and Presentation of Samples to the Panel

Preparation of Sensory Samples: Selection of sample preparation method and serving size should be determined based on project objectives and the amount of variation between and within treatments. It is critical that each panellist receive a standardized amount of each sample. Standardization of samples should be not only by weight or dimensions but also by temperature.

Trained Panel Evaluations: In order to account for the moderate to sometimes high degree of variability between and within treatments, meat samples often are cut into cubes, and each panellist receives two to three cubes from different locations within the piece of meat. For steaks, chops, and roasts, cubes that are 1.27 cm × 1.27 cm × the thickness of the cooked cut are suggested. If cooking procedures result in variation in cut thickness and charred surfaces, however, the thickness dimension should be standardized and cooked surfaces removed from cuts.

Consumer Panel Evaluations: During the normal eating experience, consumers assess the juiciness of the meat visually as well as get an initial impression of the tenderness as they cut the sample into bite-sized pieces. Therefore, when conducting consumer tests, it is best to serve samples that are large enough for the panellist to cut in order to provide a more accurate representation of the actual consumer eating experience. The serving size should be standardized. Location effects within a sub-primal should be randomized. The effect of location can then be included in the analysis of variance (ANOVA) model, usually as a random effect.

Sample Presentation: Standard presentation procedures need to be followed to ensure that all panellists receive samples at the most appropriate and consistent temperature for the attributes being measured. The minimum recommended serving temperature for meat is 60°C and under most conditions should be adhered to. At this temperature, especially for poultry, maximum volatile aromatics will be detected. For consumer studies designed with endpoint temperatures to meet the consumer preferences, however, lower serving temperatures may be needed. Under these conditions, it is recommended that meat be held at 49°C for no more than 20 minutes in a glass, covered container to not affect the sensory properties for steaks and roasts. If samples are at 49°C when cut, placed in the serving container, and served, the sample should be about 38°C. Ideally, the samples will be served immediately after being cut, with panellists receiving cubes from various locations within the steak or, in the case of larger portions for consumer panellists, the same steak portion for all treatments. If samples need to be held prior to serving, preliminary tests need to be conducted to assure that holding methods do not affect colour, tenderness, juiciness, or flavour of the samples.

Order of Sample Presentation: Every sample should be served in each serving order an equal number of times to reduce any bias related to serving position. Furthermore, every sample should be served before and after every other sample an equal number of times in order to nullify any bias related to carry-over effects. William's Square designs are one way to achieve both types of balance. For some studies with a large number of treatments, this may not be possible. Order should be randomized using a random number generator, and order should be analyzed as a random effect in the model.

For trained descriptive panels, first sample bias and variation associated with evaluations conducted on different days can be an issue. A standardized warm-up sample–usually a sample that represents the typical product being evaluated in the study–should therefore be served to the panellists at the initiation of the sensory session and then discussed. This warm-up sample will assist in standardization among panellists, improve panellists' concentration, increase panellists' confidence, and increase calibration of panellists before initiating the sensory session. The sensory leader can use the warm-up sample as a tool to address panel drift and lack of motivation, to increase panellist confidence, and to remove prior environmental factors that can influence the sensory verdict on a given day.

Number of Samples Per Session: The number of samples that should be presented in a given session depends upon product characteristics, experience of the panellists, sensory and mental fatigue and number of attributes to be measured per sample.

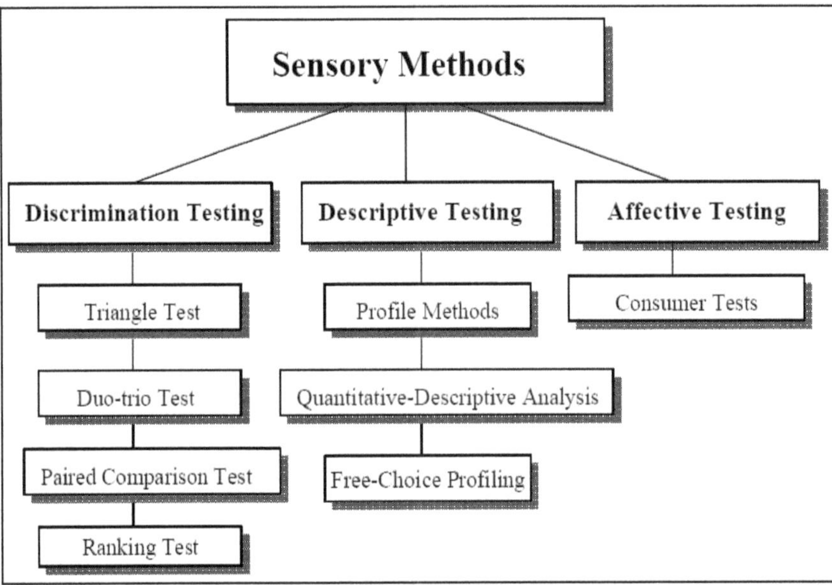

Figure 18.4. Overview of Sensory Evaluation Methods.

8. Sensory Evaluation Tests

The first step in evaluating the sensory properties of muscle foods is to determine the type of trained sensory panel needed to meet the objectives of the study. Two basic types of sensory panels are used in the evaluation of muscle foods – Difference testing and Descriptive attribute testing.

8.1 Difference Testing

Difference testing can be categorized into- overall differences tests and attribute difference tests.

8.1.1 Overall difference Tests

These tests provide an avenue to evaluate if a sensory difference exists between samples. Commonly used overall difference tests are triangle tests, two-out-of-five test, duo-trio test, simple difference test, "A"-"not A" test, difference-from-control test, sequential test and similarity tests.

Triangle Test: This test involves presenting three coded samples (using three-digit random numbers) to the sensory panellist where one sample is different from the other two samples. The panellist is asked to identify the sample that is different. The number of correct responses is counted and the probability (the probability level has to be determined) that the sensory panellists can determine a difference is identified.

Two-Out-of-Five Test: Sensory panellists are presented with five coded samples and are asked to identify the two samples that are similar. As with the triangle test, the number of correct responses is counted and the probability that the two samples differ is determined. This test is very effective when sensory fatigue is not an issue. With meat samples, sensory fatigue usually occurs fairly rapidly, as meat samples consist of moderate levels of flavour attributes, and lingering after tastes can be apparent. Therefore, the two-out-of-five test is not commonly used to discriminate flavour differences between two meat samples.

The Duo-Trio Test: It provides the sensory panellists a one in two chance of identifying the correct samples. The sensory panellists are given a reference sample to evaluate. Then two coded samples are presented and the sensory panellists are asked to identify the coded sample that is the same as the reference. The correct number of replies is determined and the probability that the two samples are difference is determined.

Simple Difference Test: Sensory panellists are given two samples and asked whether the samples are different or similar. Sensory panellists can be presented with either a single pair or any combination of the two pairs (1 then 2, 2 then 1, 1 then 1 and 2 then 2) for the two products being evaluated. The number of responses identified as the same and different are counted for matched and unmatched pairs, giving a total of four numbers. The placebo effect is compared to the treatment effect using X^2 analysis. The X^2 statistic is calculated by

$$X^2 = \frac{\sigma\,(O-E)^2}{E}$$

Where "O" is the observed value and "E" is the expected number. The determined X^2 value is compared to a X^2 table at a probability of 0.05. If the calculated value is greater than the table value, a significant difference exists between the two samples.

The Difference-From-Control Test: This test determines if the samples are different from the control and what the magnitude of the difference is in the test sample(s) from the control. Sensory panellists are presented with a control and the test samples(s). Then, they are asked to evaluate the difference in the test sample(s) from the control using a scale. The mean is calculated from the difference-from-control and the blind controls. A paired t-test can be used to compare the means if only one sample is compared, but if two or more samples are compared to the control, analysis of variance should be used to test for treatment differences.

Sequential Tests: It is a method of utilizing difference tests (such as triangle and duo-trio tests) while minimizing the number of evaluations needed to make a decision on the difference or similarity of two products.

Similarity Testing: This test should be used to test if no perceivable differences exist between two samples, instead of testing if two samples are different. The triangle, duo-trio, and two-out-of-five tests can be used in similarity testing.

8.1.2 Attribute difference Tests

This test asks if a specified attribute differs between samples. Attribute difference tests include directional difference test, pair-wise ranking test, simple ranking tests, multi sample difference test-rating approach, and multi sample difference test-BIB ranking test.

The Directional Difference Test/Paired Comparison Tests: Paired comparison tests are the most common method used to determine the differences in one attribute between two samples. The advantage of this method is that it is simple to administer and for respondents to understand. Two samples are presented to the respondents and they identify whether the samples are the same or different. There is a 50 per cent guessing rate with this test. Paired comparison tests can either be one-sided or two-sided tests. How the question is asked determines if the test is one sided or two sided. For a one-sided test, the question is whether A is more (or less) than B. For one-sided tests, the α, β, and Pd are determined using the same tables as for Duo-Trio tests. A two-sided test asks if A is different than B. When a test is two sided, the distribution is bilateral and is also defined as a Two-sided Directional Difference test. To determine the number of respondents, the α, β, and Pd should be predetermined and the number of respondents determined.

8.2 Descriptive Attribute Testing

Descriptive attribute sensory evaluation is method of quantitating specific sensory attributes of meat products. A family of attributes, called a lexicon, describes the specific sensory attributes in a muscle food that can be used to evaluate the development or change in these attributes. A lexicon of descriptive attributes can be developed for any meat product.

Descriptive attribute sensory evaluation is very useful in identifying the sensory attributes of a product and then monitoring how each attribute is altered with treatment or time. The descriptive attributes of muscle foods can be defined by a trained sensory panel through 1. Ballot development sessions where a trained panel comes to a consensus as to the attributes of a muscle food, 2. Established lexicon and 3. Free profiling where individually trained panellists develop their own descriptors for a meat product.

Meat descriptive analysis has been described and standardized by AMSA. For whole muscle meat products, the major meat descriptive attributes are: juiciness, muscle fibre tenderness, connective tissue amount, overall tenderness (the combined effect of muscle fibre tenderness and connective tissue amount) and flavour intensity.

For processed meat products meat descriptive attribute method can be used, but ballot should be modified to reflect the characteristics of the product being tested. The spectrum method, the Quantitative descriptive analysis method, or the free-choice profiling methods are more commonly used for processed meat. The data are analysed using ANOVA where treatment effects are tested for each attribute on the ballot. Some multivariate analysis, such as principle component analysis, can be used to determine the combined effect of each attribute on muscle food's palatability.

The Spectrum Method: This method is a descriptive attribute method that uses a 0-15 point universal scale where 0= none and 15= extremely intense to evaluate the sensory attributes of processed meat products. The intensity ratings are universal scale are anchored so that equal distance exists between points of the scale, and intensity for a given point does not change across attributes. The specific attributes of a product or the attributes of interest are identified either through ballot development sessions or using a standard lexicons. These data are analysed using ANOVA and common mean separation techniques. Multivariate analysis of variance also can be used to examine the effect of combinations of attributes on treatments.

The Quantitative Descriptive Analysis Method (QDA method): The quantitative descriptive analysis method was developed by the Tragon Corporation as a method to determine differences in sensory attributes of food products. From a large pool of trained panellists, the one who can discriminate differences in sensory attributes of specific products are selected. Panellists are trained as described for descriptive panellists where the product and ingredient references are used to generate descriptive terms. The panel leader acts as facilitator, not an instructor. In the QDA sensory method, a 15 cm line scale is used to quantitate each sensory attributes. Panellists are free to develop their own approach to scoring; scoring is not anchored as in spectrum method. Panellists independently evaluate each product. Panellists do not discuss data, terminology, or sample after each taste session. The only feedback on performance relative to other panel members or any differences between samples that panellist will receive is from the panel leader. Data are entered into a computer, analysed statistically, and reported in graphic representation in the form a 'spider web" where a branch form centre point is the spider web represents the intensity of that attribute. Difference in the spider web between products determines if sensory differences exists.

Free Choice Profile Method: In the free choice profile method, panellists are given the freedom to develop their own sensory descriptors for a product to use the sensory scale at their preferred range. This method eliminates the concerns of uniformly anchoring panellists to a scale and assuring that all panellists are using the scale in the same manner. Panellists training entail a short session where the sensory procedures are explained. Panellists are served the samples and they describe the sensory attributes. The sensory verdicts are analysed using the generalized Procrustes analysis. This method adjusts the data for how the panellists used different parts of the scale and it collates sensory terms that are similar. Sensor terms grouped as being similar and then evaluated by the sensor professionals and the meaning or what was being described is determined. The advantages of this method are that sensory panellists have not been biased by training and are still considered "consumers" and the time required for training, panel performance evaluation, and retraining are almost eliminated. However, the disadvantages are that the sensory professionals play a significant role in interpreting the data and biases can be introduced into the results. Free choice profiling is not commonly used in the sensory evaluation of muscle foods, but it is currently being evaluated as a viable sensory tool.

9. Evaluation of Sensory Characteristics using Consumer Survey Panels

Major factors to be considered in the determination of consumer perceptions of sensory characteristics of meat products are:

- ☆ Consumer evaluation is not a pure marketing test; it is an evaluation of the sensory properties of the muscle foods (taste, smell, sight, feel or sound). Do not train descriptive attribute terms when conducting consumer sensory tests; consumer terms can be identified in focus groups
- ☆ Strict attention must be given to ballot designing and development; it is important that the sensory test examines consumer's sensory perceptions and not if the consumer can properly interpret your directions or questions.
- ☆ Strict attention must be given to the project protocol that defines product handling, preparation, and presentation.

9.1 Consumer Sensory Evaluation Protocol

9.1.1 Scope

While use of a trained panel to assess treatment differences in a set of products is critical in determining if the treatment differences are detectable, it is also critical to know if these differences impact consumer acceptance. Are these differences large enough for consumers to detect? Are they important enough to affect overall acceptance? Consumer testing often is a key part of a study as the fate of a food product always has rested on the acceptance by the consuming public. It is only through careful planning of the consumer test, including choosing the most appropriate test method, that the researcher can get reliable, repeatable results that can be projectable to the real consumer population.

9.1.2 Determining the Test Type

The first step in setting up a consumer study is determining the most appropriate test method to use. The test method should be selected based on the study objectives and on how the results will be used. Consumer panels can be defined as either qualitative or quantitative tests.

Qualitative Tests: Qualitative tests provide a subjective response of a sample from consumers by having those consumers talk about their feelings regarding the sensory properties of a set of products (Meilgaard *et al.*, 2007). Examples of qualitative tests include focus groups or panels; mini groups; diads and triads; and one-on-one interviews.

Quantitative Tests: Quantitative tests are used to determine consumers' sensory perception of products by using a set of questions to measure preference, liking, and impressions of various sensory attributes. Quantitative tests can be conducted using central location testing (CLT) with pre recruited participants; non-prerecruited, such as a mall intercept test; or using home use testing (HUT).

9.1.3 Protocol Design

The number and type of panellists, as well as sample preparation and presentation, can significantly influence consumer responses and, therefore, significantly affect or limit interpretation of the results from the study.

Panel Size: Determining the optimal panel size is influenced by a number of factors, such as the test objectives, test method (CLT, HUT, *etc.*), and the expected difference to be detected. According to Hough *et al.* (2006), the optimum panel size is dictated by the following:

- ☆ d = size of difference to be detected, *i.e.*, 0.4 units vs. 0.8 units on a 9-point hedonic scale.
- ☆ s = standard error of the experiment.

  This value is influenced by several factors such as the amount of product variability, both within treatment and among treatments, as well as product acceptability differences driven by age, gender, and geographical location.
- ☆ α level = probability of Type I error.
- ☆ β level = probability of Type II error.

Panellists Selection: The application of results will depend extensively on the criteria used in selecting test participants. Therefore, selecting the proper panellists is key to obtaining test results that are projectable to the target population. When selecting panellists for a consumer test, the following factors should be considered:

Who is the target population? Testing with target consumers is recommended as the best way to get an accurate measure of consumer acceptance. Is it sufficient to test with company employees or is it necessary to recruit local area residents? Or, in some cases, it is sufficient to test from the general population by recruiting from the targeted user group in either one or multiple national markets in order to more accurately reflect the consumer base.

9.1.4. Ballot Development

The consumer sensory ballot, which is defined as the sensory instrument for consumer testing, is critical to ensure that the sensory professional is accurately testing consumer responses and not biasing the responses with a poorly designed ballot.

Hedonic Scales: The hedonic scale can be shown horizontally or vertically, and scales can be verbally anchored at each point along the scale with nine categories as follows: Like extremely, like very much, like moderately, like slightly, neither like nor dislike, dislike slightly, dislike moderately, dislike very much, dislike extremely.

The purpose of anchoring the scale at each point is to encourage a continuum with equal spaces between each successive increase in like/dislike or preference. These scales range from the anchored, nine-point hedonic; to the end-anchored, nine-point hedonic; to the end- and neutral-anchored, nine-point hedonic; to the non-balanced scale (more or less categories of like in relation to dislike).

10. Instrumental Measures of Tenderness and Textural Properties

Instrumental methods of measuring meat tenderness are–like trained descriptive attribute panels–objective measures of meat tenderness. Conversely, a consumer panel can provide a subjective measure of tenderness. Historically, the trained descriptive attribute sensory panel tenderness rating has been considered the gold standard to which all other measures are compared. It should be clear, however, that both trained sensory panel and instrumental measures give only measures of relative differences in tenderness. They give little indication of the acceptability of a given tenderness measure, other than that derived from associating objective measures with consumer acceptance data. Acceptability of a given level of tenderness can only be determined by the ultimate users, consumers. Historically, the lack of consumer data on meat tenderness has led many researchers to over-interpret their objective measures as indicating whether meat would be considered tough, tender, or acceptable.

11. Conclusion

Sensory evaluation of meat includes the use of either trained sensory panellists to rate eating quality differences or untrained, consumer sensory evaluation to determine consumer preference or acceptability of fresh meat and processed meat products. Sensory evaluation is a common and very useful tool in quality assessment of meat and meat products. It makes use of the senses to evaluate the general acceptability and quality attributes of the products. Many factors affect the sensory quality of meat under the dependence of various sensory factors, hygiene and toxicology and, equally, under the influence of factors related to its nutritional value, and its processing technology. The methods used for preparing and presenting samples to the panel require decisions about the testing environment, number of sessions, and physical condition of the samples. The first step in evaluating the sensory properties of muscle foods is to determine the type of trained sensory panel needed to meet the objectives of the study. Data analysis is a critical component of sensory evaluation. Sensory data are, in the true sense, multivariate. When a human, either trained or consumer, evaluates a meat sample, they utilize multiple senses during the evaluation process. Trained, descriptive attribute sensory panellists are educated to incur sensory input, segment the information into single attributes, and then rate the intensity of each attribute using an anchored, defined scale. Recognition of WOF by the consumer prior to scientific acknowledgement is very likely, as evidenced by consumers who consider reheated meats to have a more off-flavoured nature compared to when eaten just after cooking.

REFERENCES

Boyd, J. V. and Sherman, P. (1975). A study of force compression conditions associated with hardness evaluation in several foods. Journal of Texture Studies 6, 507.

Georgescu, Gh., Banu, C. (2000). Tratat de producerea, procesarea °i valorificarea cãrnii, Ed. Ceres, Bucure°ti.

Heymann, H., Hedrick, H. B., Karrasch, M. A., Eggeman, M. K., and Ellersieck, M. R. (1990). Sensory and chemical characteristics of fresh pork roasts cooked to different endpoint temperatures. Journal of Food Science, 55, 613.

Hough, G., Van Hout, D., and Kilcast, D. (2006). Workshop summary: Sensory shelf-life testing. Food Quality and Preference, 17, 640–645.

Huff-Lonergan, E. (2009). Fresh meat water-holding capacity. In J. R. Kerry, and D. Ledward (Eds.), Improving the sensory and nutritional quality of fresh meat. CRC Press: Woodhead Publishing Limited, pp. 147–160.

Hunt, M. R., Garmyn, A. J., Q Quinn, T. G., Corbin, C. H., Legako, J. F., Rathmann, R. J., Brooks, J. C., and Miller, M. F. (2014). Consumer assessment of beef palatability from four beef muscles from USDA Choice and Select graded carcasses. Meat Science, 98, 1–8.

Jeremiah, L. E., Murray, A. C. and Gibson, L. L. (1990). Meat Science, 27, 305.

Matsuishi, M., Igeta, M., Takeda, S., and Okitani, A. (2004). Sensory factors contributing to the identification of the animal species of meat. Journal of Food Science, 69, S218–S220.

McMillin KW. (2008). Where is MAP Going? A review and future potential of modified atmosphere packaging for meat. Meat Science, 80, 43–65.

Meilgaard, M., Civille, G. V., and Carr, T. T. (1991). Sensory evaluation techniques. CRC Press, Boca Raton, Florida.

Meilgaard, M., Civille, G.V. and Carr, B.T. (2007). Sensory Evaluation Techniques, fourth ed. CRC Press, pp 448.

Moeller, S. J., Miller, R. K., Edwards, K. K., Zerby, H. N., Logan, K. E., Aldredge, T. L., and Box-Steffensmeier, J. M. (2010). Consumer perceptions of pork eating quality as affected by pork quality attribute and end-point cooked temperature. Meat Science, 84(1), 14-22.

Ngapo, T. M., Riendeau, L., Laberge, C. and Fortin, J. (2013). Marbling and ageing–Part 2. Consumer perception of sensory quality. Food Research International, 51, 985–999.

Papanagiotou, P., Tzimitra-Kalogianni, I., and Melfou, K. (2013). Consumers' expected quality and intention to purchase high quality pork meat. Meat Science, 93(3), 449-454.

Chapter 19

Food Safety Issues Related to Genetically Modified (G.M.) Meat Animals

Anurag Pandey[1]* and Ankit Kumar Singh[2]

[1]Assistant Professor, Department of Livestock Products Technology,
[2]Teaching Associate, Department of Veterinary Pharmacology and Toxicology,
PGIVER, Jaipur, Rajasthan
**e-mail: dranuragonline@gmail.com*

ABSTRACT

Genetic modification is the process in which gene of interest is inserted in to a gene sequence to get desired product. Genetically modified meat animals are those animals which are used for food purpose and are genetically modified (GM) to get certain desired response e.g. cattle, pigs, goat and fish. These genetic modifications are done by mainly following two technologies, the first is the technology which causes non heritable changes and another causes heritable changes. These genetically modified foods can show very much potential to solve many of the global problems like hunger and malnutrition. The genetic modification can help to reduce the dependency of human on drugs for various diseases. Genetic modification can help the breeders to bring changes beneficial for human health. Besides lots of benefit of GM food there are certain issues regarding the consumption of GM foods. People are not aware about the GM foods, and the scientist have various doubts for the consumption of GM foods, whether the produced food is fit for the consumption or not, whether the food produced by the genetic modification will be able to solve the problem of hunger in future. There is also a large risk of health related issue associated with GM foods like allergies, toxins or any genetic hazard. The insertion of gene may give rise to new types of viruses or it can also change the enzymatic metabolic pathways.

***Keywords**: Genetically modified food, Modification, Health, Safety.*

1. Introduction

Foods that are produced from organisms in which changes is done by doing changes into their DNA using different methods or technology are known as genetically modified (GM) food products or genetically engineered foods. The genetically modified organisms (GMOs) which are present in the environment are mostly genetically modified plants, animals which constitute only a very less portion of all GMOs. The situation may change in future as there is continuous increase in the number of GM animals. Among them, there is large range of GM fish in the market. GM animals can be divided into two groups; the first group is of the genetic modification which enhances overall performance and therefore, added their importance in an agronomic and/or economic sense. The second group of GM animals has been transformed to generate particular substances in eggs, blood or milk or serve as a medical research model. In this case only specific tissues incorporate the new trait(s), whereas the other parts of the animal will remain genetically unaltered. When focusing on the food safety aspects of GM animals, two points should be considered:

1. The intentional introduction of GM animals in the food production chain
2. The unintentional introduction of GM animals

In recent years regulatory agencies have gained considerable expertise in the field of food safety assessment of GM foods. It can be stated that in recent years several proven strategies are invented that are able to minimize the adverse health effects of GMO-derived animal products on consumer health. Safety assessment strategies address different number of issues which are directly related to the genetic modifications and adverse effect of the genetic modification on the food organisms. Commonly asked information on genetically modified food organisms according to Kuiper *et al.*, 2001 are as following:

- ✰ What is the process of genetic modification?
- ✰ Regarding safety issues of genetically modified food organisms specially its allergenicity
- ✰ Rate of incidence of side effects
- ✰ Gene transfer and recombination effects
- ✰ Role of GM foods in diet

OECD (1993), FAO/WHO (2000) and the US National Research Council (NRC, 2002) have conducted a number of meetings for discussing the safety assessment of GM animal-derived food materials. In meetings it was concluded that the minimal food safety assessment of GM animals should comprise (Kleter and Kuiper, 2002)

- ✰ The molecular characterization of the inserted foreign DNA
- ✰ The safety assessment of the introduced genes and their products
- ✰ Any unintended effects of the insertion of foreign DNA in the organism
- ✰ The effect of disease resistance brought about in transgenic food animals on consumer's exposure to disease-causing agents

Experts evaluating the safety assessment of GM animals are taking advantage from the experience from GM plants as experience in the field of safety assessment of GM animals is still very much inadequate as basic approaches for GM plant materials also apply to GM food animals. GM animals can be developed by two ways: first by doing alterations which are permanent and second by non-heritable transgenic constructs. The former one is often described as "transgenic animals" while the latter techniques is known as "gene therapy". The distinction between heritable and non-heritable modifications is not dependent on the intent of the modification but it is a function of the technology selected for the intended modification.

2. Methodologies for Gene Transfer

2.1. Non-heritable Modifications

Non-heritable modifications are developed in the animal by introducing the gene of interest in a vector, targeting the somatic cells of the animal. Preferentially two types of vectors are of used: those based on viral sequences and those based on transposable elements. Viral based technologies are advantageous because of integrative properties of retroviruses and adenoviruses. Transposons (also known as jumping gene) catalyse its own movement in the animal genome. The risk scenario for both viral and transposon-based vectors also includes concern for recombining with existing viruses in the intended target animals or in humans who are exposed to them, or in other animals that may be exposed to the target animals or their wastes. Recombination increases host ranges *i.e.*, swine viruses becoming capable of infecting humans, virulence (innocuous viruses causing serious illness), and leads to development of entirely new generation, pathogenic viruses.

2.2. Heritable Modifications

GM animals of heritable modifications are produced by insertion of a stable genetic construct into the nuclear chromosomes or in mitochondrial genomes. Production of transgenic animals is done by giving injection of a DNA solution having all the information for the expression of special character to the early embryo. The expression of the character also depends on the cell's internal recombinatory enzymes. Scientists are also using viral vectors or transposon-based vectors for producing heritable traits in GM animals. Animals with transgenic line are produced commonly by two step process. In this process transgenic construct is introduced in early stage embryos and the animal produced by this process is known as mosaic transgenic animal. It is expected that the majority of the cells present in developing embryo will contain the gene of interest, including some germ cells. These animals are considered as "mosaics" as they are made up of two or more genetically distinct cells. These mosaic animals are then further breed and the progeny is tested for the presence of 100 per cent transgenic cell. The animal containing the 100 per cent transgenic cell is then selected and is further breed for the propagation of transgenic line.

3. Genetically Modified Food/Meat Product

Pork: In Canada a project is running since 1999 entitled as "Enviropig", in this project pigs are genetically modified by taking genes from mice and *E. coli*. These GM pigs are able to produce phytase enzyme by which the pigs are able to break the indigestible phosphorus. By the help of phytase enzyme the animal is now able to break phosphorus present in the cereal grains. But in scientific community there is still concern about the environmental impact of these GM pigs and food safety of these pigs for human consumption.

Salmon: salmon is a famous fish food and is a good source of protein, omega 3 fatty acid, vitamin D. The growth rate of GM salmon is more than normal salmon and it grows twice in size. Salmon is genetically modified so that it can be grown in farm also.

Cattle: Chinese scientist in 2011 produced a genetically engineered dairy cow by inserting human being genes and these cows are producing milk resembling same as human breast milk (Gray, 2011). These GM cows are similar to regular cows. Researchers from New Zealand had also developed a GM cow in 2012 producing allergy free milk.

Goat: Goats have been genetically engineered to produce milk containing strong spiderweb-like silk proteins.

4. Need of GM Foods

Until 2050 it is expected that the global population will reach around 9 billion and population of wealthier people with growing appetites is expected to rise up to 100 per cent by the mid century. The agriculture productivity which was increased over the past few years is now begun to decrease. The fast increase in the productivity of rice and wheat yield around the world is also showing signs of slowing down. The GM foods can be a better alternative to solve the problem of worlds hunger and malnutrition problem. The GM food can be also helpful in protecting and preserving environment as it can reduce the dependence on synthetic pesticide and herbicide. Beside all of these advantages there are several challenges in many areas of GM food *viz.* safety testing, regulations, policies and food labelling. GM food is a technology of huge potential which cannot be afforded to be ignored. GM foods are beneficial for both producer and consumer also that is why it is being produced and marketed. Genetic modification can be used to translate a poor quality food in terms of its nutritive and durability quality to a food of great quality.

The weather has become very much unpredictable, this is the reason the crops are also being genetically modified. The GM crops currently present in the market are aimed with an increased level of crop protection by the introduction of resistance against plant diseases caused by insects or viruses or through increased tolerance towards herbicides. Climate is also one of the major problems in the production of crops. Climate change favors the condition for occurrence of disease and insects in crops and also helps in the transfer of disease. The genetically modified crops can easily adopt the sudden changes in the environment. Genetic engineering is helping plant breeders to make more precise changes in the plants drawn from a far greater variety of genes, wild relatives or from different types of organisms.

5. Market of GM Foods

The demand of GM foods in food market over the world is increasing because of its high nutritional value for different age groups. The unusual taste of GM foods over non-GM foods and the safety concern of GM foods are interfering in the demand of GM foods. Greater nutritious content of food, increase in the production quality and lesser uses of drugs are the major factors which are playing role in the increase in the demand of GM foods. Increase in the demand of healthy food all around the year and its easy availability are also helping the demand of GM foods.

The market of GM foods is divided according to type and region. On the basis of type, it is divided as crops, vegetables, fruits and animal products. On the basis of market it is divided as North America, Latin America, Western Europe, Eastern Europe, Asia Pacific, Japan and Middle East and Africa. Few major players of GM food markets are like Monsanto US, DuPont US, Syngenta Switzerland, Bayer Crop Science Germany, Sakata Japan, BASF Gmbh and Group Limagrain France. The international companies are focusing on partnership and collaboration with other companies so that they can increase their product number and also can transport their products in different parts of the world. The global companies are collaborating with the Advanced Research Institutes (ARIs) in the industrial countries such as Brazil, Argentina, China, India, Malaysia and the Philippines for significant R&D program in biotechnology and transgenic crops. In order to fulfil the increasing demand of GM foods companies are also increasing their R&D in various countries such as South Africa, Kenya, Zimbabwe, Mali, Nigeria, Egypt, Uganda *etc.*

6. Human Health Concern

There are lots of issues of concern for GM food on human health but the issues which are mainly discussed during debates are allergenicity, gene transfer and outcrossing.

6.1. Allergenicity

Foods developed either by conventional methods or by biotechnology are good source of allergens. All food allergies are mediated by antigen-specific IgE and are characteristic of type-I reactions. In the case of new proteins being expressed in the GM animal, the allergenic potential of the protein will need to be established. In the case of production of specific well-characterized (medicinal) proteins by the GM animal, it needs to be established whether the post-translational modifications are comparable to the same substances being produced by more traditional sources in order to assess potential altered toxicological or allergenic properties of the newly synthesized proteins (Dyck *et al.*, 2003). The strategies used to address allergenicity rely on following parameter (Metcalfe *et al.*, 1996; FAO/WHO, 2001).

- ☆ Gene source
- ☆ Homology of the sequence
- ☆ Testing of patients' serum to know the allergenicity to the source organism or to sources distantly related

- ☆ Pepsin resistance
- ☆ Prevalence of the trait and
- ☆ Animal models

6.2 Gene Transfer

Large number of endogenous viruses and virus-like sequences are present in host cell (Chakraborty *et al.*, 1994). The DNA construct used to change the genetic make-up of the animal should be considered within an assessment especially if the gene or its promoter is obtained by a viral source since horizontal transfer or recombination may occur. Additionally, bacterial host-derived materials may contain additional sequence fragments unrelated to the target gene. Because such fragments can be heterogeneous in size and sequence, they are difficult to detect, this problem is found with retroviral vectors. The unwanted introduction of gene into the animals can not only cause unintended damage to the genetic makeup of animal but it can also cause generation of new infectious viruses. The horizontally transferred gene by food ingestion may adversely affect the human health, as the transferred DNA may not be completely degraded by the GIT.

6.3 Outcrossing

Transfer of gene from GM animals to the non GM animals in the wild is known as outcrossing this mixing of gene from GM animal to non GM animals may affect the human food safety. Several countries are adopting certain strategies to stop the mixing of gene between GM and non GM animals.

7. Consumer Awareness

In the part of consumer awareness it is very important for consumer to understand what GM food means, as consumers are taking the food through the technology that they barely know. Consumer must be aware about the beneficial and harmful effects of the GM foods. According to some direct consumer surveys it is interpreted that consumer knowledge is very low about the GM food. In a very recent survey done by Rutgers University found that 43 per cent people know that GM products are sold in supermarkets, 26 per cent are believing that they have probably eaten a GM food, 54 per cent people knew very little or nothing at all, and 25 per cent people had admitted that they had never heard of them (Shahla Wunderlich and Gatto, 2015). The results of these studies are showing that less than half of the population knows about the GM food. Awareness of GM foods also varies by country. A cross cultural survey comparing the knowledge of consumers in the United States, Japan, and Italy showed that US consumers were more likely to be at least somewhat familiar with GMOs (40.9 per cent reported being somewhat or very familiar) compared with Italian (just 28.0 per cent) and Japanese (33.3 per cent) consumers (McGarry *et al.*, 2012). This shows that US consumers are more readily accepting the GM foods than other part of the world. There are also some misconceptions about the taste and texture of GM foods like, insertion of a fish gene would not make a tomato taste fishy, eating a GM food may cause toxic or allergic reactions.

8. Safety Issues for GM Food

Several controversies are present at different levels for GM foods, which includes whether the produced food is safe, whether it should be labelled or not and if yes than how, whether this technology is capable of solving world hunger and malnutrition problem now or in the future, and more specifically with respect to its intellectual property and market dynamics and about its environmental effects. Genetic modification increases the risk of "tampering the Mother Nature" which can cause an increase in the number of health related issues. The naturally viruses can recombine with the viral fragment present in the genetically modified organism and there can be formation of new virus (Steinbrecher, 1996).

Safety issues regarding GM foods are related with produced toxins, allergies and genetic hazard. Food hazard issues are divided mainly in three categories according to Conner and Jacobs, 1999 which are as following; gene inserted and their expression in products, pleiotropic effects of gene expression and mutation because of insertion of genes. Regarding the first category it is not always the transferred gene is responsible for the health issues but expression of inserted gene and its product are also important. The new synthesized protein can also be responsible for the allergic reaction.

Genes are responsible for encoding an enzyme for different biochemical pathway. If the change in gene resulted in any pleiotropic effect it will cause changes in various biochemical pathways. Also, the presence of a new enzyme could cause depletion in the enzymatic substrate and subsequent build up of the enzymatic product. The newly expressed enzyme may also lead to change in the direction of one metabolic pathway to another. Changes in the direction of metabolism can cause generation of toxins or it can also cause increase in the concentration of toxins. Mutation can disturb or cause change in the expression of existing genes in a host. Random insertion can cause inactivation of endogenous genes. Many of the genes created produces nonsense products which are either eliminated or selected. However most concern should be given upon activation or up regulation of silent or low expressed genes, because it is possible to activate "genes that encode enzymes in biochemical pathways toward the production of toxic secondary compounds" (Conner and Jacobs, 1999).

Attention should be given towards the GM foods which are commonly causing allergies such as such as milk, eggs, nuts, wheat, legumes, fish, molluscs and crustacean (Maryanski, 1997). The effect of the product should be previously identified before public consumption as the transgenic property of the food is already known to the scientists. Any probable risk like immunological, allergic, toxic or genetic hazardous should be documented if any health concern issue arises. There is a lot of variation between animal's experimental group and it is quite difficult to attain a transgenic food dose in animal providing result comparable to humans (Butler and Reichhardt, 1999). This shows that the initial step for evaluation of food safety will be a Community supported agriculture (CSA) of the GM animal with its traditional counterpart, if available. This approach identifies potential differences between the GM animal-derived food products and its traditional counterpart as the first phase, and then the next phase will be to evaluate the toxicological and nutritional relevance of genetically modified food. As every GM (founder) animal at this moment will have a different genetic constitution with respect to the integration of the genetic construct, the safety evaluation should be carried out on a case-by-case basis, even if the same genetic construct was used for the genetic modification. If homologous recombination will reduce the likelihood of insertion effects in the future, it may become more feasible to come to more harmonized approaches for the safety assessment of GM animals and products thereof. Application of the idea of substantial equivalence allows for analysis of intended and unintended alterations in the GMO and is central to the CSA. The intended changes can be evaluated with knowledge of: the nature and gene construct used in the genetic modification, process of genetic modification, in situ characterization of the genetic modification in the animal, information on animal breeding and propagation of the GM animal, the amino acid sequence of expressed product from the gene construct, the expression rates in different tissues of the expressed product, and traditional toxicological testing of the expressed product.

Depending on the identity of differences detected, further toxicological testing may be necessary to assess the safety and nutritional impact of the observed differences. A few major differences are also observed when comparing the GM animal to the GM plant situation. Firstly, the numbers of GM animals derived from a single GM founder animal will in general be much lower compared to GM plant genetic modification events and numbers available in subsequent plant generations. This will result in less animals being available for the comparative safety assessment. This will have major influence on the reliability of the results of the comparative safety assessment. Knowledge on the natural background variation in animal tissue constituents will even be more important compared to the plant situation as it will be less feasible to obtain statistically significant results from analysis of the GM animals versus the conventional counterpart.

An extra difference is the presence of natural toxins in plant products and the very few cases of animal products that have proved to contain anti-nutritional substances for the consumer. The fact that the physiology of animals has major resemblance to our own physiology may in some aspects make the assessment of (GM) animal-derived food products 'easier'. On the other hand animal-derived food products form an important part of the human diet. Relatively small compositional changes may therefore have considerable effects on the nutritional status of the consumer. With increasing numbers of genetically altered plant and animal derived food products the nutritional aspects, beside the safety aspects, will increasingly gain weight. The recent development in the field of GM animals necessities our approach to maintain our current standard for a safe and nutritious food supply in the light of growing numbers of different (GMO-derived) foods and food ingredients and increasingly complex food supply chains.

REFERENCES

Butler, T. and Reichhardt, T. (1999). Long-term effect of GM crops serves up food for thought. Nature, 398(6729),651–653.

Chakraborty, A.K., Zink, M.A. and Hodgson, C.P. (1994). Transmission of endogenous VL30 retrotransposons by helper cells used in gene therapy. Cancer Gene Therapy, 1,113-8.

Conner AJ, Jacobs JME. (1999). Genetic engineering of crops as potential source of genetic hazard in the human diet. Mutation Research- Genetic Toxicology and Environmental Mutagenesis., 443,223–234.

Dyck M.K., Lacroix D., Pothier F., and Sirard M.A. (2003) Making recombinant proteins in animals – different systems, different applications. Trends in Biotechnology, 21(9),394-399.

FAO/WHO (2000). Safety aspects of genetically modified foods of plant origin. Report of a Joint FAO/WHO Expert Consultation on Foods Derived from Biotechnology, Geneva, Switzerland. Food and Agriculture Organisation of the United Nations, Rome.

FAO/WHO. (2001). Evaluation of Allergenicity of Genetically Modified Foods. Report of a Joint FAO/WHO Expert Consultation on Allergenicity of Foods Derived from Biotechnology 22 – 25 January 2001. http: //www.fao.org/fileadmin/templates/agns/pdf/topics/ec_jan2001.pdf

Gray,R. (2011). "Genetically modified cows produce 'human' milk". The Daily Telegraph, April 2, 2011

Kleter, G.A. and Kuiper H.A. (2002). Considerations for the assessment of the safety of genetically modified animals used for human food or animal feed. Livestock Production Science, 74,275-285

Kuiper, H.A., Kleter, G.A. and Kok, E.J. (2001). Assessment of the food safety issues related to genetically modified foods. Plant Journal, 27, 503-528

Maryanski, J.H. (1997). Bioengineered foods: will they cause allergic reactions? NY: U.S. Food and Drug Administration (FDA)/Centre for Food Safety and Applied Nutrition (CFSAN).

McGarry, W.M., Bertolini, P., Shikama, I. and Berger, A. (2012). A comparison of attitudes toward food and biotechnology in the U.S., Japan, and Italy. Journal of Food Distribution Research, 43: 103–10.

Metcalfe, D.D., Astwood, J.D., Townsend, R., Sampson, H.A., Taylor, S.L. and Fuchs, R.L. (1996). Assessment of the allergenic potential of foods derived from genetically engineered crop plants. Critical Reviews in Food Science and Nutrition, 36,165–186.

National Research Council of the National Academies (U.S.), Committee on Defining Sciencebased Concerns Associated with Products of Animal Biotechnology. (2002) Animal biotechnology: science-based concerns. The National Academies Press, Washington D.C., U.S.

OECD (1993). Safety Evaluation of Foods Derived by Modern Biotechnology: Concepts and Principles. Organisation for Economic Co-operation and Development, Paris, http: //www.oecd.org/pdf/M00034000/M00034525.pdf

Shahla Wunderlich,S and Gatto, K.A. (2015).Consumer Perception of Genetically Modified Organisms and Sources of Information. Advances in Nutrition, 6(6),842-851.

Steinbrecher, R.A. (1996). From green to gene evolution: the environmental risks of genetically engineered crops. Ecologist, 26,273–281

Chapter 20

Regulations and Standards for the Quality and Safety of Meat and Meat Products

H.V. Mohan[1]*, S. Wilfred Ruban[2] and R. Narendra Babu[3]

[1]Assistant Professor, Department of Veterinary Public Health and Epidemiology,
[2]Associate Professor, Department of Livestock Products Technology, Veterinary College, KVAFSU, Bengaluru, Karnataka
[3]Professor, Department of Livestock Products Technology,
Madras Veterinary College, TANUVAS, Chennai, Tamil Nadu
**e-mail: mohanhv@gmail.com*

ABSTRACT

The growing meat industry needs certain guideline in form of regulations and standards so that the industry can follow and the consumer gets access of good quality, safe and wholesome meat. Unless there is a uniformity in standards available, it is difficult for the industry to follow it and also difficult for the monitoring agency to check the quality and standard of meat produced by the industry. Hence, the regulatory body has made the standards following scientific discussions and these are made available for microbial limit, residues etc. Further the role of regulatory body is that the regulations should be followed and be supported by the law so that if the standards are not met then the regulatory agencies can reject the meat consignment with appropriate fine or punishment. The ultimate aim for the whole standards, regulation is to ensure safe and wholesome meat to consumer.

1. Introduction

Safety is a growing concern with the industrialization and globalization of food and food products. In recent years, food safety of animal products is receiving greater attention and importance. Efforts to assure food safety has consequently grown throughout the food supply chain. The effects of population growth, income increase and urbanization will fuel the growth in demand for animal products which would further demand increased emphasis on food safety issues. The role of animal products in ensuring food and nutritional security is important in Indian context. Food-borne diseases and hazards have greatly increased with industrialization of food production and processing units.

The intensive rearing of food animals and poultry production and the increasing trade in raw materials for the feed industry as well as livestock products, demands more attention on risk management by all countries. This calls for effective monitoring and control of undesirable substances in the products which also depends on monitoring and controls applied in primary production, processing and utilization of all ingredients and products at several stages in the production chain. Thus, effective intervention measures and risk management strategies to protect the consumer against the food borne hazards have become important and the role of standards for contaminants in the ingredients and food is important to assure consumer protection. Improved traceability of components of food throughout the food production chain is important though it involves practical difficulties and heavier costs.

Toxic and genetically modified foods need strict laws, however locally processed meat in small retail outlets cannot be subjected to industrial regulations, because they are not centralized producers needing centralized regulation.

Due to these risks in the food market, now it is must for us to know about the food laws, standards and regulations in different countries and the way in which food safety is handled; and in addition the steps being taken in our country to meet the present day needs.

2. WTO

World Trade Organization (WTO) is the international organization dealing with the global trade rules between nations. The mandate of WTO covers trade in goods, services, trade-related investment measures and trade related intellectual property rights. WTO is committed to establish an open and liberal global environment free from any barriers or restrictions encouraging participation of developed and developing countries. Among several WTO agreements, those that are relevant to agricultural products including livestock sector, dairying and animal products are:

- ✰ Agreement on Agriculture (AOA)
- ✰ Sanitary and Phytosanitary (SPS) Agreement
- ✰ Technical Barriers to Trade (TBT) Agreement

3. OIE

Office International des Epizooties (OIE- World Organization for Animal Health) was founded in 1924 in Paris with 28 founding members as a response to a Rinderpest outbreak in Europe arising from cattle undergoing international shipment. The OIE develops the International Animal Health Code which sets out animal health conditions for trade, including definitions of the criteria by which individual animals, herds or areas may be considered free from a particular disease.

With the establishment of WTO, the work of the World Organization for Animal Health has assumed a new distinction through recognition of its role in providing standards, guidelines and recommendations for livestock health and zoonoses within the Sanitary and Phytosanitary Agreement (SPS) of the WTO.

4. Codex Standards

The term Codex Alimentarius is taken from latin and means food code- a code of food standards for all nations. It was established in 1962, when the FAO and WHO of the United Nations recognized the need for international standards for foods to guide the World's growing food industry, food safety and to protect the health of consumers. Codex develops *commodity standards* which are product specific and *general standards* which have across the board application to all foods and are not product specific. A number of codex standards have dealt with safety aspects of animal products. Some important standards with a brief mention of scope and content are as follows:

4.1 List of Some Codex Standards

1. General principles of food hygiene (CAC/RCP 1-1969, Rev. 4-20031)
2. Code of hygienic practice for meat (CAC/RCP 58-2005)
3. Codex standard for luncheon meat (CODEX STAN 89-1981)
4. Codex standard for cooked cured chopped meat (CODEX STAN 98-1981)
5. Recommended international code of hygienic practice for processed meat and poultry products (CAC/RCP 13-1976, Rev 1 (1985)
6. Code of practice on good animal feeding (CAC/RCP 54-2000)
7. Principles for the establishment and application of microbiological criteria for foods (CAC/GL 21-1997)
8. Principles and guidelines for the conduct of microbiological risk assessment (CAC/GL- 30 (1999)
9. General principles for the use of food additives 1 (CAC/MISC I-1972)
10. Code of practice for source directed measures to reduce contamination of food with chemicals (CAC/RCP 49-2001)
11. Recommended international code of practice for control of the use of veterinary drugs (CAC/RCP 38-1993)
12. Guidelines on the judgment of equivalence of sanitary measures associated with food inspection and certification systems (CAC/GL 53-2003)
13. Guidelines on good laboratory practice in pesticide residue analysis (CAC/GL 401)
14. Codex general standard for contaminants and toxins in food (Codex Stan 193-1995, Rev.1-1997)

5. Sanitary and Phytosanitary Measures (SPS)

The agreement on the application of sanitary and phytosanitary measures is approved at the conclusion of the Uruguay round of the multilateral trade negotiations. It applies to all sanitary (relating to animals) and phytosanitary (relating to plants) measures that may have a direct or indirect impact on international trade. The sanitary and phytosanitary (SPS) agreement includes a series of understandings (trade disciplines) on how sanitary and phytosanitary measures will be established and used by countries when they establish, revise and apply their domestic laws and regulations. In this sanitary and phytosanitary (SPS) agreement, countries maintain the sovereign right to provide the level of health protection they deem appropriate, but agree that this right will not be misused for protectionist purposes nor result in unnecessary trade barriers. A rule of equivalency rather than equality applies to the use of SPS measures.

6. BRC-Global Standard-Food

BRC Global Standard has been prepared at the initiative of British Retail Consortium (BRC) which is the leading trade association for UK retailing. The British Retail Consortium Global Standard is a food safety and quality management protocol based on Hazard Analysis and Critical Control Points (HACCP) and designed for manufacturers of all types of foods and food products. Primary producers (including farmers) are not concerned since a company with primary product preparation activities, must exclude these activities from the scope of the certification programme. Furthermore, the standard does not apply to wholesale, import, distribution and storage activities. The BRC standard deals with quality management systems and HACCP, but also establishes general good manufacturing practices for food safety.

7. International Food Standard (IFS)

The International Food Standard (IFS) was set up in 2002 by Hauptverband des Deutschen Einzelhandels (HDE), the German retail association. In 2003, French retailers and wholesalers from the FCD (Federation des enterprises du Commerce et de la Distribution) joined the International Food Standard Working Group and contributed to the development of the current version of the normative document. The IFS is a food safety and quality management standard based on HACCP that is designed for producers of all types of food products. This standard was set up specifically in view of needs of retailers. In fact, it is dedicated to the manufacturing suppliers for private label products, in order to ensure food safety at all levels of the manufacturing supply chain of the products under retailer's responsibility (that is their private label goods). Like the BRC Global Standard-Food, the IFS is not dedicated to primary producers, it comes into force when the product is handled. As with the BRC standard, IFS deals with quality management system and HACCP, but also establishes general good manufacturing practices related to food safety.

8. Safe Quality Food (SQF) Code

The safe quality food standards were originally established by the Western Australian Department of Agriculture in 1996, in response to the demands of the farming and small food manufacturing sectors for a quality assurance system that enabled their businesses to meet regulatory food safety and commercial food quality criteria.

The SQF programme is intended to deal with complete food safety management systems; however, in comparison to the BRC or IFS standards, it only specifies requirements on quality management systems and does not specify good practices nor HACCP plans (though it demands them). It is designed for all types of food products and for all types of suppliers (the SQF 1000 Code for primary producers, the SQF 2000 Code for food industries).

9. Dutch HACCP Code

The standard was set up in 1996 by the Dutch National Board of Experts-HACCP, (a board composed by representatives of all parties involved in the food chain (*i.e.* National Bureau for the Provision Trades, certification bodies, consumer associations, food production and industry). The standard focuses on all operators along the food chain (concerned with preparation, processing, manufacturing, packaging, storage, transportation, handling, distribution for sale or supply), but not on suppliers or service companies to food business (eg. suppliers of packaging materials, food equipment, industrial cleaning services). Primary producers are neither explicitly included in, nor excluded from the scope of the standard. It establishes requirements on quality management systems and HACCP systems, but not on good practices.

10. EurepGAP

The Euro-Retailer Produce Association (EUREP) was created in 1997 by large European retail chains; and was subsequently joined by large fresh produce suppliers and producers. There are also associate members from the input and service side of agriculture (mainly suppliers of agrochemicals, consultancy firms and certification bodies), who can participate in meetings but are not part of the decision-making process. The EUREP has developed different certification programmes, named EUREP Good Agriculture Practices (EurepGAP), to promote good agricultural practices and regain consumer confidence in food safety, environmental protection, animal welfare and worker welfare.

In contrast to other international food safety standards, EurepGAP standards are specific to fresh products and therefore concern farm companies directly.

11. ISO 22000

ISO 22000 is a Food Safety Management System that can be applied to any organization in the food chain, farm to fork. The objective of the ISO 22000 is precisely to establish a single internationally- recognized standard for food safety management systems. The International Organization for Standardization (ISO) is a network of 163 national standards bodies, one per member country, coordinated by a Central Secretariat in Geneva, Switzerland. ISO is a non-governmental organization, occupies a special position between the public and private sectors. Although it is not an intergovernmental organization, the WTO Agreement on Technical Barriers to Trade (TBT) explicitly recognizes ISO as providing internationally accepted standards.

ISO has established standards on quality management systems, the ISO 9000 series that became an international reference for quality management requirements in business-to-business dealings. However, this standard is not specific to the food industry and does not take into account food process specificities, especially in regard to food safety issues. In this context, ISO recently developed the ISO 22000 standard ("Food safety management systems – Requirements for any organization in the food chain") specifically for food safety management systems. However, the ISO 22000 standard only specifies requirements on quality management systems and HACCP systems, and not on good practices. ISO 22000 applies to all food operators (feed producers, primary producers, transport and storage operators, retail and food service outlets and related organizations such as producers of equipment, packaging material, cleaning agents, ingredients or additives *etc.*). Thus, primary producers are included in the scope of the standard.

12. ISO Standards

A brief description of ISO standards related to food safety is given below:

67.120.1 Meat, meat products and other animal produce including frozen products.

67.120.10.1 Meat and meat products

65.120 Animal feeding stuffs Microbiology of animal feeding stuffs.

12.1. ISO 9000 and ISO 14000 Importance

The **ISO 9000** family addresses **"quality management"**. This means what the organization does to fulfill:

- ☆ The customer's quality requirements
- ☆ Applicable regulatory requirements
- ☆ Enhance customer satisfaction
- ☆ Achieve continual improvement of its performance.

The **ISO 14000** family addresses **"environmental management"**.

- ☆ To minimize harmful effects on the environment caused by its activities.
- ☆ To achieve continual improvement of its environmental performance.

ISO codes for meat and meat products are given in Table 20.1. Other related codes are available at www.iso.org.

Table 20.1. ISO Codes for Meat and Meat Products

Sl.No.	*Code*	*Details*
1	ISO 936: 1998	Determination of total ash
2	ISO 937: 1978	Determination of nitrogen content
3	ISO 1442: 1997	Determination of moisture content
4	ISO 1443: 1973	Determination of total fat content
5	ISO 1444: 1996	Determination of free fat content
6	ISO 1841-1: 1996	Determination of chloride contnt – Part 1: Volhard method
7	ISO 1841-2: 1996	Determination of chloride content – Part 2: Potentiometric method
8	ISO 2294: 1974	Determination of total phosphorus content
9	ISO 2917: 1999	Measurement of pH
10	ISO 2918: 1975	Determination of nitrite content
11	ISO 3091: 1975	Determination of nitrite content
12	ISO 3496: 1994	Determination of hydroxyproline content
13	ISO 4134: 1999	Determination of L- (+) glutamic acid content
14	ISO 5553: 1980	Detection of polyphosphates
15	ISO 5554: 1978	Determination of starch content
16	ISO 13493: 1998	Determination of chloramphenicol content- method using liquid chromatography
17	ISO 13496: 2000	Detection of colouring agents – Method using thin layer chromatography
18	ISO/CD 13720	Enumeration of *Pseudomonas* spp.
19	ISO 13730: 1996	Determination of total phosphorus content – spectrometric method
20	ISO 13965: 1998	Determination of starch and glucose contents – enzymatic method

13. European Union (EU) Regulations

The European Food Safety Authority (EFSA) is an independent European agency funded by the European Union budget that operates separately from the European Commission, European Parliament and Member States of European Union. The European Food Safety Authority (EFSA) was set up in January 2002, as an independent source of scientific advice and communication on risks associated with the food chain. EFSA was created as a part of comprehensive programme to improve European Union food safety and ensure a high level of consumer protection and restore and maintain confidence in the EU food supply.

14. USDA Regulations

In the US Department of Agriculture implements regulations related to animal products safety through Food Safety and Inspection Service (FSIS). FSIS stepped up its research studies to apply the Hazard Analysis and Critical Control Point (HACCP) system to meat and poultry inspection, setting the stage for the most significant change in regulatory philosophy and food inspection programs. It focuses on the prevention and reduction of microbial pathogens on raw products that can cause illness. HACCP clarifies the respective roles of industry and government.

Industry is accountable for producing safe and wholesome food. Government is responsible for setting up of appropriate food safety standards, maintaining vigorous inspection oversight to ensure those standards are met, and maintaining a strong enforcement program to deal with food plants that do not meet regulatory standards.

15. Indian Regulations in Livestock Products

1. The Prevention of Cruelty to Animals Act, 1960.
2. The Prevention of Food Adulteration Act, 1954.
3. Meat Food Products Order, 1973.
4. Water (Prevention and Control of Pollution) Act, 1974.
5. Air (Prevention and Control of Pollution) Act, 1981.
6. Environment (Protection) Act, 1986.
7. Export (Quality Control and Inspection) Act, 1963.
8. Export (Quality Control and Inspection) Rules, 1964.
9. Export of Raw Meat (Chilled/Frozen) (Quality Control and Inspection) Rules, 1992.
10. Export of Processed Meat (Quality Control and Inspection) Rules, 1995.
11. Foreign Trade (Development and Regulation) Act, 1992.
12. Food Safety and Standards Act, 2006 (Act No. 34 of 2006).

16. Food Safety and Standards Act, 2006

Food Safety and Standards Authority of India (FSSAI) is an autonomous body established under the Ministry of Health and Family Welfare, Government of India. The FSSAI has been established under the Food Safety and Standards Act, 2006 which is a consolidating statute related to food safety, standards and regulations in India. FSSAI is responsible for protecting and promoting public health through the regulation and supervision of food safety. The promulgation of Food Safety and Standards Act, 2006 repeals the Prevention of Food Adulteration Act, 1954 and the following orders mentioned in the second schedule to the act.

1. The Fruit Products Order, 1955
2. The MEAT FOOD PRODUCTS ORDER, 1973
3. The Vegetable Oil Products (Control) Order, 1998
4. The Solvent Extracted Oil, De-oiled Meal and Edible Flour (Control) Order, 1967
5. The Milk and Milk Products Order,1992 (this order has been deemed by section 99 of the Act to have been issued by the Food Authority under the Act from the date of commencement of the Act), and
6. Any other order issued under the Essential Commodities Act, 1955 (10 of 1955) relating to food. (Section 97(1) of the Act).

Food safety is also regulated under other state acts such as The Andhra Pradesh (Andhra Area) Public Health Act, 1939; The Punjab Municipal Act, 1911; The Bihar and Orissa Municipal Act, 1922 *etc.*

17. Bureau of Indian Standards (BIS)

The Indian Standards Institution gave the nation the standards it needed for nationalization, industrial and commercial growth, competitive efficiency and quality production. BIS is engaged in formulation of Indian Standards for the different sectors including food sector and agricultural sector. Each of these sectors has division council to oversee and supervise the work. Food and Agriculture Division Council oversee and supervise Standardization in the field of food and agriculture including food processing, agricultural machinery and agricultural inputs. Apart from Indian Standards, BIS is also working with international organizations. BIS as a founder member of **International Organization for Standardization (ISO)** and continues to take part in international standardization activities. BIS as a member of ISO -participates in its policy making bodies like Committee on Developing Country Matters (DEVCO), Committee on Conformity Assessment (CASCO), Committee on Information (INFCO) and Committee on Consumer Policy (COPOLCO).

The BIS standards for various meat and meat related products are given in Table 20.2. Other food related standards are available on www.bis.org.in

Table 20.2. The BIS Standards for Various Meat and Meat Related Products

Sl.No.	*Document Number*	*Standard Title*	*Status*
1.	IS 1743: 1973	Specification for Mutton and Goat Meat Canned in Brine	Active
2.	IS 1982: 1971	Code of practice for ante-mortem and post-mortem inspection of meat animals	Active
3.	IS 2536: 1995	Meat and Meat Products - Mutton and Goat Meat (Chevon) - Fresh, Chilled and Frozen - Technical Requirements	Active
4.	IS 2537: 1995	Meat and Meat Products - Beef and Buffalo Meat - Fresh, Chilled and Frozen - Technical Requirements	Active
5.	IS 3044: 1973	Specification for Mutton and Goat Meat, Curried and Canned	Active
6.	IS 3545: 1982	Specification for Meat Choppers	Active
7.	IS 4352: 1967	Specification for Pork Luncheon Meat, Canned	Active
8.	IS 5065: 1986	Specification for Meat Meal as Livestock Feed Ingredient	Active
9.	IS 5960: Part 1: 1996	Methods of test for meat and meat products: Part 1 Determination of nitrogen content	Active
10.	IS 5960: Part 2: 2000	Meat and Meat Products - Methods of Test - Part 2: Determination of Total Ash	Active
11.	IS 5960: Part 3: 1970	Methods of test for meat and meat products: Part 3 Determination of total fat content	Active
12.	IS 5960: Part 4: 1997	Meat and meat products - Method of test: Part 4 Determination of free fat content	Active
13.	IS 5960: Part 5: 2001	Meat and Meat Products - Methods of Test - Part 5: Determination of Moisture Content (Reference Method)	Active

Sl.No.	Document Number	Standard Title	Status
14.	IS 5960: Part 6: Sec I: 1997	Meat and meat products - Method of test Part 6 Determination of chloride content Section 1 Volhard method	Active
15.	IS 5960: Part 6: Sec 2: 1997	Meat and meat products - Method of Test Part 6 Determination of Chloride content Section 2 Potentiometric method	Active
16.	IS 5960: Part 7: 1996	Methods of test for meat and meat products: Part 7 Determination of nitrite content	Active
17.	IS 5960: Part 8: 1996	Methods of test for meat and meat products: Part 8 Determination of nitrate content	Active
18.	IS 5960: Part 9: 1988	Meat and Meat Products - Methods of Test - Part 9: Determination of Total Phosphorus Content	Active
19.	IS 5960: Part 10: 2001	Meat and Meat Products - Methods of Test - Part 10: Measurement of pH - Reference Method	Active
20.	IS 5960: Part 11: 1988	Method of test for meat and meat products: Part 11 Determination of glucone-delta-lactone content	Active
21.	IS 5960: Part 12: 2001	Meat and Meat Products - Methods of Test - Part 12: Determination of L-(+)-Glutamic Acid Content - Reference Method	Active
22.	IS 5960: Part 13: 1988	Methods of test for meat and meat products: Part 13 Determination of polyphosphates	Active
23.	IS 5960: Part 14: 1988	Methods of Test for Meat and Meat Products - Part 14: Determination of Starch Content	Active
24.	IS 5960: Part 15: 2000	Meat and Meat Products - Methods of Test - Part 15: Determination of L (-) Hydroxyproline Content	Active
25.	IS 5960: Part 16: 2004	Meat and Meat Products - Methods of Test - Part 16: Determination of Total Phosphorus Content - Spectrometric Method	Active
26.	IS 5960: Part 17: 2004	Meat and Meat Products - Methods of Test - Part 17: Determination of Starch and Glucose Content - Enzymatic Method	Active
27.	IS 6851: 1973	Specification for Meat Extract, Microbiological Grade	Active
28.	IS 7053: 1996	Meat and Meat Products - Basic Requirements for a Stall for Sale of Meat of Small and Large Animals	Active
29.	IS 7143: 1973	Specification for Crab Meat Canned in Brine	Active
30.	IS 7582: 1975	Specification for Crab Meat, Solid Packed	Active
31.	IS 7802: 1975	Code of hygienic conditions for sweetmeat shop	Active
32.	IS 8182: 1976	Code for hygienic conditions for processed meat products	Active
33.	IS 8539: Part 1: 1977	Terminology of meat products and meat animals: Part 1 Poultry	Active
34.	IS 8700: 1977	Basic requirements for a stall for sale of meat of large animals	Withdrawn
35.	IS 9800: 1993	Meat and Meat Products - Day-Old Chicks (layers and broilers) Basic Requirements	Active
36.	IS 10380: 1982	Methods of test for printing ink permeation of paper (castor oil test)	Active

Sl.No.	Document Number	Standard Title	Status
37.	IS 10726: 1983	Specification for Knife, Metal Skin Incision	Active
38.	IS 10763: 1983	Specification for Frozen Minced Fish Meat	Active
39.	IS 10974: Part 3: 1984	Code for hygienic condition for production, transport, storage and distribution of indigenous milk products: Part 3 Coagulated products CHHANA and CHHANA based sweetmeats	Active
40.	IS 11748: 1986	Specification for Meat Extract (Beef), Food Grade	Active
41.	IS 12486: 1988	Specification for Meat Inspection Table	Active
42.	IS 12541: 1988	Meat and Meat Products - Poultry - Chicken Curry, Canned - Specification	Active
43.	IS 12542: 1988	Meat and Meat Products - Canned Ham, Minced - Specification	Active
44.	IS 12543: 1988	Meat and Meat Products - Poultry Products - Canned Egg Curry - Specification	Active
45.	IS 13059: 1991	Meat and meat products - enumeration of micro-organisms colony count technique at 30°C (reference method)	Withdrawn
46.	IS 13060: 1991	Meat and meat products - detection and enumeration of presumptive coliform bacteria and presumptive *Escherichia coli* (reference method)	Withdrawn
47.	IS 13061: 1991	Meat and meat products - detection of salmonellae (reference method)	Withdrawn
48.	IS 13165: 1991	Meat and Meat Products - Mutton Biryani (Canned) - Specification	Active
49.	IS 13400: 1992	Meat and Meat Products - Chicken Sausages - Specification	Active
50.	IS 13401: 1992	Meat and Meat Products - Determination of Thiobarbituric Acid Value in Meat - Test Method	Active
51.	IS 14514: 1998	Clam Meat - Frozen - Specification	Active
52.	IS 14843: 2000	Meat and Meat Products - Enumeration of *Pseudomonas* Spp.	Active
53.	IS 14844: 2000	Meat and Meat Products - Enumeration of Lactic Acid Bacteria - Colony-Count Technique at 30C	Active
54.	IS 14920: 2001	Meat and Meat Products - Enumeration of Yeasts and Moulds - Colony-count Technique	Active
55.	IS 15463: 2004	Meat and Meat Products - Enumeration of *Escherichia coli* - Colony-Count Technique at 44°C Using Membranes	Active
56.	IS 15478: Part 2: 2004	Meat and Meat Products - Sampling and Preparation of Test Samples - Part 2: Preparation of Test Samples for Microbiological Examination	Active
57.	IS/ISO 3100-1: 1991	Meat and Meat Products - Method of Sampling	Active
58.	IS 8700: 1977	Basic requirements for a stall for sale of meat of large animals	Withdrawn
59.	IS 8539: Part 1: 1977	Terminology of meat products and meat animals: Part 1 Poultry	Active

Sl.No.	Document Number	Standard Title	Status
60.	IS 7053: 1996	Meat and Meat Products - Basic Requirements for a Stall for Sale of Meat of Small and Large Animals	Active
61.	IS 14988: Part 2: 2002	Microbiology of Food and Animal Feeding Stuffs - Horizontal Method for the Detection and Enumeration of *Listeria monocytogenes* - Part 2: Enumeration Method	Active
62.	IS 7909: 1993	Slaughter House Equipment - Electrical Stunning Tongs for Pigs - Specification	Active
63.	IS 8895: 1978	Guidelines for handling, storage and transport of slaughter house by-products	Active

18. APEDA

The Agricultural and Processed Food Products Export Development Authority (APEDA) was established by the Government of India under the Agricultural and Processed Food Products Export Development Authority Act passed by the Parliament in December, 1985. APEDA is mandated with the responsibility of export promotion.

19. National Regulations: Constitutional Provisions and Slaughter Regulations

Slaughter of animals is a state subject and state legislatures have exclusive power to legislate as per Entry 15 of List II in the Seventh Schedule of the Constitution which reads as under: preservation, protection and improvement of stock and prevention of animal diseases, veterinary training and practice. States have to apply Directive Principles which are fundamental in the governance of the country while making laws. Articles 47, 48 and 51 A(g) under the Directive Principles of the Constitution concern with level of nutrition, standard of living, public health, agriculture and animal husbandry and protection of natural environment. Clearances from Pollution Control Board, Airport Authority and Defence Department are required for establishing slaughter houses. Meat hygiene is regulated as per bye-laws of the local body and meat as food is also covered under the Prevention of Food Adulteration Act, 1954.

Water (prevention and control of pollution) Act, 1974; Air (Prevention and control of pollution) Act, 1981 and Environment (protection) Act, 1986 stipulate requirements for abattoirs. Motor Vehicles Transport Act and The Prevention of Cruelty to Animals Act, 1960 are also relevant in meat sector. Meat Food Products Order, 1973 regulates meat food products in the country. Meat export is regulated as per Export (quality control and inspection) Act, 1963 and the Export (Quality Control and Inspection) Rules, 1964. Export of fresh and frozen meat is regulated as per Raw Meat (chilled/frozen) (Quality Control and Inspection) Rules 1992 and meat products under Processed Meat (Quality control and Inspection) Rules, 1995. Export of animal casings is regulated as per Plant Registration and Animal Casings Rules, 1995.

A pragmatic approach is necessary with respect to slaughter and utilization of livestock so as to sustain their production and prevent depletion.

20. Quality Standards of Meat and Meat Products

20.1 Sampling Plans and Recommended Microbiological Limits for Raw Meat (ICMSF, Microorganisms in Foods.Vol.2, 1986)

Sl.No.	Product	Test	n	c	m	M
1.	Carcass meat, before chilling	APC	5	3	10^5	10^6
2.	Carcass meat, chilled	APC	5	3	10^6	10^7
3.	Edible offal, chilled	APC	5	3	10^6	10^7
4.	Carcass meat, frozen	APC	5	3	$5x10^5$	10^7
5.	Boneless meat, frozen (beef, pork, mutton)	APC	5	3	$5x10^5$	10^7
6.	Comminuted meat, frozen	APC	5	3	10^6	10^7
7.	Edible offal, frozen	APC	5	3	$5x10^5$	10^7
8.	Raw chicken, fresh or frozen	APC	5	3	$5x10^7$	10^7

n: Number of samples to be taken from the lot; c: Number of samples permitted to fail; m: Microbial count below which the sample is considered to be satisfactory; M: Microbial count above which the sample is considered unsatisfactory..

20.2 Microbiological Standards Prescribed by Government of India: (Ref: Export of Raw Meat (Chilled/frozen) Quality Control and Inspection Rule, 1990 and Raw Meat (Chilled/frozen) Grading and Marking Rules, 1991)

<table>
<tr><th>Sl.No.</th><th>Products</th><th>Grades</th><th>Microbial Standards*</th></tr>
<tr><td>(a)</td><td>Buffalo meat</td><td>Choice good commercial grade X</td><td rowspan="4">Minimum 5 samples should be drawn and tested.
(1) TPC: Out of 5 samples, 3 should have count not exceeding 10^6/g and the remaining 2 samples can have upto 10^7/g.
(2) Escherichia coli: out of 5 samples 3 should have E.coli counts not exceeding 10/g and the remaining 2 samples can have E.coli counts upto 100/g.
(3) Salmonella should be absent in all the samples</td></tr>
<tr><td>(b)</td><td>Veal</td><td>Choice Standard Grade X</td></tr>
<tr><td>(c)</td><td>Mutton</td><td>Choice Standard Grade X</td></tr>
<tr><td>(d)</td><td>Minced meat (Keema)</td><td>Grade 1
Grade X</td></tr>
</table>

*: Microbiological standards are same for all the products and grades.

20.3 Meat Food Products Order (1973) and Meat Food Products Order Amendment (1993)

Sl.No.	Name of Poisonous Metal	Parts per million by Weight
1.	Lead	2.5
2.	Copper	20
3.	Arsenic	1
4.	Tin	250
5.	Zinc	50

Sl.No.	Name of Preservative	Parts per million by Weight
1.	Sulphur Dioxide	450
2.	Sodium and Potassium Nitrite (Calculated as Sodium Nitrite)	200
3.	Commercial Saltpetre (Calculated As Sodium Nitrite)	500

Sl.No.	Name of Insecticide	Tolerance Limit mg/kg (ppm)
1.	Aldrin Dieldrin	0.2
2.	D.D.T.	7.0
3.	Fenitrothion	0.03
4.	Lindane	2.0
5.	Chlorfenvinphos	0.2
6.	Chlorpyrifos	0.1
7.	2,4-D	0.05
8.	Ethion	0.2
9.	Monocrotophos	0.02
10.	Trichlorfon	0.1
11.	Carbendazim	0.10
12.	Benomyl	0.10
13.	Carbofuran	0.10
14.	Cypermethrin	0.20
15.	Edifenphos	0.02
16.	Fenthion	2.00
17.	Fenvalerate	1.00
18.	Phenthoate	0.05
19.	Phorate	0.05
20.	Pirimiphos- Methyl	0.05

20.4 Extract from Gazette of India: Part II – Sect 3, Sub-Sect (ii)

Ministry of Commerce (1993) order for development of the export trade of India, regarding raw meat (chilled/frozen) buffalo beef/buffalo veal/sheep and goat meat bacteriological standards as per raw meat (chilled or frozen) grading and marking rules (1991)

Specification for Buffalo Beef or Meat

1.	Total plate count	not exceeding 10^7 microorganisms per gram
2.	*Escherichia coli*	not exceeding 100 per gram
3.	*Salmonella*	absent

Specification for Buffalo Calf Meat or Veal

1.	Total plate count	not exceeding 10^7 microorganisms per gram
2.	*Escherichia coli*	not exceeding 100 per gram
3.	*Salmonella*	Absent in 25 gm sample

Specification for Mutton (Sheep) and Chevon (Goat meat)

1.	Total plate count	not exceeding 10^7 microorganisms per gram
2.	*Escherichia coli*	not exceeding 100 per gram
3.	*Salmonella*	Absent in 25 gm sample

Specification for Minced Meat (keema)

1.	Total plate count	not exceeding 10^7 microorganisms per gram
2.	*Escherichia coli*	not exceeding 100 per gram
3.	*Salmonella*	Absent in 25 gm sample

Method of Sampling and Testing for Confirmation of Quality of Meat in Cartons

No. of Cartons in each Lot	*No. of Cartons to be Selected in Routine Inspection*	*No. of Cartons to be Selected for Revalidation*
1- 10	4	8
101- 200	5	10
201- 500	6	12
501- 800	7	14
801- 1200	8	16
1201- 3200	9	18
3201- 8000	10	20
8001- and above	12	24

20.5 Fixation of Maximum Residue Limits in India: Maximum Residue Limits of Pesticides in Livestock Products

Name of the Pesticide	*MRL mg/kg*		
	Meat and Poultry	*Milk and Milk Products*	*Eggs*
Aldrin, Dieldrin	0.2*	0.15	
Benomyl	0.10	0.10	0.10
Carbendazim	0.10	0.10	0.10
Carbofuran	0.1	0.05	
Chlordane		0.05	
Chlorfenvinphos	0.2		
Chorpyriphos	0.1	0.01	

Name of the Pesticide	*MRL mg/kg*		
	Meat and Poultry	*Milk and Milk Products*	*Eggs*
Cypermethrin	0.20	0.01	
2,4- D	0.05	0.5	0.05
Ddt	7.0	1.25	0.5
Edifenphos	0.02	0.01	
Ethion	0.2	0.5	0.2
Fenitrothion	0.03*	0.05	
Fenthion	2.00	0.05	
Fenvalerate	1.00	0.01	
Heptachlor		0.15	
Lindane	2.0	0.2	0.1
Monocrotophos	0.02		0.02
Phenthioate	0.05	0.01	0.05
Phorate	0.05	0.05	0.05
Primiphos- Methyl	0.05	0.05	0.05
Trichlorofon	0.1		

*: Meat alone.

Source: Pesticide Residues Significance, Management and Analysis by Handa *et al.* (1999).

20.6 Ministry of Commerce and Industry Order (2002) for Development of the Export Trade of India Regarding Poultry Meat and Poultry Meat Products

Tolerance Levels of Residues for Poultry Standard

Residues	*Tolerance Level By EU (Microgms/Kg)*
Tetracycline Residue (Total) (Including Tetracycline Hydrochloride, Oxytetracycline Hydrochloride, Chlorotetracycline)	100
Pencillins	
A. Benzylpencillin	50
b. Ampicillin	50
c. Amoxicillin	50
d. Oxacillin	300
e. Cloxacillin	300
f. Dicloxacillin	300
Tylosin	100
Amprolium **Streptomycin sulphate** **Streptomycin hydrochloride**	Limits to be fixed

20.7 Microbial Standards for Meat and Meat Products

Microbiological Standards for Meat and Meat Products: Process Hygiene Criteria

Sl. No.	Product Category	Aerobic Plate Count				Yeast and Mold Count				Escherichia coli				Staphylococcus aureus (Coagulase +ve)			
		Sampling Plan		Limits (cfu/g)		Sampling Plan		Limits (cfu/g)		Sampling Plan		Limits (cfu/g)		Sampling Plan		Limits (cfu/g)	
		n	*c*	*m*	*M*	*n*	*c*	*m*	*M*	*n*	*c*	*m*	*M*	*n*	*c*	*m*	*M*
(1)	Fresh meat/ Chilled meat	5	3	$1x10^6$	$5x10^6$	5	2	$1x10^4$	$5x10^4$	5	2	$1x10^2$	$1x10^3$	5	2	$1x10^2$	$1x10^3$
(2)	Frozen meat	5	2	$1x10^5$	$5x10^6$	5	2	$1x10^3$	$1x10^4$	5	2	1x10	$1x10^2$	5	2	10	$1x10^2$
(3)	Raw marinated/ minced/ comminuted meat	5	2	$5x10^5$	$5x10^6$	5	2	$1x10^2$	$1x10^3$	5	2	$1x10^2$	$1x10^3$	5	2	$1x10^2$	$1x10^3$
(4)	Semi-cooked/ Smoked Meat/ Meat food product	5	2	$1x10^4$	$1x10^5$	5	2	10	$1x10^2$	5	2	10	$1x10^2$	5	2	1x10	$1x10^2$
(5)	Cured/Pickled meat	5	2	$5x10^2$	$5x10^3$	5	2	$1x10^2$	$1x10^3$	5	2	10	$1x10^2$	5	1	$1x10^2$	$1x10^3$
(6)	Fermented meat products	NA	NA	NA	NA	NA	NA	NA	NA	5	2	10	$1x10^2$	5	1	$1x10^2$	$1x10^3$
(7)	Dried/ dehydrated meat product	5	2	$1x10^3$	$1x10^4$	5	2	$1x10^2$	$1x10^3$	5	2	10	$1x10^2$	5	1	10	$1x10^2$
(8)	Cooked Meat Products	5	2	$1x10^3$	$1x10^4$	5	1	10	$1x10^2$	5	2	10	$1x10^2$	5	1	10	$1x10^2$

Sl. No.	*Product Category*	*Aerobic Plate Count*				*Yeast and Mold Count*				*Escherichia coli*				*Staphylococcus aureus (Coagulase +ve)*			
		Sampling Plan		*Limits (cfu/g)*		*Sampling Plan*		*Limits (cfu/g)*		*Sampling Plan*		*Limits (cfu/g)*		*Sampling Plan*		*Limits (cfu/g)*	
		n	*c*	*m*	*M*	*n*	*c*	*m*	*M*	*n*	*c*	*m*	*M*	*n*	*c*	*m*	*M*
(9)	Canned/Retort pouch Meat Products	NA	NA	NA	NA	NA	NA	NA	NA	5	0	Absent	NA	5	0	Absent	NA
	Test Methods	IS: 5402/ISO 4833				IS: 5403/ISO 21527				IS: 5887 Part 1 or ISO 16649-2				IS 5887: Part 2 or IS 5887 Part 8 (Sec 1)/ISO: 6888-1 or IS 5887 Part 8 (Sec 2)/ISO 6888-2			

NA: Not applicable.

Source: The Gazette of India, Ministry of Health and Family Welfare (Food Safety and Standards Authority of India), Notification, New Delhi, the 10th October, 2016.

Microbiological Standards for Meat and Meat Products: Food Safety Criteria

Sl. No.	Product Category	*Salmonella*			*Listeria monocytogenes*			*Sulphite Reducing Clostridia*				*Clostridium botulinum*				*Campylobacter Spp.*			
		Sampling Plan		*Limits (cfu/g)*	*Sampling Plan*		*Limits (cfu/g)*	*Sampling Plan*		*Limits (cfu/g)*		*Sampling Plan*		*Limits (cfu/g)*		*Sampling Plan*		*Limits (cfu/g)*	
		n	*c*	*m M*	*n*	*c*	*m M*	*n*	*c*	*m*	*M*	*n*	*c*	*m*	*M*	*n*	*c*	*m*	*M*
(1)	Fresh meat/Chilled meat	5	0	Absent	NA	NA	NA	NA	NA	NA	NA	NA	NA	NA	NA	NA	NA	NA	NA
(2)	Frozen meat	5	0	Absent	NA	NA	NA	NA	NA	NA	NA	NA	NA	NA	NA	NA	NA	NA	NA
(3)	Raw marinated/ minced/ comminuted meat	5	0	Absent	NA	NA	NA	NA	NA	NA	NA	NA	NA	NA	NA	NA	NA	NA	NA
(4)	Semi-cooked/ Smoked Meat/Meat food product	5	0	Absent	NA	NA	NA	NA	NA	NA	NA	NA	NA	NA	NA	5	0	Absent	
(5)	Cured/Pickled meat	5	0	Absent	5	0	Absent	5	2	5×10^2	5×10^3	NA	NA	NA	NA	NA	NA	NA	NA
(6)	Fermented meat products	5	0	Absent	5	0	Absent	5	2	5×10^2	5×10^3	NA	NA	NA	NA	NA	NA	NA	NA
(7)	Dried/dehydrated meat product	5	0	Absent	5	0	Absent	5	2	5×10^2	5×10^3	NA	NA	NA	NA	NA	NA	NA	NA
(8)	Cooked Meat Products	5	0	Absent	5	0	Absent	5	1	1×10^2	1×10^3	NA	NA	NA	NA	5	0	Absent	
(9)	Canned/Retort pouch Meat Products	5	0	Absent	5	0	Absent	5	0	Absent		5	0	Absent		5	0	Absent	
	Test Methods	IS: 5887 Part 3/ ISO 6579			IS: 14988, Part 1 and 2/ISO 11290-1 and 2			ISO 15213				IS 5887: Part 4 or ISO 17919				ISO 10272-1 and 2			

NA: Not Applicable.

Source: The Gazette of India, Ministry of Health and Family Welfare (Food Safety and Standards Authority of India), Notification, New Delhi, the 10th October, 2016.

20.8 Extract From Gazette of India: Part II – Sec 3, Sub- Sec (II)

Ministry of Commerce (1995) Order for Development of the Export Trade of India, Regarding Canned Processed Meat Products

Selection of Cans for Testing

Number of Cans in the Lot (N)	*Number of Cans to be Selected (n)*
Upto 500	6
500- 1000	7
1001- 5000	8
5001- 10000	9
Above 10000	10

Opening of Packing Cases

Number of Packing Cases in the Lot (N)	*Number of Packing Cases to be Opened (n)*
Upto 100	2
101- 500	3
501- 1000	4
1001- 5000	5
Above 5000	6

21. Conclusion

In the present scenario, the food safety and quality awareness of the consumers in developed countries triggered the retailers in the food market chain to motivate and monitor the producers/processors to adhere to the standards fixed to satisfy the consumers. Now our consumers have more chances to know about the changes happening throughout the globe due to globalization so our food law approved by the Indian parliament has to be implemented with the global standards in all the quality aspects of the food commodities to enter into the global trade and also to suit domestic consumers.

REFERENCES

CAC (Codex Alimentarius Commission) (1997). Principles for the establishment and application of microbiological criteria for foods, CAC/GL-21. FAO, Rome.

CAC (Codex Alimentarius Commission) (2007a). Principles and guidelines for the conduct of Microbiological Risk Management (MRM). CAC/GL-63. FAO, Rome.

CAC (Codex Alimentarius Commission) (2007b). Joint FAO/WHO Food Standards Programme. Procedural Manual. Seventeenth edition. ISSN 1020-8070.

FAO/WHO (Food and Agriculture Organisation, World Health Organisation) (2006). The Use of Microbiological Risk Assessment Outputs to Develop

Practical Risk Management Strategies: Metrics to improve food safety. Report of a Joint Expert Meeting, Kiel, Germany, 3 – 7 April 2006. FAO, Rome.

Gazette notification on FSS (Food Product Standards and Food Additives) Tenth amendment regulation, 2016 related to Microbiological Standards for Milk and Milk Products and Meat and Meat Products.

Handa, S.K., Agnihotri, N.P. and Kulshrestha, G. (1999). Pesticide Residues Significance, Management and Analysis. Research Periodicals and Books Publishing House, pp 226.

http: //www.oie.int/.

ICMSF (International Commission on Microbiological Specifications for Foods) (1986). Microorganisms in Foods 2: Sampling for Microbiological Analysis: Principles and Specific Applications 2nd ed. University of Toronto Press, Toronto.

ICMSF (International Commission on Microbiological Specifications for Foods) (1988). Microorganisms in Foods 4: Application of the Hazard Analysis Critical Control Point (HACCP) System to Ensure Microbiological Safety and Quality. Blackwell Scientific Publications, Oxford.

ICMSF (International Commission on Microbiological Specifications for Foods) (1994). Choice of sampling plan and criteria for Listeria monocytogenes. International Journal of Food Microbiology, 22 (2 – 3), 89-96.

ICMSF (International Commission on Microbiological Specifications for Foods) (2002). Microorganisms in Food 7. Microbiological Testing in Food Safety Management. Kluwer Academic/Plenum, NY.

Legan, J.D., Vandeven, M.H., Dahms, S. and Cole, M.B. (2001). Determining the concentration of microorganisms controlled by attributes sampling plans. Food Control, 12(3), 137-147.

WTO (World Trade Organisation) (1994). Agreement on the Application of Sanitary and Phytosanitary Measures. pp 16.

Chapter 21

Modern Techniques to Assess the Nutrition and Freshness of Meat and Meat Products

B.K. Ojha

College of Veterinary Science and Animal Husbandry, Rewa, Madhya Pradesh
(Nanaji Deshmukh Veterinary Science University, Jabalpur, M.P.)
e-mail: drbrijesh.ojha@gmail.com

ABSTRACT

The objective of evaluating the quality of meat is to present the consumer a nutritious, delicious and safe product at rational cost. It is an important part for producing good quality meat products. The meat obtained from livestock is highly perishable in nature. For assuring quality of such products nowadays modern techniques such as rapid and nondestructive methods are used. Prospective modern techniques, which can add or replace many of traditional long destructive methods, include high performance liquid chromatography, colour and computer image analysis, near-infrared spectroscopy, ultrasound, and biosensors etc. These techniques are useful in speedy test of huge number of meat samples, maintaining quality and value of meat products. With recent advancements in the basic science applied to these non-destructive methods, it is anticipated that these methods can be used extensively and commercially for determining nutritional quality of meat and meat products at large scale in coming future. Keeping this objective in centre these methods are briefly described and reviewed here.

Keywords: *Modern techniques, Nutritional quality, Meat, Products.*

1. Introduction

Quality control and monitoring is crucial component of meat industry. Consumers' hopes for lesser cost and reliable quality, demand the need for expansion of dependable instruments for both assessing worth and pricing. Consumer evaluates the meat quality by inspecting factors such as colour, tenderness, juiciness, and flavour of meat. Proper quality of meat is deciding element in purchase of the products. Meat quality and constancy are significant in ensuring customer pleasure. Quality of meat is affected by the genetic predilection of the livestock, rearing management, and the nutritional condition during production (Narsaiah and Jha, 2012). Hoffman (1990) defined meat quality as the 'addition of the entire meat quality factors such as sensory, nutritional, hygienic and meat technological properties.' From the user perception, quality is attached to "efficient" distinctiveness that range from physical appearance and sensory characteristics to price, keeping quality and distribution criteria. In the present text main stress will be on nutritive quality of meat and its evaluation along with some other quality parameters. The baseline characteristics concern meat content in terms of nutrients, chemical residues, toxic elements, pathogenic microbes, and any other contaminants which may be harmful for health. For measuring these characteristics we have to use recent non-destructive techniques. With the help of these techniques we can assure the meat quality and wholesomeness by using simple probes derived from the lab techniques. Industry people with proper use of these probes can maintain the quality control of their products and ensure their customers with certified product.

2. Nutritional Characteristics

Nutritional characteristics are high on the list of consumer interests. Here, we describe some of the more sophisticated methods for assessing the main encouraging or depressing nutritional aspects of meat/meat products.

2.1 High Performance Liquid Chromatography

This technique is used to measure protein quality of meat by analyzing the type and profile of the essential amino acids. For determining nutritional quality of meat it is essential to analyze the composition of essential amino acids present in the meat. Thus, analyzing the amino acid composition of the proteins is necessary to assess the nutritional value of the meat products in terms of protein quality. The recent methods used for the partition and quantitative measurement of the free amino acids are high performance liquid chromatography (HPLC) (Fontaine, 2003; Gilani *et al.*, 2008), capillary electrophoresis (Peace *et al.*, 2005), near infra-red spectrometry (Fontaine, 2003; Kryeziu *et al.*, 2007).

By HPLC primary and secondary amino acids can be analyzed rapidly and accurately. The amino acid composition of proteins is not only used to measure meat quality as well as can be used to determine the origin of meat products also and thus helps to detect adulteration of meat (Primer, 2001). HPLC along with automated online derivatization is now a well-established technique for detecting amino acids owing to its rapid and moderately simple sample preparation procedure. Hydrolyzation by hydrochloric acid or enzyme is used to break down protein bonds.

The HPLC method can also efficiently be used for the measurement of the fatty acid profile of meat. This method is followed by hydrolysis with hot KOH/methanol and online derivatization with bromophenacyl bromide (Primer, 2001). To find out the fatty acid pattern of a meat fat, firstly free fatty acids are obtained through hydrolysis, and then derivatization is performed to introduce a chromophore, which enables analysis of the fatty acids via HPLC and UV-visible detection. Liquid-liquid extraction and on- and offline solid-phase extraction methods are used in the analysis of liquid samples or extracts from solid meat samples. Liquid-liquid extraction is the most commonly used extraction method. It requires a suitable solvent and a separating funnel, or a continuous or counter-current distribution device.

Vitamins are biologically active trace organic factors that act as controlling agents for an organism's normal health and growth. The level of vitamins in meat may be as low as a few micrograms per 100 g. Vitamins frequently are accompanied by a glut of compounds with similar chemical properties. Thus not only quantification but also identification is mandatory for the detection of vitamins in meat. Vitamins generally are very sensitive and heat labile compounds that should not be exposed to high temperatures, light, or oxygen. HPLC efficiently separates and detects these compounds at room temperature and blocks oxygen and light. With the use of spectral information, UV-visible diode-array detection yields qualitative as well as quantitative data (Primer, 2001). Another extremely sensitive and selective HPLC method for detecting vitamins is electrochemical detection. Vitamins such as A, D, and E (fat soluble vitamins) and water-soluble vitamins, such as vitamins B_1, B_2, B_6, B_{12}, and C, have been analyzed with this technique. Different food matrices require different extraction procedures. Water-soluble vitamins are extracted with water in an ultrasonic bath after homogenization of the meat test sample. While alkaline hydrolysis, enzymatic hydrolysis, alcoholysis, direct solvent extraction, and supercritical fluid extraction of the total lipid content of meat sample are the procedures used for fat soluble vitamins (Primer, 2001).

HPLC analytes can be recognized on the basis of their retention times and either their UV-visible or mass spectra. Compounds, alternatively, are identified primarily according to the degree of agreement between retention times recorded using calibration standards and those obtained from the sample. Unluckily, co-eluting peaks can falsify results obtained with samples containing unknowns, especially for food matrices such as meat, vegetables, or beverages. In such cases, samples frequently can be identified using UV-visible spectral information.

2.2 X-Ray Measurements

The principle of x-ray is to obtain a measurement of the attenuation of penetrating energy. Different materials have different attenuation properties, and accordingly depending on the level of penetrating energy delivered, it must be feasible to obtain quantitative measurements particularly for bone, lean meat, and fat. A broad panel of technology tools has been developed by means of x-ray beams at diverse energy levels, making it feasible to differentiate bone, fat, and lean meat on the basis of attenuation of energy measured.

A novel x-ray technology dubbed DXA (Dual-energy X-ray absorptiometry) offers useful capabilities for meat fat assessment. Absorption at low x-ray energies (*e.g.*, 62 keV) is dependent on fat content and sample density both, while absorption at higher energies (*e.g.*, 120 keV) is dependent on only density essentially. Superposing the two measurements and subtracting one from another gives the fat content (Brienne *et al.*, 2001; Hansen *et al.*, 2003) with excellent exactness compared to chemical analysis (*R*2 values from 0.7 to 0.97). Brienne *et al.* (2001) and Mercier *et al.* (2006) reported that, even though DXA is too slow for use commercially, it may be used in carcass composition research/studies as a reference method.

Fat content and marbling percentage appearance of meat product/muscle tissue also constitute an important factor for consumers. As explained by Swatland (1989), video image analysis can easily measure meat cut quantification and can potentially be mechanized for in-line use. Pearce *et al.* (2009) showed that dual-energy x-ray absorption (DXA) is a precise apparatus for measuring carcass composition in both live and dead animals/carcasses. An additional potentially important technique is x-ray computed tomography, which has given extremely precise predictions of meat cut composition, marbling percentage (intramuscular fat), and fatty acid in primal meat cuts (Prieto *et al.*, 2010), but is extreme less precise for technical and sensory characters.

2.3 NMR (Nuclear Magnetic Resonance) Spectroscopy and MRI (Magnetic Resonance Imaging)

NMR micro-imaging on meat samples may be used to quantitatively characterize lipids. Fat content analysis by NMR involved selecting a sequence of impulses (Inversion-Recovery, Spin Echo, or others) to be applied to the product. The results highlight the versatility and practicability of the technique, since the equipment involved is compact and the method can be equally well deployed for controlling food composition as for checking food quality. Moreover, diffusion-weighted imaging of muscle is another MRI method which makes use of the large difference between apparent diffusion coefficients of myofibers and lipids (intramuscular fat and frying oil). Using a sufficiently large *b*-value, the water signal is attenuated so that only the lipid signal becomes detectable (Clerjon *et al.*, 2010; Clerjon and Bonny, 2011).

2.4 Atomic Absorption Spectrophotometer (AAS)

Meat is a source of many valuable nutrients for human health, out of which protein and fat possibly being the most important nutrients. However, beef and other meat are also an important source of mineral ions including the trace minerals as well as selenium. AAS was developed in mid 60s and it is considered to be one of the most modern and precise technique for estimation of trace minerals in fresh and processed meat. Primarily it operates on the absorption of radiant energy by free atom in their ground state; these then tend to absorb characteristic radiation identical to those which the same substance can emit. It is an automatic PC based machine. It uses a double beam with a wave length of 190-900 nm range. Separate hollow UV lamps for each element are used (Nunes *et al.*, 2011). Air/acetylene flame is used as fuel. Sample analysis can be done by attached computer and concentration of mineral samples should expressed in parts per million (ppm).

Principle of AAS follows Lambert's Beers law. A liquid sample (extract of meat) to be aspirated, aerolized and mixed with flammable gases like acetylene and air or acetylene and nitrous oxide. Sample solution or sample is burned in a flame (2100 to 2800°C) or heated in a tube and the individual atoms of the sample are released (Paul *et al.*, 2014). After absorbing UV light at specific wavelengths the ground state mineral atoms present in the sample are transitioned to higher electronic state, consequently reducing its intensity. The instrument measures the change in intensity. An attached computer system converts the change in intensity into an absorbance. Absorption is directly related to sample concentration and data. Liquid sample is turned into an atomic gas all the way through desolations, vaporization, atomization. There are numerous atomization techniques *i.e.* flame, graphite furnace and hydride generation. Hollow cathode lamp is regularly used as a light source. Since AAS is extremely sensitive method, sample preparation and storage is very important and sample should be free from any contaminants. The concentration of trace mineral is expressed in ppm or ppb level. There is number of sample preparation techniques such as acid digestion, fusion, dry ashing, microwave digestion and pressure dissolution *etc.* A set of standard solution is required for making a calibration curve. Unknown sample concentration can be determined from the calibration curve (Paul *et al.*, 2014).

2.5 Ultrasound

Analyzing the acoustic parameters of waves propagating in a medium, makes it promising to assess and characterize the propagation medium. Research discerning meat samples depending on fat and collagen contents have reported superior results than those obtained by basic analysis of chemical and mechanical properties of meat sample (Abouelkaram *et al.*, 2000; Morlein *et al.*, 2005). Fat content of sample has been reported to be associated with speed of ultrasound propagation, with fat and lean muscle showing reverse temperature dependencies on sound speed (Abouelkaram *et al.*, 2000; Benedito *et al.*, 2001).

2.6 Optical Spectroscopy

Raman spectroscopy was initially used in biomedical study for determining degree of saturation in fatty acids. Due to this extremely significant characteristic of nutrition this is now recommended to use in study of meat science. Furthermore, it has been established that this method can provide structural information on the lipid changes (Beattie *et al.*, 2006) that occur in muscle food during spoilage. This spectroscopy technique offers several advantages over traditional methods, since it is a straight and noninvasive technique that requires only small sample portions. Rapid nondestructive methods to determine fatty acid composition are also being investigated as a solution for assaying the saturated, monounsaturated, and polyunsaturated fatty acids that influence human health. Near infrared reflectance spectroscopy (NIRS) is used to rapidly evaluate fatty acid composition in dry-cured meat product (Fernandez-Cabanas *et al.*, 2011) and is close to deployment at industrial level (Pérez-Juan *et al.*, 2010).

2.7 Macroscopic and Microscopic Imaging

Image analysis is used to analyze the texture of images produced from meat cuts at different wavelengths in meat science research. Collagen and lipid structures of muscular tissue can be clearly highlighted with the use of this imaging process.

Marbling score, which is an extent of fat allocation density in the rib–eye region, is considered a texture pattern.

Multispectral image analysis on images found at diverse wavelengths in the UV-visible range was used to correlate meat components with physical and sensory properties, providing better results in terms of predicting lipid content ($R2 = 0.87$) (Abouelkaram *et al.*, 2006). The two key components of connective tissue that is, fat and collagen-vary depending on breed, muscle, and age, and both are evidently observable on photographable images. Histology may be utilized to improve lipid imaging: for example, Thakur *et al.* (2003) researched on the effect of composition and deposition of Sudan dye-stained lipids on fish meat texture.

2.8 NIR (Near Infra Red) Spectroscopy

The near-infrared and visual spectroscopy is a rapid and often nondestructive method used for determining the composition of biological samples. It is based on the principle of absorption, reflection, transmission and/or scattering of light in or through a meat/food material following the Beer Lambert law. Now different NIR spectrometers are accessible and are being used commercially. Absorption or reflectance of light in known range of wavelengths are measured and correlated with different quality parameters of meat samples. NIR spectroscopy in combination with chemo metrics can be used to determine all constituents (proteins, fat, sugars *etc.*) of meat products at the same time.

Prieto *et al.* (2009) described in detail about use of NIR spectroscopy for determining chemical composition, technical and sensory value parameters of meat and meat products. Prieto *et al.* (2006) showed that numerous nutritional parameters such as crude protein (CP), gross energy (GE), ether extract (EE) and dry matter (DM) in beef meat have been calibrated effectively in NIR wavelength range but myoglobin, collagen and total ash percentage showed a reduced predictability. Hoving-Bolink *et al.* (2005) reported with the help of NIRS positive correlation of intra-muscular fat content in early post mortem can be measured whereas drip loss prediction was not that successful. Prieto *et al.* (2008) effectively differentiated between meat of young and adult cattle through scanning the homogenized beef meat in 1100–2500 nm wavelengths. Their results were primarily on the basis of differences in the intramuscular fat and water content. Gaitán-Jurado *et al.* (2008) found that the better result of proximate analysis of meat sample with the use of NIR spectroscopy may be obtained in homogenized samples compared to intact one principally due to homogeneous division of fat and protein after homogenization. So for chemical analysis of meat sample we should use homogenized sample for performing NIR spectroscopy. Grimes *et al.* (2008) reported that with the use of hyper spectral imaging apparatus tenderness of meat can be predicted. Accuracy is more if predicted about two tenderness categories (acceptable, tough) and comparatively less if predicted three tenderness categories (tender, intermediate, tough). Andrés *et al.* (2007) also used NIR spectroscopy to predict the in general acceptability or texture report of meat samples and it is useful for quality control for consumer satisfaction along with helpful in price fixation of the product.

Chao *et al.* (2001) reported that with the proper use of optical spectral reflectance and multi–spectral image analysis, chicken hearts disease diagnosis can be done effectively so it is useful for differentiating diseased meat too. Dey *et al.* (2003) recognized a method for differentiating normal and septicemia/toxemia affected chicken using spectral measurements and neural network classification of the spectral data after principal component analysis. Hsieh *et al.* (2002) also reported that the accuracy of using spectroscopy to differentiate septicemic livers from normal ones is very high upto 94-98 per cent. The NIR spectroscopy could be easily used in meat industry for analysis of proximate composition, texture and diseased condition of meat. It is also very accurate method to screen large number of meat samples in very short time.

3. Colour Measurement and Computer Image Analysis

Colour of meat is very important and unique property while used for analytical evaluation. Colour is mainly an exterior character recognized to the spectral distribution of light and, in a way, is connected to various source of radiant energy (the illuminant), to the object to which the colour is attributed, and to the eye of the spectator. Colour does not exist in dark *i.e.* without light. The quality of an object that gives it a distinguishing colour is its light-absorptive ability. Consumer commonly used colour as an indicator of meat quality before taking decision of purchasing. They commonly take discolouration as a condition of deterioration of freshness and integrity of meat sample and do not like to purchase such meat.

Mancini and Hunt (2005) used computer vision with the help of colorimeters and spectrophotometers as tools for measurement of meat colour. Colorimeter is the most common instrument for measuring colour. A colorimeter produces a measurement in three numbers called tristimulus values. The values are the quantity of primary colours in the coating red (X), green (Y), and blue (Z). Colour can also be measured in terms of L (lightness), a (redness) and b (blueness) values and/or Hue (H), Intensity (I) and Saturation (S). All systems of colour measurement are scientifically interconnected.

Rosenvold and Andersen (2003) experienced that the early post mortem temperature rise plays vital role in colour and colour constancy. It has been seen that when pigs were slaughtered in stress condition there were increase in the early post mortem temperature and the pH decline observed. But when pigs were slaughtered after feeding the experimental diet these effects were absent. Mancini *et al.* (2005) proposed that elimination of oxygen and the adding of low concentration of carbon-mono-oxide minimized discoloration of beef marrow as compared to high oxygen modified ambiance packaging.

Computer vision expertise offers immense advantages in assessment of meat colour as the image can be segmented with software and other soft computing methods such as neural network models can be collective in study and analysis of this extremely subjective data. Lu *et al.* (2000) evaluated fresh pork loin colour by using image processing and back propagation algorithm of neural networks. Du and Sun (2004) and Brosnan and Sun (2004) reviewed developments in image processing techniques by means of computer vision for meat/food quality assessment. Faucitano *et al.* (2005) estimated marbling fat percentage quantitatively and evaluated effect of marbling variation on eating quality of pork with the use of computer image analysis. A device with a computer vision structure set to scan or analysis by acquiring dimensional, volume, shape, or meat quality or grading information for packing meat cuts to divert the meat cut to respective packing station was patented by Koke and Steele (2002). In India ICAR and CSIR laboratories are studying the use of colour, computer vision and image analysis for evaluation of food/meat quality. They have commercialized some techniques based on computer image analysis for fruits gradation and finding maturity status of fruits. The work on meat quality estimation and use of computer image processing to assess the meat cuts is in a developmental stage and with the electronics and communication advancement will come into a regular practice.

4. Biosensors

Biosensors are a comparatively new and promising techniques used for rapid detection of meat borne pathogen, chemical intoxicants and residues in edible items. This technique can also be handy for determining the freshness of some livestock products. For meat science maximum work is focused in detection of pathogenic organisms for determining meat safety and optical biosensors are suitable for pathogen detection because of their sensitivity and rapid response. Bosch *et al.* (2007) and Passaro *et al.* (2007) listed the optical bio-sensing transducers which could be based on different optical phenomenon such as absorption, reflection, surface plasmon resonance, fluorescence, and interference. Many of above biosensor are tried for testing with pure microbial culture or toxins; however only a few were ready to be used as a large scale sampling as a lot of laboratory work is required before validating the technique for commercial utilization.

Yano *et al.* (1995) tested both bacterial spoiling and aging of stored meat using two-line flow injection analysis biosensor. They used putrescine oxidase immobilized electrode for measuring putrescine, cadaverine for bacterial spoilage and a xanthine oxidase immobilized electrode as detectors for measuring hypoxanthine and xanthine for aging. Biosensors can also be used potentially for assessment of meat freshness. Watanabe *et al.* (2005) used xanthine oxidase based biosensor for assessing freshness of sashimi fish via measuring colour change of thiazole blue. Fillit *et al.* (2008) and Mitsubayashi *et al.* (2004) reported that flavin-containing monooxygenase type-3 based biosensors can estimate high-quality quantitative indications of trimethyl amine with high sensitivity. Frébort *et al.* (2000) used amine oxidase based flow biosensor for assessing freshness of fish in terms of putrescine and histamine measurement. Ghosh *et al.* (1998) also used biosensors containing electrodes immobilized with xanthine oxidase, nucleoside phosphorylase and nucleotidase enzymes for determining ATP degradation products, hypoxanthine, inosine and inosine monophosphate metabolites in meat/fish products. By using different enzymes/chemical based biosensors we can rapidly detect the pathogens, meat aging, freshness and different targeted meat qualities very efficiently.

5. Electronic Nose

Meat flavours are very important criteria for selection of meat by the consumers. For quality assessment and meat safety point of view this is also an important sensory parameter. An "electronic or artificial nose" is an instrument, which comprises a sampling system, an array of chemical gas sensors with differing selectivity, and a computer with an appropriate pattern-classification algorithm, capable of qualitative and/or quantitative analysis of simple or complex gases, vapors, or odours. The word "electronic nose" is used to specify artificial olfaction, since many current electronic noses are constructed with more than one class of sensors in them. Electronic nose can substitute several traditional methods used for evaluation of fresh meat quality based on costly and comparatively individual taste panels and time-consuming and invasive chemical tests. So, as it is a modern technique and very limited number of research regarding application of electronic nose for meat evaluation is available; a few of them are discussed in brief here.

Descalzo *et al.* (2007) used electronic nose for differentiating the antioxidant status of pasture or grain fed beef on the basis of their odour profile. Vestergaard *et al.* (2006) have worked in swine and established the prospective of electronic/artificial nose based on ion mobility to distinguish boar taint of male pigs. O'Sullivan *et al.* (2003) worked on cooked meat to see the sensitivity of consumer towards over cooked flavour. They used data of over cooking flavour in pork by electronic nose and correlated well with data of sensory evaluation. Then they grind this meat and used this as an ingredient to prepare various products. During the process of meat grinding meat comes in the contact of oxygen into close proximity of oxidisable fats which causes rancidity in meat and meat products. Cooked pork products prepared from even slightly rancid meat taste cardboardy and painty and are rejected by consumers. Braggins *et al.* (1999) reported that electronic nose is very sensitive instrument as compared to panelists to rapidly assess the onset of rancidity in frozen ground meat/beef. Meat odours or volatile profile is very sensitive sensory parameter and depends upon animal of different origin and odours also vary with the diet of animal. Panigrahi *et al.* (2006) reported that they used commercially available electronic nose integrated with neural networks to classify the different beef/meat on the basis of volatile compounds released due to amplified microbial load and got cent per cent accuracy of classification of beef.

6. Conclusion

Modern techniques to assess the nutritional quality of meat which have revealed huge prospective are non-invasive rapid techniques such as HPLC, NMR and NIR spectroscopy, AAS, ultrasound, x-ray method and biosensors. HPLC can analyze protein, fats, carbohydrates, vitamins and minerals present in the meat. AAS can measure trace minerals rapidly. Similarly, colour and computer image analysis correlate well with meat integrity, marbling percentage of meat, tenderness, meat disorders/pathogens and effects of processing and storage time. The method is rapid and highly accurate. These methods are very efficient for quality determination of meat than traditional proximate analysis methods. These non-destructive techniques are rapid, more sensitive and able to screen large number of sample in comparatively less time. With the help of these techniques we can save time, money and can provide consumer quality labled meat. The different methods can be used for nutritional analysis, intramuscular fat, microbial spoilage and antioxidant status of meat.

REFERENCES

Abouelkaram, S., Chauvet, S., Strydom, P., Bertrand, D., and Damez, J. L. (2006). Muscle study with multi- spectral image analysis. 52nd International Congress of Meat Science and Technology. Dublin, Ireland. pp. 669-670.

Abouelkaram, S., Suchorski, K., Buquet, B., Berge, P., Culioli, J., Delachartre, P., and Basset, O. (2000). Effects of muscle texture on ultrasonic measurements. Food Chemistry, 69, 447-455.

Andrés, S., Murray, I., Navajas, E. A., Fisher, A. V., Lambe, N. R., and Bünger, L. (2007). Prediction of sensory characteristics of lamb meat samples by near infrared reflectance spectroscopy. Meat Science, 76, 509-516.

Beattie, R. J., Bell, S. J., Borggaard, C., and Fearon, A. (2006). Prediction of adipose tissue composition using Raman spectroscopy: Average properties and individual fatty acids. Lipids, 41, 287-294.

Benedito, J., Carcel, J. A., Rossello, C., and Mulet, A. (2001). Composition assessment of raw meat mixtures using ultrasonics. Meat Science, 57(4), 365-370.

Bosch, M. E., Sánchez, A. J. R., Rojas, F. S., and Ojeda, C. B. (2007). Recent development in optical fiber biosensors. Sensors, 7, 797-859.

Braggins, T. J., Frost, D. A., Agnew, M. P., and Farouk, M. M. (1999). Application of an electronic nose in the meat industry. In: Hurst W. J., editor. Electronic noses and sensor array based systems design and applications. Lancaster: Technomic Publishing Company, pp. 51-82.

Brienne, J. P., Denoyelle, C., Baussart, H., and Daudin, J. D. (2001). Assessment of meat fat content using dual energy x-ray absorption. Meat Science, 57(3), 235–244.

Brosnan, T., and Sun, D. W. (2004). Improving quality inspection of food products by computer vision-a review. J Food Eng., 61, 3-16.

Chao, K., Chen, Y. R., Hruschka, W. R., and Park, B. (2001). Chicken heart disease characterization by multi-spectral imaging. Appl Eng Agric, 17, 99-106.

Clerjon, S., and Bonny, J. M. (2011). Diffusion-Weighted NMR Micro-Imaging of Lipids: Application to food products. In: J. P. Renou, P. S. Belton, and G. A. Webb (Eds.). Magnetic Resonance in Food Science. An Exciting Future. Cambridge: The Royal Society of Chemistry. pp.182-189.

Clerjon, S., Kondjoyan, A., Bauchart, D., and Bonny, J. M. (2010). Diffusion-weighted NMR micro-imaging (DWI) for frying oil penetration evaluation. International Conference on the Applications of Magnetic Resonance in Food Science. Clermont Ferrand, France.

Descalzo, A. M., Rossetti, L., Grigioni, G., Irurueta, M., Sancho, A. M., Carrete, J., and Pensel, N. A. (2007). Antioxidant status and odour profile in fresh beef from pasture or grain-fed cattle. Meat Science, 75, 299-307.

Dey, B. P., Chen, Y. R., Hsieh, C., and Chan, D. E. (2003). Detection of septicemia in chicken livers by spectroscopy. Poultry Science, 82, 199-206.

Du, C., and Sun, D. W. (2004). Recent developments in the applications of image processing techniques for food quality evaluation. Trends Food Sci Technol., 15, 230-249.

Faucitano, L., Huff, P., Teuscher, F., Gariepy, C., and Wegner, J. (2005). Application of computer image analysis to measure pork marbling characteristics. Meat Science, 69, 537-543.

Fernandez-Cabanas, V. M., Polvillo, O., Rodriguez-Acuna, R., Botella, B., and Horcada, A. (2011). Rapid determination of the fatty acid profile in pork dry-cured sausages by NIR spectroscopy. Food Chemistry, 124(1), 373-378.

Fillit, C., Jaffrezic-renault, N., Bessueille, F., Leonard, D., Mitsubayashi, K., and Ttardy, J. (2008). Development of a microconductometric biosniffer for detection of trimethylamine. Mater Sci Eng, C., 28, 781-786.

Fontaine, J. (2003). Amino acid analysis of feeds, amino acids in animal nutrition. Second Edition; J.P.F D'Mello, formerly of the Scottish Agricultural College, Edinburgh, UK

Frébort, I., Skoupá, L., and Peè, P. (2000). Amine oxidase-based flow biosensor for the assessment of fish freshness. Food Control, 11, 13-18.

Gaitán-jurado, A. J., Ortiz-somovilla, V., España-españa, F., Pérez-aparicio, J., and Pedro-sanz, E. J. D. (2008). Quantitative analysis of pork dry-cured sausages to quality control by NIR spectroscopy. Meat Science, 78, 391-399.

Ghosh, H. S., Sarker, D., and Misra, T. N. (1998). Development of an amperometric enzyme electrode biosensor for fish freshness detection. Sens Actuators, B., 53, 58-62.

Gilani, S. G., Xiao, C., Lee, and N. (2008). Need for accurate and standardized determination of amino acids and bioactive peptides for evaluating protein quality and potential health effects of foods and dietary supplements. Journal of AOAC International, 91, 4.

Grimes, L. M., Naganathan, G. K., Subbiah, J., and Calkins C. R. (2008). Predicting aged beef tenderness with a hyperspectral imaging system. Nebraska Beef Report. 138-139.

Hansen, P. W., Tholl, I., Christensen, C., Jehg, H. C., Borg, J., Nielsen, O., Ostergaard, B., Nygaard, J., and Andersen, O. (2003). Batch accuracy of on-line fat determination. Meat Science, 64(2), 141-147.

Hoffman, K. (1990). Definition and measurement of meat quality. Proceedings of 36th Ann. Int. Cong. of Meat Science and Technology, Cuba, pp, 941.

Hoving-Bolink, A. H., Vedder, H. W., Merks, J. W. M., De Klein, W. J. H., Reimert, H. G. M., Frankhuizen, R., Van den Broek, W., and En Lambooij, E. (2005). Perspective of NIRS measurements early post mortem for prediction of pork quality. Meat Science, 69, 417-423.

Hsieh, C., Chen, Y. R., Dey, B. P., and Chan, D. E. (2002). Separating septicemic and normal chicken livers by visible/near-infrared spectroscopy and back-propagation neural networks. Trans ASAE. 45, 459-469.

Koke, J. P., and Steele, C. B. (2002). Apparatus and method for use in packing meat cuts. Sealed Air (NZ) Ltd. PCT Patent No. WO 2002/016210.

Kryeziu, A., Bakalli, R. I., Kamberi, M. A., Kastrati, R., Mestani, N. (2007). Comparison of commercial near-infrared reflectance spectroscopy (NIRS) calibrations and standard chemical assay procedures for prediction of crude protein levels in poultry feed ingredients. 16th European Symposium on Poultry Nutrition., August 26 - 30, 2007 Strasbourg, France.

Lu, J., Tan, J., Shatdal, P., and Gerrard, D. E. (2000). Evaluation of pork colour by using computer vision. Meat Science, 56, 57-60.

Mancini, R. A., and Hunt, M. C. (2005). Current research in meat colour. Meat Science, 71, 100-121.

Mancini, R. A., Hunt, M. C., Hachmeister, K. A., Kropf, D. H., and Johnson D. E. (2005). Exclusion of oxygen from modified atmosphere packages limits beef rib and lumbar vertebrae marrow discolouration during display and storage. Meat Science, 2005, 69, 493-500.

Mercier, J., Pomar, C., Marcoux, M., Goulet, F., Theriault, M., and Castonguay, F. W. (2006). The use of dual-energy x-ray absorptiometry to estimate the dissected composition of lamb carcasses. Meat Science, 73(2), 249-257.

Mitsubayashi, K., Kubotera, Y., Yano, K., Hashimoto, Y., Kon, T., Nakakura, S., Nishi, Y., and Endo, H. (2004). Trimethylamine biosensor with flavin-containing monooxygenase type 3 (FMO3) for fish-freshness analysis. Sens Actuators, B., 103, 463-467.

Morlein, D., Rosner, F., Brand, S., Jenderka, K. V., and Wicke, M. (2005). Non-destructive estimation of the intramuscular fat content of the longissimus muscle of pigs by means of spectral analysis of ultrasound echo signals. Meat Science, 69(2), 187-199.

Narsaiah, K., and Jha, S. N. (2012). Nondestructive methods for quality evaluation of livestock products. Journal of Food Science and Technology, 49(3), 342-348.

Nunes, A. M., Acunha, T. S., Oreste, E. Q., Lepri, F. G., Vieira, M. A., Curtius, A. J., and Ribeiro, A. S. (2011). Determination of Ca, Cu, Fe and Mg in fresh and processed meat treated with tetramethylammonium hydroxide by atomic absorption spectrometry. Journal of the Brazilian Chemical Society, 22(10), 1850-1857.

O'sullivan, M. G., Byrne, D. V., Jensen, M. T., Andersen, H. J., and Vestergaard, J. (2003). A comparison of warmed-over flavour in pork by sensory analysis, GC/MS and the electronic nose. Meat Science, 65, 1125-1138.

Panigrahi, S., Balasubramanian, S., Gu, H., Logue, C. M., and Marchello, M. (2006). Neural-network-integrated electronic nose system for identification of spoiled beef. LWT Food Sci Technol., 39, 135-145.

Passaro, V. M. N., Dell'olio, F., Casamassima, B., and de Lleonardis, F. (2007). Guided-wave optical biosensors. Sensors. 7, 508-536.

Paul, B. N., Chanda, S., Das, S., Singh, P., Pandey, B. K. and Giri, S. S. (2014). Mineral Assay in Atomic Absorption Spectroscopy. The Beats of Natural Sciences. 1(4): 1-17.

Peace, R. W., and Gilani G. S. (2005). Chromatografic determination of amino acids in foods. Journal of AOAC International, 88, 3.

Pearce, K. L., Ferguson, M., Gardner, G., Smith, N., Greef, J., and Pethick, D. W. (2009). Dual x-ray absorptiometry accurately predicts carcass composition from live sheep and chemical composition of live and dead sheep. Meat Science, 81(1), 285-293.

Pérez-Juan, M., Afseth, N. K., Gonzalez, J., Diaz, I., Gispert, M., Furnols, M. F., Oliver, M. A., and Realini, C. E. (2010). Prediction of fatty acid composition using a NIRS fibre optics probe at two different locations of ham subcutaneous fat. Food Research International, 43(5), 1416-1422.

Prieto, N., Andrés, S., Giráldez, F. J., Mantecón, A. R., and Lavín, P. (2006). Potential use of near infrared reflectance spectroscopy (NIRS) for the estimation of chemical composition of oxen meat samples. Meat Science,74, 487-496.

Prieto, N., Andrés, S., Giráldez, F. J., Mantecón, A. R., and Lavín, P. (2008). Discrimination of adult steers (oxen) and young cattle ground meat samples by near infrared reflectance spectroscopy (NIRS). Meat Science, 79, 198-201.

Prieto, N., Navajas, E. A., Richardson, R. I., Ross, D. W., Hyslop, J. J., Simm, G., and Roehe, R. (2010). Predicting beef cuts composition, fatty acids and meat quality characteristics by spiral computed tomography. Meat Science, 86(3), 770-779.

Prieto, N., Roehe, R., Lavín, P., Batten, G., and Andrés, S. (2009). Application of of near infrared reflectance spectroscopy to predict meat and meat products quality: a review. Meat Science, 83, 175-186.

Primer, A. (2001). HPLC for food analysis. First edition; Agilent technologies company, Germany.

Rosenvold, K., and Andersen, H. J. (2003). The significance of preslaughter stress and diet on colour and colour stability of pork. Meat Science, 63, 199-209.

Swatland, H. J., Irving, T. C., and Millman, B. M. (1989). Fluid distribution in pork, measured by x-ray-diffraction, interference microscopy and centrifugation compared to paleness measured by fiber optics. Journal of Animal Science, 67(6), 1465-1470.

Thakur, D. P., Morioka, K., Itoh, Y., and Obatake, A. (2003). Lipid composition and deposition of cultured yellowtail Seriola quinqueradiata muscle at different anatomical locations in relation to meat texture. Fisheries Science, 69(3), 487-494.

Vestergaard, J. S., Haugen, J., and Byrne, D. V. (2006). Application of an electronic nose for measurements of boar taint in entire male pigs. Meat Science, 74, 564-577.

Watanabe, E., Tamada, Y., and Hamada-Sato, N. (2005). Development of quality evaluation sensor for fish freshness control based on K_I value. Biosens Bioelectron, 21, 534-538.

Yano, Y., Miyaguchi, N., Watanabe, M., Nkamura, T., Youdou, T., Miyai, J., Numata, M., and Aasano, Y. (1995). Monitoring of beef aging using a two-line flow injection analysis biosensor consisting of putrescine and xanthine electrodes. Food Research International, 28, 611-618.

Index

C

D

E

F

M

N

O

P

Q

R

S

T

U

V

W

X